W0018278

Aeronautics and Astronautics

Aeronautics and Astronautics

Editor

Nicolaos Sabella

Aeronautics and Astronautics
Edited by **Nicolaos Sabella**

ISBN: 978-1-68117-233-0
Library of Congress Control Number: 2016934765

© 2017 by
SCITUS Academics LLC,
www.scitusacademics.com
Box No. 4766, 616 Corporate Way,
Suite 2, Valley Cottage,
NY 10989

This book contains information obtained from highly regarded resources. Copyright for individual articles remains with the authors as indicated. All chapters are distributed under the terms of the Creative Commons Attribution License, which permits unrestricted use, distribution, and reproduction in any medium, provided the original author and source are credited.

Notice

Reasonable efforts have been made to publish reliable data and views articulated in the chapters are those of the individual contributors, and not necessarily those of the editors or publishers. Editors or publishers are not responsible for the accuracy of the information in the published chapters or consequences of their use. The publisher believes no responsibility for any damage or grievance to the persons or property arising out of the use of any materials, instructions, methods or thoughts in the book. The editors and the publisher have attempted to trace the copyright holders of all material reproduced in this publication and apologize to copyright holders if permission has not been obtained. If any copyright holder has not been acknowledged, please write to us so we may rectify.

Preface

Aerospace engineering is the primary field of engineering concerned with the development of aircraft and spacecraft. It is divided into two major and overlapping branches: aeronautical engineering and astronautical engineering. Aeronautics is the study of the science of flight. Aeronautics is the method of designing an airplane or other flying machine. Astronautics is often referred to as astronomical engineering. It is the science and technology of space flight and is a field of aerospace engineering that deals with machinery designed to work beyond the Earth's atmosphere. Astronautics works by applying scientific principles as well as engineering techniques to aviation technology.There are four basic areas that aeronautical engineers must understand in order to be able to design planes. To design a plane, engineers must understand all of these elements. Aeronautical engineers apply principles and concepts of engineering to create new aircraft containing the most current and sophisticated technologies.Airplanes, gliders and helicopters all stay to the principles within the flight science called aeronautics. These include physics concepts like Newton's laws of motion; aerodynamics, the motion of air; and Bernoulli's principle of lift. Aeronautical engineering is the design and construction of airplanes and other aircraft that fly within the atmosphere. Aerospace engineering encompasses aeronautics as well as the design and construction of spacecraft. Aeronautical engineering and aerospace engineering degree programs typically contain many of the same courses. Principles and theories that are common to aeronautical engineering are also applied in the construction of buildings and evaluation of the performance of space vehicles within the earth's atmosphere. This book entitled Aeronautics and Astronautics encompasses the study, design, and manufacture of airplanes or other aviation vehicles. It also focuses on the theory and practice of navigation beyond Earth's atmosphere.

Table of Contents

CHAPTER 1

Rotorcraft Design for Maximized Performance at Minimized Vibratory Loads

Marilena D. Pavel[1]

[1] *Faculty of Aerospace Engineering, Delft University of Technology, The Netherlands*

1. INTRODUCTION

Rotorcraft (helicopters and tiltrotors) are generally reliable flying machines capable of fulfilling missions impossible with fixed-wing aircraft, most notably rescue operations. These missions, however, often lead to high and sometimes excessive pilot workload. Although high standards in terms of safety are imposed in helicopter design, studies show that "*it is ten times more likely to be involved in an accident in a helicopter than in a fixed-wing aircraft*" (Iseler et. al. 2001). According to World Aircraft Accident Summary (WAAS, 2002), nearly 45 percent of all accidents of single-piston helicopters is attributed to pilot loss of control, where - because of various causes, often involving vibrations, high workload, and bad weather - a pilot loses control of the helicopter and crashes, sometimes with fatal consequences. Quoting the Royal Netherlands Air Force ('Veilig Vliegen' magazine, 2003), "*for helicopters there is a considerable number of inexplicable incidents (...) which involved piloting loss of control*". The situation is likely to get worse, as rotorcraft missions are becoming more difficult, demanding high agility and rapid manoeuvring, and producing more violent vibrations. (Kufeld & Bousman, 1995; Hansford & Vorwald, 1996; Datta & Chopra, 2002)

The primary cause of pilot control difficulties and high-workload situations is that even modern helicopters often have poor Handling Qualities (HQs) (Padfield, 1998). Cooper and Harper (Cooper & Harper, 1969), pioneers in this subject, defined these as: "*those qualities or characteristics of an aircraft that govern the ease and precision with which a pilot is able to perform a mission*". Below, the current practice in rotorcraft handling qualities assessment will be discussed, introducing the key problem addressed in this chapter.

1.1. State-of-the-Art in Rotorcraft Handling Qualities – The Aeronautical Design Standard Ads-33

Helicopter handling qualities used to be assessed with requirements defined for fixed-wing aircraft, as stated in the FAR (civil) and MIL (military) standards. In the 1960's, however, it became clear that these standards were not sufficient (Key, 1982). Helicopters have *strong cross-coupling effects* between longitudinal and directional controls, their behaviour is *highly non-linear* and requires *more degrees of freedom* in modelling than the rigid-body models used for aircraft. Therefore, the MIL-H-8501A standard (MIL-H-8501A, 1962) was developed. This standard was used up until mid 1980's. From a safety perspective, these requirements were merely 'good minimums', and a new standard was developed in the 1970's, that is used up until today, the Aeronautical Design Standard ADS-33 (ADS-33, 2000)

The crucial point, understood by ADS-33, is that helicopter HQ requirements need to be related to the mission executed, as this will determine the needed pilot effort. E.g., a shipboard landing at night and in high sea with strong ship motions demands more precision of control from the pilot than when flying in daytime and good weather.2 ADS-33 introduced handling qualities *metrics* (HQM), a *combination* of flight parameters such as rate of climb, turn rate, etc., that reflect how much manoeuvre-capability the pilot has *per specific mission*. These metrics are then mapped into handling qualities criteria (HQC) that yield boundaries between 'good' (Level 1), 'satisfactory' (Level 2) and 'poor' (Level 3) HQs.[1] -

Despite their importance for the helicopter safety, its operators and, above all, the helicopter pilots, achieving good handling qualities is still mainly a secondary goal in helicopter design. The first phase in helicopter design is the 'conceptual design' phase in which the main rotor and fuselage parameters are established, based on desired performance and, to some extent, vibration criteria.20 Only in the following phase, that of preliminary design, are 'high-fidelity' simulation models developed and the handling qualities considered. The high fidelity models allow an analysis of helicopter behaviour for various flight conditions. Applying the ADS-33 metrics/criteria to these models results in predicted levels of HQs. When these are known, the *experimental* HQ assessment begins, illustrated in Fig. 1.

In this process, first databases of missions and environments are defined, according to customer requirements. Manoeuvrability, i.e., how easy can pilots guide the helicopter, and agility, i.e., how quickly can they change its flight direction, are critical performance demands. Second, experienced test pilots are asked to fly the missions in a flight simulator. Through insisting pilots to execute the mission very thoroughly, one aims to expose deficient handling qualities. Pilots rate the HQs they experience in each manoeuvre flown, generally using the Cooper-Harper rating scale.[2] - Third, Operational Flight Envelopes (OFEs) and Service Flight Envelopes (SFEs) are defined, based on a mapping of the HQ ratings. OFEs represent the limits within which the helicopter must be able to operate in order to accomplish the operational missions. SFEs stem from the helicopter limits and are expressed in terms of any parameters believed necessary to ensure safety, see Fig. 1.

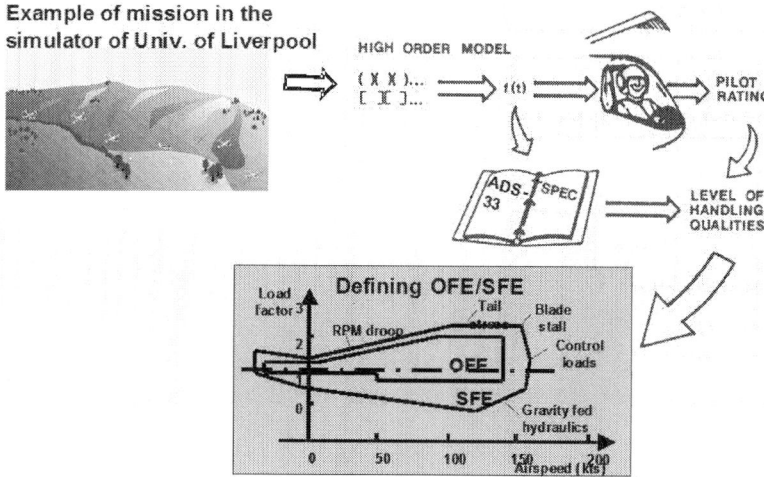

Figure 1. Experimental assessment of helicopter handling qualities

In this first experimental assessment of the helicopter's handling qualities, the first problems arise, as more often than not, large differences arise between the theoretical predictions and the experimentally-determined pilot judgments. The gaps that occur are bridged by applying optimisation techniques using the simulation models developed in preliminary design, to improve the designs of helicopter, load alleviation system and flight control system (Celi, 1991; Celi, 1999; Celi, 2000; Fusato & Celi, 2001, Fusato & Celi, 2002a,2002b; Ganduli, 2004; Sahasrabudhe & Celi, 1997). Computational approaches have three important disadvantages, however. First, many designs that "roll out" of the procedure are unfeasible, the optimisation "pushing" the solution along the boundaries of the problem and not inside of the feasible region. Second, optimising for ADS-33 requires calculations of the helicopter time-domain responses, and the numerical methods become computationally very intensive (Sahasrabudhe & Celi, 1997; Tischler et. al., 1997) Third, and most important, the lack of quantitative, validated helicopter pilot models, capable of accurately predicting the effects of helicopter vibrations on pilot control behaviour, prevents the proper inclusion of pilot-centred considerations in mathematical optimization techniques (Mitchell et. al., 2004; Tischler et. al., 1996)

Whereas the first and second disadvantages are common in multi-dimensional design problems, the lack of knowledge on how helicopter vibrations affect pilot performance is a typical and fundamental problem of modern helicopter design (Mitchell et. al., 2004, Padfield, 1998). Although ADS-33 proposes criteria and missions regarding helicopter limits, these only characterize a helicopter's performance, and do not require an adequate knowledge of helicopter vibratory loads (Kolwey, 1996;Tischler et. al., 1996). This shortcoming stems from the fact that, when the ADS-33 criteria were defined, helicopter missions were not so demanding, and the vibratory loads associated with them were low. In the last twenty years, however, ever-increasing performance requirements and extended flight envelopes were defined, for reasons of heavy competition, demanding manoeuvres that impose heavy vibrations on both structure and pilot. These vibrations, combined with cross-coupling effects, rapidly lead to pilot overload and degradation in performance (Padfield, 2007).

Key Problem: The current practice of assessing rotorcraft handling qualities reveals significant gaps in the ADS-33 HQ criteria, especially regarding the effects of vibrations.

1.2. Goal

The design of high-performance rotorcraft has become an arduous process, regularly leading to surprises, demanding 'patches' to safety-critical systems, and needing more iterations than expected, all contributing in very high costs. Unmistakably, the helicopter and flight control system design teams do not have up-to-date criteria to adequately assess the effects of helicopter vibrations on its handling qualities. There is an urgent need for a much more fundamental understanding of how helicopter vibrations affect pilot control behaviour (Mitchell et. al., 2004) and for new tools to incorporate this knowledge as early as possible in the design of both helicopter and flight control system. The goal of this chapter is to portray a novel approach to rotorcraft handling qualities (HQs) assessment by defining a set of consistent, complementary metrics for agility and structural loads pertaining to vertical manoeuvres in forward flight. These metrics can be used by the designer for making trade—offs between agility and vibrational/load suppression. The emphasis of the chapter will be on agility characteristics in the pitch axis applied to helicopter and tiltrotor. Especially in such new configurations, the proposed approach could be particularly useful as the performance tools for fixed-wing mode and helicopter mode must merge together within new criteria (Padfield, 2008).

The chapter is structured as follows: The second section will present an overview of traditional metrics for measuring pitch agility; The third section will present some alternative metrics proposed in the 90's for better capturing the transient characteristics of the agility; Then, based on the rational developments of the metrics from the previous two sections, fourth section will propose the new approach that can better quantify the agility from the designer point of view. Finally, general conclusions and potential extension of this work will be discussed.

2. TRADITIONALLY DESIGN OF AIRCRAFT FOR PITCH AGILITY

One of the most important flying quality concepts defining the upper limits of performance is the so-called "agility". Generally, it is well known that the level of performance achieved by the pilot depends on the task complexity. Fig. 2 presents generically this situation, showing that there is a line of saturation up to which the pilot is able to perform optimally the specified mission; increasing the task difficulty above this line leads quickly to stress, panic and even incapacity to cope anymore with the task complexity and blocking, sometimes with fatal consequences.

It is difficult to point precisely to the origins of the concept of agility but probably these go back to the moment when it was realized that, in a combat, a "medium performance" fighter could win over its superior opponent if the first aircraft possesses the potential for faster transient motions, i.e. superior agility. In its most general sense, the concept of agility is defined with respect to the overall combat effectiveness in the so-called "Operational agility". Operational agility according to measures the '*ability to adapt and respond rapidly and precisely, with safety and poise, to maximize mission effectiveness*'(McKay, 1994). In the mid 80's a strong wave of interest arose in seeking metrics and criteria that could quantify the aircraft agility (Mazza, 1990; McKay, 1994). However, there have been developed almost as many criteria of agility as there were investigators in the field. The problem was partially due to the lack of coordination in the research studies performed but also due to a disagreement on the most fundamental level: there simply was very little agreement on what agility was.

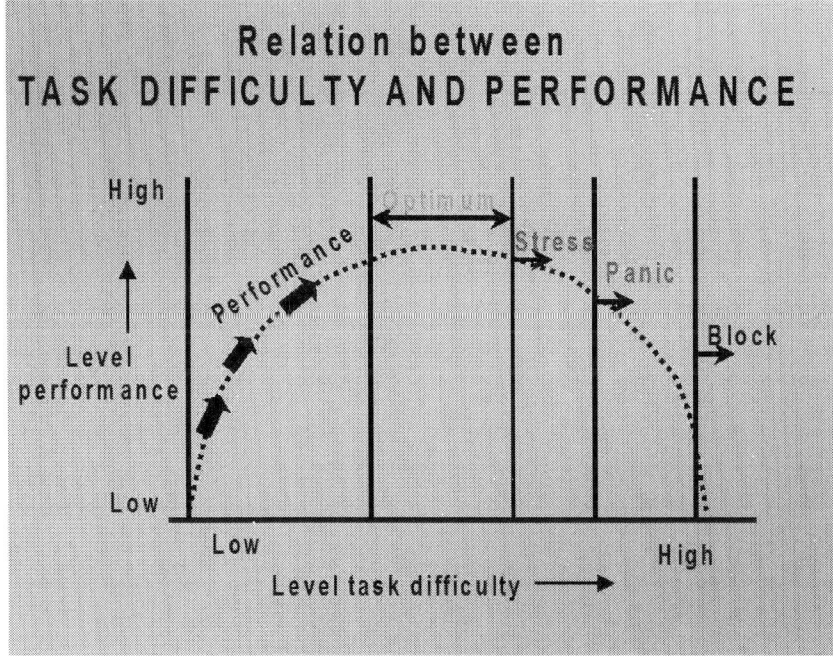

Figure 2. Correlation between task difficulty and performance

Within the framework of operational agility one can see agility as a function of the airframe, avionics, weapons and pilot. Airframe agility is probably the most crucial component in the operational agility as it is designed in from the onset and cannot be added later. The present chapter focuses on airframe agility and within this, the chapter will relate to the airframe agility in the pitch axis.

A large number of agility metrics have been proposed during the years to determine the aircraft realm of agility. The AGARD Working Group 19 on Operational Agility (McKay, 1994) put together all the different metrics and criteria existing on agility and fit them into a generalizable framework for further agility evaluations. The present section presents the traditional approach on pitch agility using as example a tiltrotor aircraft. This specific aircraft combines the properties of both fixed and rotary-wing aircraft and can be used to define a unified approach in the agility requirements at both fixed and rotary-wings.

The tiltrotor example to be investigated in this study is the Bel XV-15 aircraft. Next, as vehicle model, the FLIGHTLAB model of the Bell XV-15 aircraft as developed by the University of Liverpool (this model is designated as FXV-15) will be used. For a complete description of this model and the assumptions made the reader is referred to (Manimala et. al., 2003). For the tiltrotor in helicopter mode, the pilot's controls command pitch through longitudinal cyclic, roll through differential collective (lateral cyclic is also provided for trimming), yaw through differential longitudinal cyclic and heave through combined collective. In airplane mode, the pilot controls command conventional elevator, aileron and rudder (a small proportion of differential collective is also included).

Pitch agility is the ability to move, rapidly and precisely, the aircraft nose in the longitudinal plane and complete with easiness that movement. This implies that in the agility analysis one has to search for sample manoeuvres to be carried out by the flight vehicle dominated by high flight path changes and high rate of change of longitudinal acceleration which can give a good picture of the agility characteristics. As starting point in this discussion on pitch agility we will consider the kinematics of a sharp pitch manoeuvre – a simple example of this type is the tiltrotor trying to fly over an obstacle (see Fig. 3). Assume that the manoeuvre is executing starting from different forward speeds (helicopter mode 60 kts and 120kts; airplane mode 120 kts and 300 kts) and the manoeuvre aggressiveness is varied by varying the pulse duration (from 1 to 5 sec). The pilot flies the manoeuvre by giving a pulse input in the longitudinal cyclic stick of 1 inch amplitude.

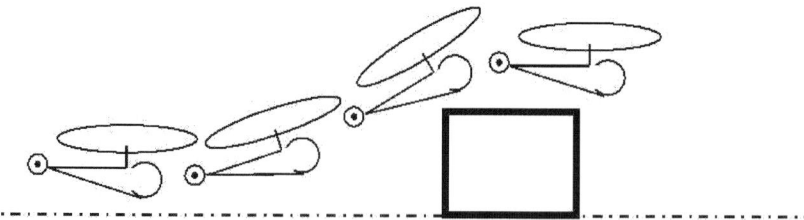

Figure 3. Executing an obstacle-avoid manoeuvre in the pitch axis

2.1. Transient Metrics

The first class of metrics developed to quantify the agility corresponds to the so-called "transient metrics". The transient class contains metrics which can be calculated at any moment for any manoeuvre. For pitch agility these metrics are pitch rate (entitled attitude manoeuvrability metric) and accelerations along the axes a_x, a_y, a_z (entitled manoeuvrability of the flight path). These metrics are next studied for the pull up manoeuvres flown with the FXV-15 in a 1 second pulse given from the initial trim at 120kts in helicopter mode and 300 kts in airplane mode. The presentation of the transient metric information is best achieved through a time history plot. Fig. 4 presents the transient metrics parameters of pitch rate q and vertical acceleration n_z (in the form of normal load factor). Looking at Fig. 4 one may see local maxima in the metric parameters q and n_z illustrating peak events in the agility characteristics. This clearly demonstrates that in a "real" manoeuvre sequence, the agility characteristics occur at key moments, depending on the manoeuvre.

2.2. Experimental Metrics

The above conclusion gave the idea to develop a new class of agility metrics, the so-called "experimental metrics" formulated as discrete parameters during a real manoeuvre sequence. These metrics are actually the basic building blocks for understanding the agility and can be related to flying qualities and aircraft design. The metrics describing pitch agility during aggressive manoeuvring in vertical plane were defined by (Murphy et. al., 1991) and are described in the next section. They referred to the ability of an aircraft to pint the nose in at an opponent and commented that what was not clear in such manoeuvres was the behaviour of the flight path. Was the nose pointing w.r.t. the velocity vector or did it include the flight path bending or perhaps both? The authors noted that longitudinal stick displacements would be expected to command the flight path in addition to the aircraft nose pointing pitch angle for agile aircraft. The study pointed out that current aircraft behave differently in the high speeds and slow speed regimes. In the high speed case the flight path displaced as per the nose pointing displacement. The low speed case exhibited no flight path or even opposite flight path displacements.

2.2.1. Peak and Time to Peak Pitch Rates

The peak and time to peak pitch rates metrics were proposed by (Murphy et. al., 1991) for fixed wing aircraft. These metrics measure the time to reach peak pitch rate and the corresponding pitch rate. Fig. 5 presents charts of peak pitch rate and time to reach this peak as a function of the velocity for the tiltrotor flying pull-up manoeuvres of increasing pulse duration. The pull-up manoeuvres are executed gradually increasing the velocity and the nacelle angle from the helicopter mode (90deg nacelle, hover and 60 kts) to conversion (60deg nacelle

120 kts) and ending in airplane mode (0deg nacelle 200 kts). Looking at these figures one can see that as the velocity increases the pilot is able to achieve higher pitch rates, the time to achieve these peaks being and faster especially if the pulse duration is short. As attributes, the peak and time to peak pitch rates metrics have the advantage that can be related to design.

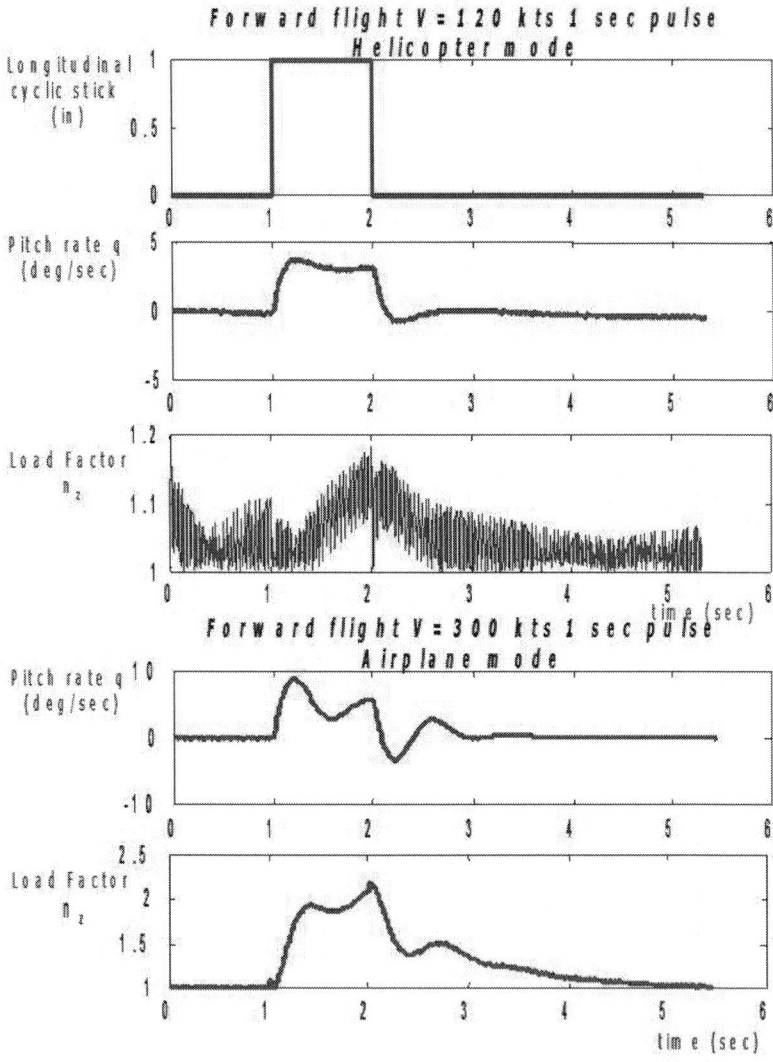

Figure 4. Transient agility metrics for pull-up manoeuvres with the tiltrotor

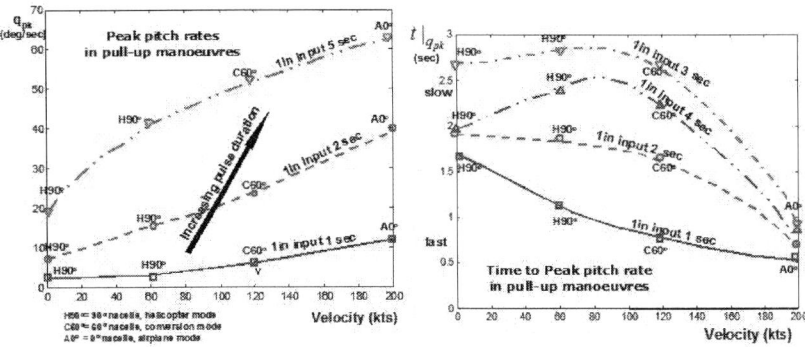

Figure 5. Peak and time to peak pitch rates in pull-up manoeuvres

2.2.2. Peak and Time to Peak Pitch Accelerations

(Murphy et. al., 1991) considered the so-called peak and timer to peak pitch accelerations as the primary metrics for pitch motion agility. The time to peak acceleration provides insight into the jerk characteristics of pitch motion: if it is too slow, then the pilot may complain that the aircraft is too sluggish for tracking-type tasks; if it is too fast, then the pilot may complain of jerkiness or over-sensitivity. Fig. 6 presents charts of peak and time to leak pitch acceleration as a function of velocity when flying pull-ups manoeuvres. One can see that as the velocity increases the pilot is able to obtain higher pitch accelerations but as is passing from the helicopter to aircraft mode this capability diminishes. For fixed wing aircraft, (Murphy et. al., 1991) commented on the differences in the data for the peak accelerations in the body and wind axes. This effect has implications on the pilot selection of flight path or nose pointing control during manoeuvring.

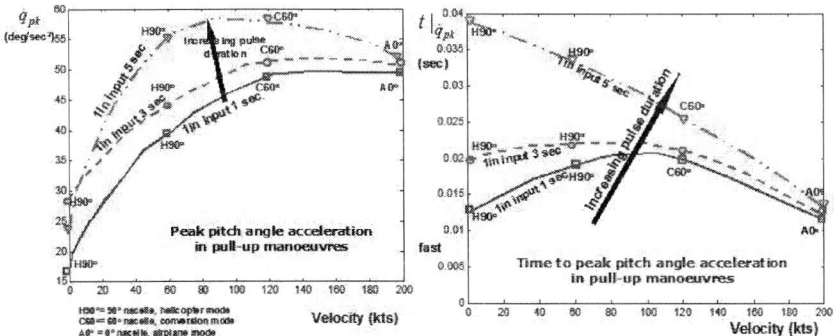

Figure 6. Peak and time to peak pitch angle acceleration in pull-ups with the tiltrotor

2.2.3. Peak and Time to Peak Load Factor

Peak and time to peak load factor metrics describe the peak and the transition time to the peak normal load factor during a manoeuvre in pitch axis. They can be used at best to determine the flight path bending capability of an aircraft. Fig. 7 presents these two metrics as a function of the velocity for the tiltrotor example. One may see that as the velocity increases the pilot is able to pull more g's as going from the airplane to helicopter mode.

2.2.4 Pitch Attitude Quickness Parameter

One of the most important pitch agility metrics introduced by ADS-33 helicopter standard (ADS-33, 2000) is the so-called "pitch attitude quickness" parameter and is defined as the ratio of the peak pitch rate to the pitch angle change:

$$Q_\theta \overset{def}{=} \frac{q_{pk}}{\Delta\theta} \left(\sec^{-1}\right)$$

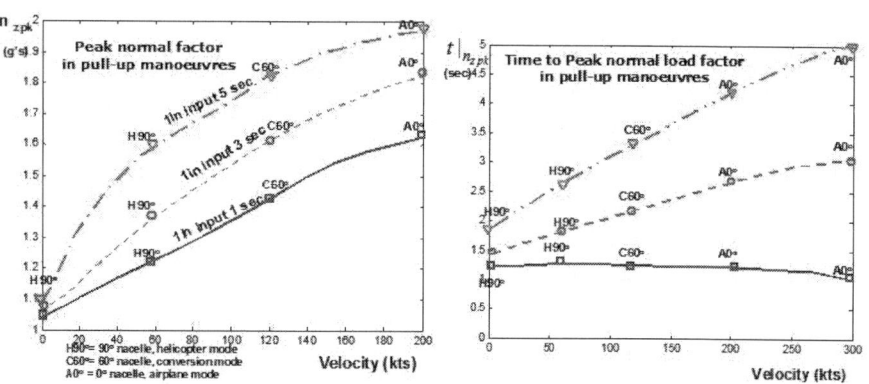

Figure 7. Peak and time to peak normal load factor

The advantage of this parameter is that it was linked to handling qualities so that potential bounds for agility could be identified. In this sense, ADS-33 presents HQs boundaries for the pitch quickness parameter as a function of the minimum pitch angular change $\Delta\theta_{min}$ (considered as the pitch angle corresponding to a 10% decay from q_{pk}). These boundaries are defined to separate different quality levels, but because they relate too to an agility metric, they become now boundaries of available agility. Fig. 8 illustrates the attitude quickness charts for the tiltrotor executing pull-up maneuvers of 1 to 5 sec 1in amplitude input at 60, 120 and 300 kts in helicopter and airplane mode. The figure shows also the Level 1/2 boundaries as defined by 1) ADS-33 for a general mission task element, low speed helicopter flight (<45kts) and 2) MIL STD 1797A for fixed wing aircraft. One may see that whereas in helicopter mode FXV-15 hardly meets Level 1 performance in ADS-33 standard, being mostly at Level 2 performance, in airplane mode FXV-15 meets Level 1 performance in AHS-33 but exhibits Level 2 performance according to the MIL standard for airplanes (MIL HDBK-1797, 1997).

3. FLYING QUALITIES METRICS FOR AGILITY DESIGNING

Linking the agility to flying qualities raised up a new question: is agility limited by pilot handling parameters or in other words what are the upper limits to agility set by flying qualities considerations? Flying qualities considerations do limit agility according to (Padfield, 1998). In this sense, in a series of flight and simulation trials research conducted at DERA (now Qinetics) the pilots were asked to fly maneuvers with increasing tempo until either performance or safety limit was reached. The results showed that in all cases the safety limit came first, thus the agility was constraint by safety.

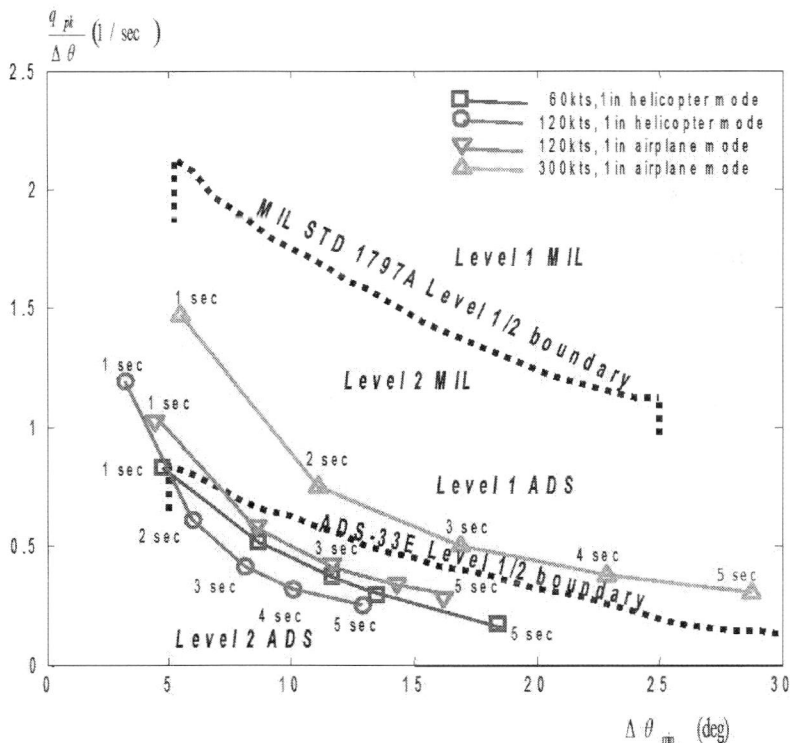

Figure 8. Pitch quickness for the tiltrotor

3.1. Agility Factor

A new metric was therefore introduced as a measure of performance margin (Padfield & Hodkinson, 1993), the so-called agility factor A_f, defined as the ratio of used to usable performance. For the simple case of the pull-up maneuver this metric can be easily calculated as the ratio of ideal task time T_i to actual task time T_a.

$$A_f \overset{def}{=} \frac{T_i}{T_a} = \frac{\omega_m \Delta t}{\omega_m \Delta t - \ln(0.1)}$$

where $T_i = \Delta t$ is the control pulse duration (1 to 5 sec), Ta is the time to reduce the pitch angle to 10% of the peak value achieved and ωm is the fundamental first-order break frequency or pitch damping which for this simple case represents the maximum achievable value of quickness. Fig. 9 illustrates the variation of Af with $\omega_m \Delta t$ -thus the quickness. The values considered for ω_m were: ωm=1.81 rad/s in hover helicopter mode, ωm=2.6 rad/s at 60kts helicopter mode, ωm=3.6 rad/s at 120kts 60deg conversion mode. Fig. 9 underlines an important aspect of the link between handling and agility: the higher the quickness, the higher the agility but when this agility is connected to Fig. 8 one may see that at the highest agility poor Level 2 ratings are awarded, i.e. the performance degrades rather than improves. This shows that actually, in practice, the closer the pilot flies to the performance boundary the more difficult it becomes to control the maneuver and thus the higher the agility the worse the HQs. In conclusion, handling qualities considerations do limit the agility.

3.2. Control Anticipation Parameter

The discussion on the experimental metrics suggests that on the one side the best metric for pitch motion agility is the peak pitch acceleration and on the other side the best metric for determining the aircraft flight path bending capability is the peak load factor. In order to capture both the transients of the maneuver and the precision achieved in flight path control, MIL standard on fixed-wing aircraft (MIL-HDBK-1797, 1997) introduced as metric a combination between these two metrics, the so-called 'control anticipation parameter CAP'. CAP is defined as the ratio of the initial pitch acceleration to the steady state load factor (effectively pitch rate) after a step-type control input:

$$CAP \overset{def}{=} \frac{\dot{q}(0)}{n_z^{qs}}$$

MIL standard defines CAP boundaries for fixed-wing aircraft. Fig. 10 presents the agility of the FXV-15 CAP as a function of speed (60 kts, 120 kts and 200 kts) in the MIL boundaries. Looking at this figure one can see the tiltrotor meets Level 1 MIL performance and some degradation to Level 2 is seen when flying at high speeds in airplane mode.

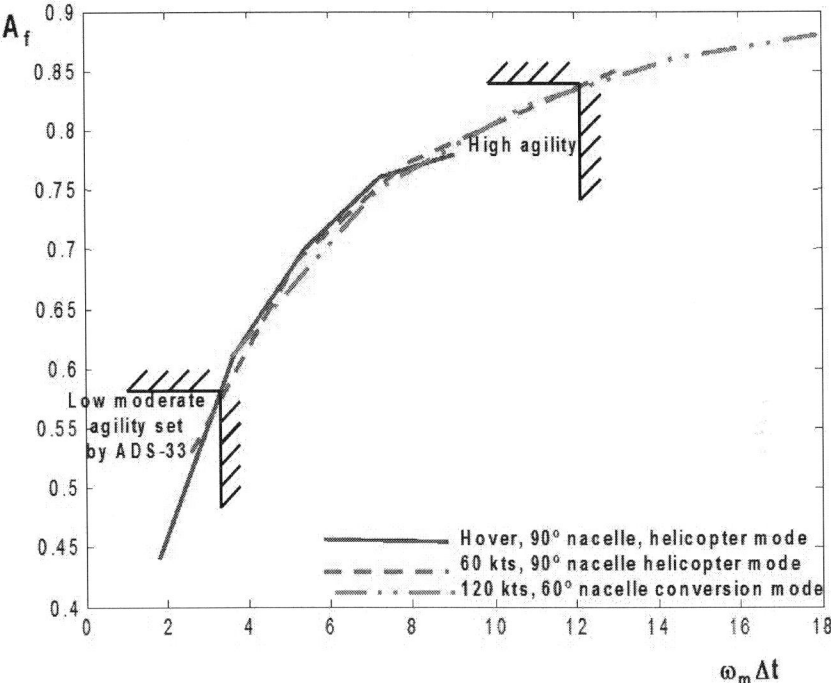

Figure 9. Agility factor as a function of quickness

3.3. Rate Pitch Quickness

For helicopters a similar metric to CAP was introduced by (Padfield & Hodkinson, 1993). This metric was called 'rate pitch quickness' and was defined as the ratio of pitch acceleration to the pitch angle change:

$$\dot{Q}_\theta \overset{def}{=} \frac{\dot{q}_{pk}}{\Delta\theta} \left(sec^{-2}\right)$$

and can be used to determine upper limits to agility based on maneuver acceleration. Fig. 11 plotted the rate quickness in the normalized form as a function of acceleration time constant $\omega_m t_{pk}$ (where t_{pk} is the time to peak acceleration).

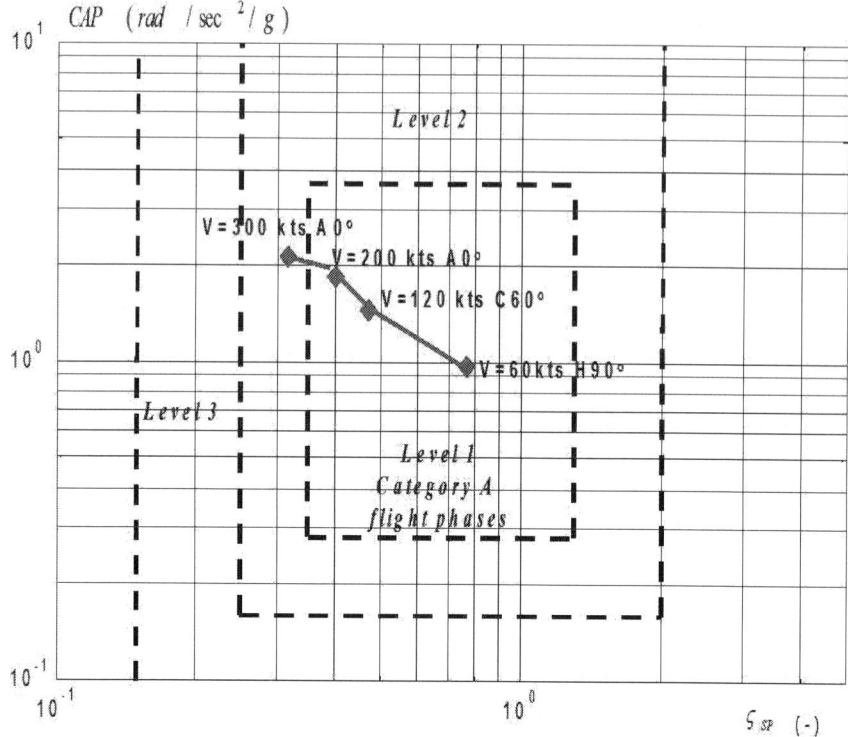

Figure 10. CAP boundaries for the tiltrotor

One can see that as the rate quickness increases the time to peak that rate is decreasing, so the agility is increasing. However, (Padfield & Hodkinson, 1993) commented that simply increasing the agility in terms of acceleration rates would lead to over-responsiveness and thus decreasing in operational capability since an over-responsive vehicle would not be controllable. In this sense, also CAP was quoted as an example of a criterion defining over-responsiveness. Unfortunately, there were no boundaries defined in this chart, although it was mentioned that intuitively there are likely to be upper and lower bounds for this metric. '*Hard and fast may be as unacceptable as soft and slow, both leading to low agility factors*' (Padfield & Hodkinson, 1993).

4. A RATIONAL DEVELOPMENT OF A MULTI-DISCIPLINARY APPROACH TO AGILITY

Combining equation (Eq. 3) for CAP with equation (Eq. 4) for rate quickness it follows that:

$$\dot{Q}_\theta = CAP \cdot \frac{n_z}{\Delta\theta}$$

Equation (5) gives the idea that rate quickness and CAP can be related to each other through a new metric which will be presented in the next paragraph.

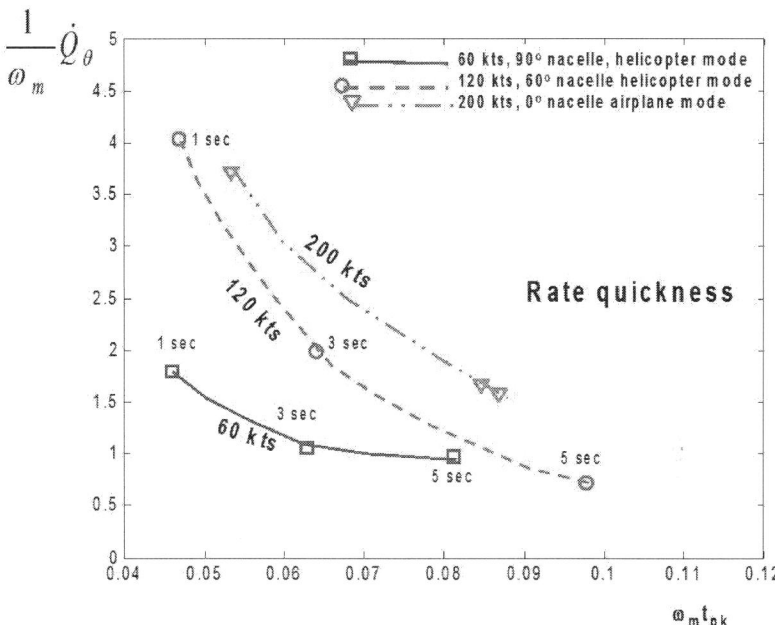

Figure 11. Rate quickness as a function of time peak acceleration

4.1. Agility Quickness Metric as a Measurement of Performance

As a potential successful metric for agility, (Pavel & Padfield, 2002) proposed a new metric for characterizing agility, the so-called 'agility quickness' defined as the ratio of peak quasi-steady normal acceleration $n_{z\,pk}^{qs}$ in g units corresponding to a step change in flight path angle $\Delta\gamma$:

$$Q_\gamma \overset{def}{=} \frac{n_{z\,pk}^{qs}}{\Delta\gamma} \quad \left(\frac{g's}{deg} \right)$$

Observe that the pitch angle from (Eq. 5) was substituted by the flight path angle, this has been done because actually during vertical axis maneuvering agility is more related to how quickly the flight path can be changed, the pilot being in reality more interested in the flight path angle change than in the pitch change. Furthermore, (Pavel & Padfield, 2003) proposed a Level 1/2 performance boundary for agility quickness by flying yo-yo maneuvers in the full motion simulator at the University of Liverpool the UH-60A model. Fig. 12 presents the example of tiltrotor on the agility quickness charts as determined in (Pavel & Padfield, 2003).

One can see that the tiltrotor is mostly at Level 2 performance in helicopter and airplane modes. (Pavel & Padfield, 2003) derived a relation between CAP and Q_γ and (Padfield & Meyer, 2003; Cameron & Padfield, 2010) connected CAP to other flying qualities parameters.

One of the reasons the attitude quickness criterion has gained large acceptance was due to its physical interpretation (in the limiting case gives the time constant of the aircraft as a function of the time constant of the maneuver). It can be demonstrated that agility quickness has also a physical interpretation, in the limiting case for small-amplitude maneuvers giving the heave damping, for large amplitudes giving the attitude quickness (Pavel & Padfield, 2002).

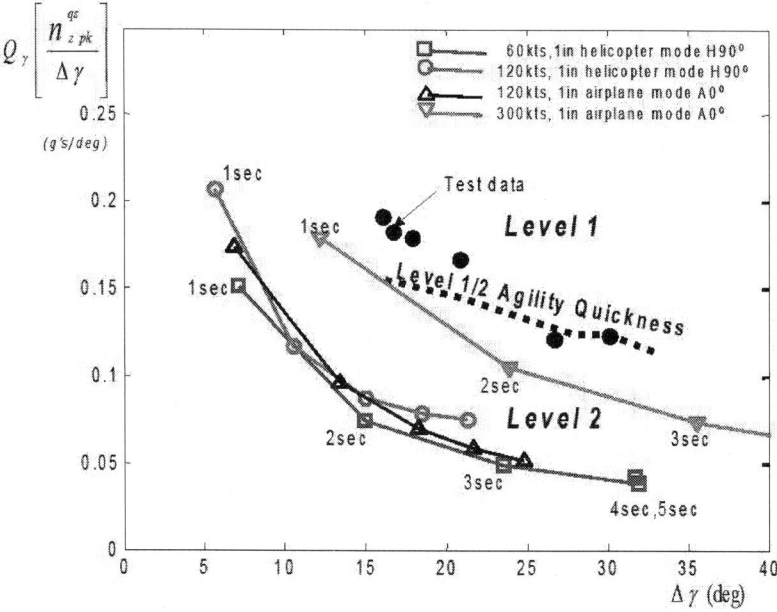

Figure 12. Tiltrotor on Agility quickness chart

4.2. Vibratory Quickness Metric as a Measurement of Vibratory Activity

The advantage of the agility quickness metric as above defined is that it could be linked to a complementary vibratory for structural alleviation. In this way, the designer is able to optimize in parallel both the performance and the vibratory loads in maneuvering flight. This novel approach may reform the methods presently used to accomplish agility of a new design as it is well known that, in practice, performance is itself often compromised by the need to control and minimize vibration.

(Pavel and Padfield, 2002) define an parallel metric, so-called 'vibratory load quickness' quantifying the buildup of loads in the rotor during maneuvering flight:

$$Q_l \overset{def}{=} \frac{F_{pk}^{vib}}{W\Delta\gamma}\left(\frac{1}{deg}\right); \quad Q_l \overset{def}{=} \frac{M_{pk}^{vib}}{\Delta\gamma}\left(\frac{lbf \cdot ft}{deg}\right)$$

Where F_{pk}^{vib}, M_{pk}^{vib} represent the peak amplitudes in the critical vibratory components for respectively hub shears and hub moments corresponding to a change $\Delta\gamma$ in flight path angle. The peak load amplitude can be calculated by using the FFT and time representations of the hub shears ($F_{x\ hub}$, $F_{y\ hub}$or $F_{z\ hub}$) and/or moments ($M_{x\ hub}$, $M_{y\ hub}$, $M_{z\ hub}$) during a maneuver flown and determining the critical loads (i.e. the loads achieving the highest peaks) during the maneuver.

For example, for the pull-up maneuver flown with the tiltrotor it was found that when flying in helicopter mode at 60 and 120 kts the critical loads developed were the 3/rev vibratory component of the hub vertical shear, the 1/rev and 2/rev components of the blade inplane moment and the 1/rev component of the blade flapping moment. When flying in the airplane mode at 120 and 300 kts, the critical loads measured by the FXV-15 were the 2/rev and 3/rev components of the vertical shear, the 1/rev and 2/rev components of the blade inplane moment.

Fig. 13 presents the equivalent vibratory quickness charts for the critical 3/rev component of the hub vertical shear in helicopter mode and 2/rev and 3/rev components of hub vertical shear in airplane mode when flying respectively at 60 and 120 kts and 120 and 300 kts, giving an 1 in input in longitudinal cyclic and varying the pulse duration (1 to 5 seconds). Each of these vibratory chart can be associated with an equivalent agility quickness chart as plotted in Fig. 12.

For helicopter mode it was plotted a presumable vibratory quickness boundary as derived in (Pavel & Padfield, 2003) when flying piloted yo-yo's in the simulator with the UH-60A helicopter. It was there showed that actually, increasing the flight path change enables the pilot to pull more g's of course but also increases the vibratory activity in the rotor. The presumable vibratory boundary would mean then that the structural designer would aim for a boundary that slope in a similar direction to the agility quickness boundary as plotted in Fig. 13. Looking at this figure one may see that as the pulse duration is increasing the vibratory quickness is decreasing. This is because the vibratory activity in the hub reaches its absolute peak rather quickly, depending mainly on the initial velocity, the input amplitude (which is a measure of the level of aggressiveness in executing the maneuver) and not on the pulse duration. However, for very aggressive maneuvering, (Pavel & Padfield, 2002) showed that it might appear the situation in which, increasing the flight path change enabled the pilot to pull more g's (so, increased performance) but also increased the vibratory activity in the rotor. The goal of the structural designer would be then to alleviate these high peak loads to lower levels and reduce the sensitivity of the vibratory loads to flight path angle.

Fig. 14 presents other critical load for the tiltrotor, namely the blade inplane moment. Both, for the helicopter and airplane mode, the critical components measured during the simulation of the pull-up maneuvers with the FXV-15 were corresponding to the 1/rev and 2/rev vibratory components.

Looking at Fig. 14 one may see again that the vibratory quickness parameter Q_l varies approximatively inversely with the flight path change. This means that the vibratory activity in the blade in the inplane direction reaches its absolute peak rather quickly, depending on the aggressiveness of the pulse (pulse amplitude) and not on pulse duration.

5. CONCLUSION

This chapter has presented a rational development of key metrics and criteria used to design for airframe agility. Concentrating on the agility in the pitch axis (vertical-plane maneuvers) and taking as case study the unique example of a tiltrotor aircraft, the chapter demonstrated how, starting from the more traditional way of quantifying the agility, the designer can develop new agility metrics that do a better job of capturing the aircraft transient motion characteristics. This chapter discussed on the many correspondences that exist between the study of agility and the study of flying qualities, emphasizing the fact that flying qualities do limit agility. In this sense, providing the pilot with a high level of maneuverability, without a high level of controllability, will reduce agility. However, especially for the tiltrotor, higher agility cannot be achieved without increasing the vibratory loads on the rotor, which means also an increasing in pilot workload. The chapter proposed therefore a unique approach by presenting a first set of complementary metrics capable of being applied to both agility and structural load analysis.

Subsequent phases of this study will include the expansion of this new approach and studying it in axial, turning (horizontal-plane maneuvers) and roll (torsion) axes. It is hoped that in this way a more unified set of design criteria will be developed enhancing multi-disciplinary design optimization.

6. ACKNOWLEDGEMENTS

The subject of this research was initiated as collaboration between The University of Liverpool and Delft University of Technology. Sincere thanks for all support given by the University of Liverpool and Prof. Padfield for using FLIGHTLAB FXV-15 model.

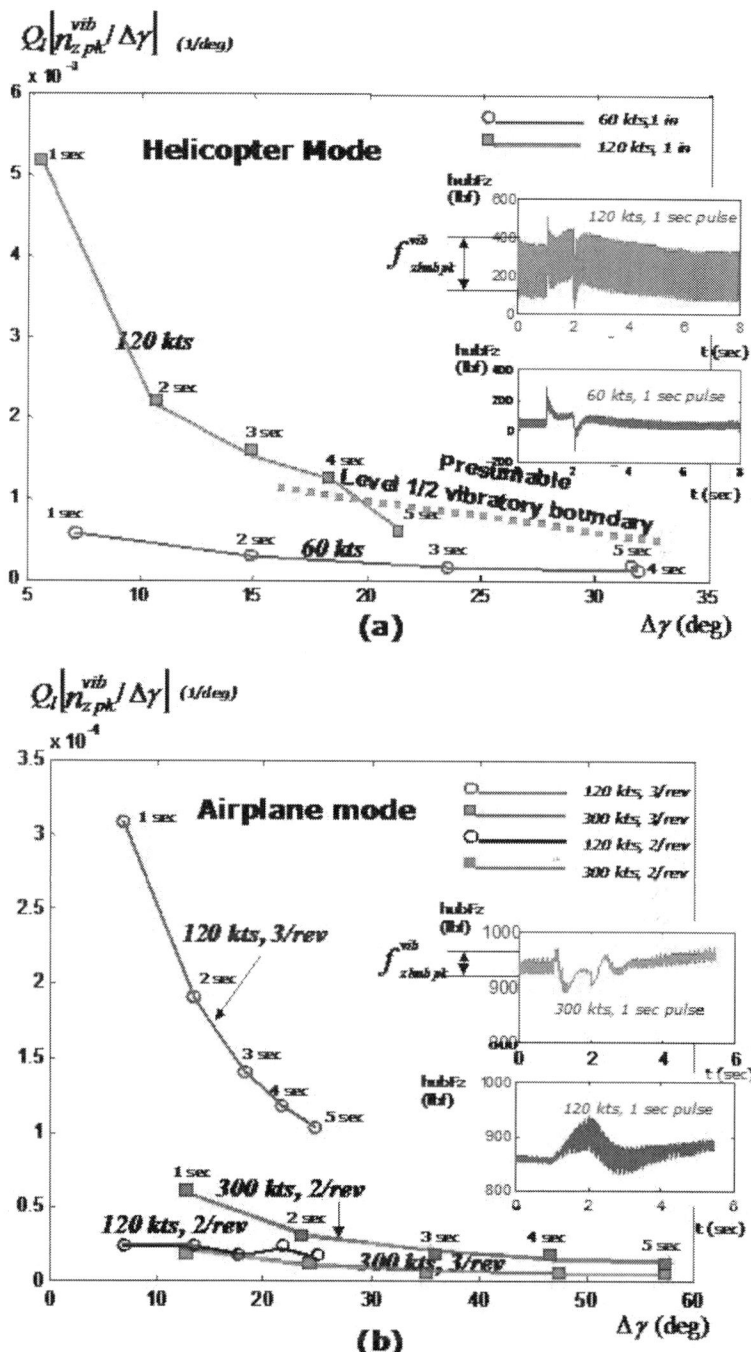

Figure 13. Vibratory quickness envelopes for the critical components of the hub vertical shear during a pull-up maneuver.

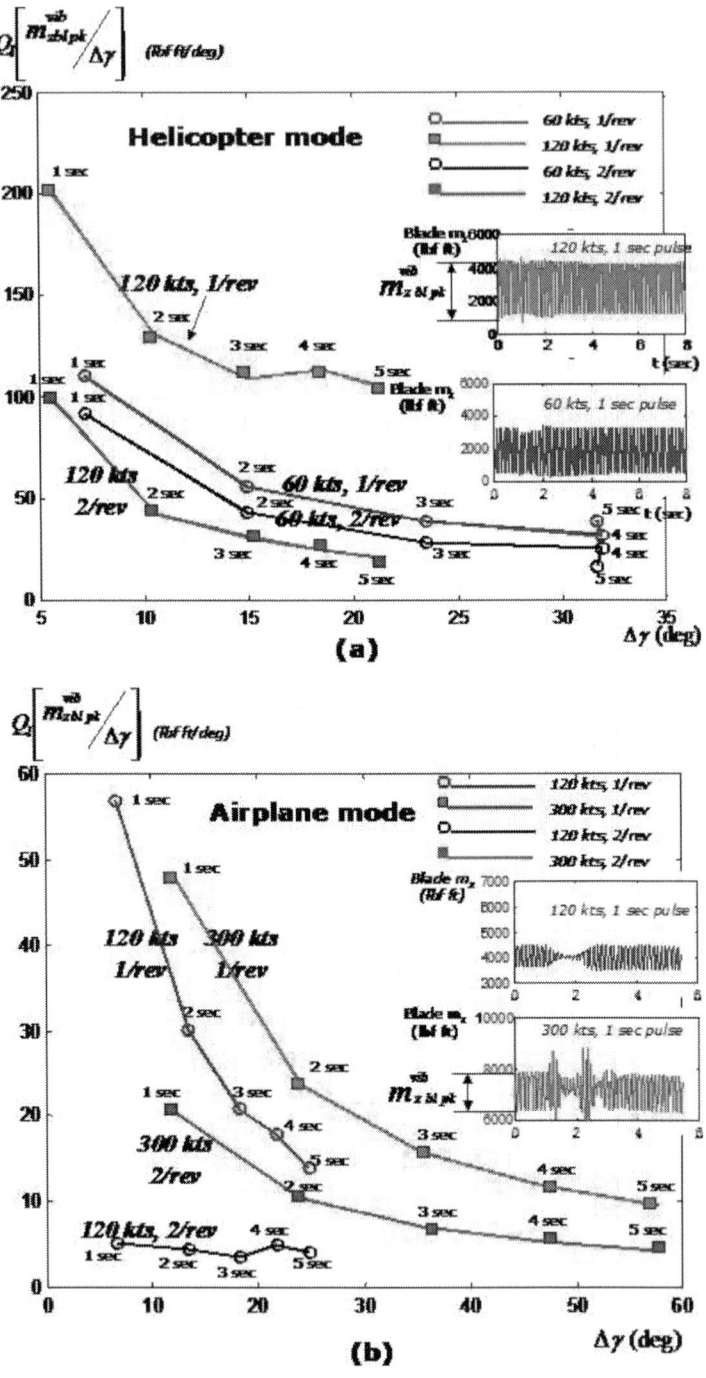

Figure 14. Vibratory quickness envelopes for the critical components of the hub vertical shear during a pull-up maneuver

NOTES

1) Level 1 HQs means that the rotorcraft is satisfactory without improvements required from the pilot's perspective. Level 2 HQs means that pilots can achieve adequate performance, but with compensation; at the extreme of Level 2, the mission is flyable, but pilots have little capacity for other duties and can not sustain flying for longer periods without the danger of pilot error. Level 3 HQs is unacceptable, it describes rotorcraft behaviour in extreme situations, like the loss of critical flight control systems.

2) The Cooper-Harper rating scale runs from 1 to 10. Rates from 1 to 3 1/2 correspond to Level 1 HQs; rates from 31/2 to 61/2 correspond to Level 2 HQs, and rates from 61/2 up to 10 correspond to Level 3 HQs.

REFERENCES

1. ADS-332000Aeronautical Design Standard-33E-PRF, Performance Specification, Handling Qualities Requirements for Military Rotorcraft, US Army AMCOM, Redstone, Alabama

2. Neil. Cameron, Gareth. D. Padfield, 2010Tilt Rotor Pitch/Flight-Path Handling QualitiesJ. Of the American Helicopter Society, 55(4), 042008042001to 042008-13

3. R. Celi, 1991Optimum Aeroelastic Design of Helicopter Rotors for Longitudinal Handling Qualities ImprovementJ. of Aircraft, 28(1), 4957

4. R. Celi, 1999Recent Applications of Design Optimization to Rotorcraft- A Survey, J. of Aircraft, 36 (1), pp.

5. R. Celi, 2000Optimization Based Inverse Simulation of a Helicopter Slalom Maneuver, J. of Guidance, Control and Dynamics, 23(2), 289297

6. G. E. Cooper, R. P. Harper, Jr , 1969The use of Pilot Ratings in the Evaluation of Aircraft Handling Qualities, NASA TM D-5133

7. A. Datta, I. Chopra, 2002Validation and Understanding of UH-60A Vibration Loads in Steady Flight, 58th American Helicopter Society Conference, 1114June, Montreal

8. D. Fusato, R. Celi, 2001Formulation of a Design-Oriented Helicopter Flight Dynamic Simulation Programth American Helicopter Society Forum, Alexandria, VA

9. D. Fusato, R. Celi, 2002Design Sensitivity Analysis for ADS-33 Quickness Criteria and Maneuver Loadsth American Helicopter Society Forum, 1114June, Montreal

10. D. Fusato, R. Celi, 2002Multidisciplinary Design Optimization for Aeromechanics and Handling Qualitiesth European Rotorcraft Forum, 1720September, Bristol, UK

11. R. Ganduli, 2004Survey of Recent Developments in Rotorcraft Design Optimization", J. of Aircraft, 41(3), 493510

12. R. E. Hansford, J. Vorwald, 1996Dynamic Workshop on Rotor Vibratory Loads. 52nd Forum of the American Helicopter Society, June 46Washington D.C.

13. Iseler, Laura; De Maio, Joe & Rutkowski, Michael2002Analysis of US Civil Rotorcraft Accidents by Cost & Injury (1990-1996), Army/NASA Rotorcraft Division Ames Research Center, USA, NASA-TM-2002209615

14. D. L. Key, 1982The status of Military Handling Qualities Criteria, AGARD CP-333

15. H. G. Kolwey, 1996The New ADS-33 Process: Cautions for ImplementationTechnical Note, J. of the American Helicopter Society, 41(1), 36

16. R. M. Kufeld, W. G. Bousman, 1995High Load Conditions measured on a UH-60A in Manoeuvring Flight, 51st Forum of the American Helicopter Society, May 911Texas

17. K. Mc Kay, 1994Operational Agility: An Overview of AGARD Flight Mechanics Working Group 19, In: Technologies for Highly Manoervrable Aircraft, AGARD-CP-548

18. Binoy. Manimala, Mauro. . Naddei, Philippe. Rollet, 2003Load Alleviation in Tiltrotor Aircraft through Active Control; Modelling and Control Concepts, 59th American Helicopter Society Conference, Phoenix, Arizona, May 6-8, 2003

19. C. J. Mazza, 1990Agility: A Rational Development of Fundamental metrics and their relationship to Flying Qualities. Proceedings of Flight Mechanics Panel Symposium, Quebec City, Canada, October 15-18, 1990, Published as AGARD-CP-508

20. MIL-H-8501A1962Military Specification, General Requirements for Helicopter Flying and Ground Handling Qualities MIL-H-8501A, Sept. 1961, through Amendment 1, April 1962 (superseding MIL-H-8501, Nov. 1952).

21. MIL-HDBK-17971997Flying Qualities of Piloted Aircraft, MIL-HDBK-1797, US Department of Defense, USA

22. D. G. Mitchell, et al.2004Evolution, Revolution and Challenges of Handling Qualities, J. of Guidance, Control and Dynamics, 27(1), 1228

23. P. Murphy, M. Bailey, Ostroff, 1991A. Candidate Control Design Metrics for an Agile Fighter, NASA-TM-4238

24. Gareth. D. Padfield, 1998The Making of Helicopter Flying Qualities; A Requirements Perspective. The Aeronautical Journal of Royal Aeronautical Society, 1021018409437

25. G. D. Padfield, 2007Helicopter Flight Dynamics, Blackwell Science LTD

26. G. D. Padfield, 2008Capturing requirements for tiltrotor handling qualities-case studies in virtual engineering, The Aeronautical Journal, 112Aug. 2008, 433448

27. Gareth. D. Padfield, John. Hodkinson, The Influence of Flying Qualities on Operational Agility", AGARD Flight Mechanics Panel Symposium Technologies for Highly Manoeuvrable Aircraft, Annapolis, Maryland, Oct. 1993Published as AGARD-CP-548

28. Marilena. D. Pavel, Gareth. D. Padfield, 2002Defining Consistent ADS-33Metrics for Agility Enhancement and Structural Loads Alleviation, 58th American Helicopter Society Conference, Montreal, Canada, 11-14 June 2002

29. Marilena. D. Pavel, Gareth. D. Padfield, 2003Progress in the Development of Complementary Handling and Loading Metrics for ADS-33 Maneuvers", 59th American Helicopter Society Conference, Phoenix, Arizona, 68May 2003

30. Gareth. D. Padfield, Michael. Meyer, 2003Progress in Civil Tilt-Rotor Handling Qualities", 29th European Rotorcraft Forum, Friedrichshafen, Germany, 1517October 2003.

31. V. Sahasrabudhe, R. Celi, 1997Efficient Treatment of Moderate Amplitude Constraints for Helicopter Handling Qualities Design Optimization, J. of Aircraft, 34(6), 730739

32. M. B. Tischler, et.al, 1997CONDUIT- A New Multidisciplinary Integration Environment for Flight Control Development, NASA TM-112203

33. M. B. Tischler, J. N. Williams, Ham, J. A. , 1996Comments on The New ADS-33 Process: Cautions for Implementation, Technical Note, J. of the American Helicopter Society, 41194195

34. Veilig Vliegen magazine2003Hoe de zaken liepen in 2002' ('How the things evolved in 2002'), May 2003, 133138

35. WAAS2002World Aircraft Accident Summary- 19902001Index-Cap 479, Civil Aviation Authority, Airclaims Limited

CHAPTER 2

Plasma Flow Control

Ying-hong Li[1], Yun Wu[1], Hui-min Song[1], Hua Liang[1] and Min Jia[1]

[1] Air Force Engineering University of China, People's Republic of China

1. INTRODUCTION

Plasma flow control, based on the plasma aerodynamic actuation, is a novel active flow control technique to improve aircrafts' aerodynamic characteristics and propulsion efficiency. Plasma flow control has drawn considerable attention and been used in boundary layer acceleration, airfoil separation control, forebody separation control, turbine blade separation control, axial compressor stability extension, heat transfer and high speed jet control. Plasma aerodynamic actuator has many advantageous features including robustness, simplicity, low power consumption and ability for real-time control at high frequency.

In this chapter, the principle of plasma aerodynamic actuation and its application in subsonic and supersonic flow control was summarized. In order to better understand the underlying physical mechanism of plasma flow control and optimize the geometric configuration of the actuator, the characteristics of the plasma aerodynamic actuation, including gas temperature, electron density and temperature, induced body force, velocity and vorticity were investigated. Both wind tunnel experiments and computations were performed to investigate the flow control capability of plasma aerodynamic actuation for airfoil separation control, corner separation control, axial compressor stability extension and shock wave control.

2. PLASMA AERODYNAMIC ACTUATION

The mechanism for plasma flow control can be summarized as momentum effect, shock effect, and chemical effect. Momentum effect induces near-surface flow velocity of 1-6 m/s. Shock effect induces local gas pressure or temperature rise near the electrode. Chemical effect adds new particles, such as ions, electrons, and excited particles, into the flow field. Surface dielectric barrier discharge plasma aerodynamic actuation is typically used in subsonic plasma flow control, while arc discharge plasma aerodynamic actuation is mostly used in supersonic and hypersonic plasma flow control. Plasma aerodynamic actuation generated by corona discharge, radiofrequency discharge, microwave discharge are also widely investigated.

2.1. Surface Dielectric Barrier Discharge Plasma Aerodynamic Actuation

A schematic of the dielectric barrier discharge plasma aerodynamic actuator is shown in Fig. 1. Most of the previous experimental and simulation studies mainly focused on the asymmetric dielectric barrier discharge plasma aerodynamic actuation excited by sinusoidal or sawtooth voltage waveforms at amplitude of 2-20 kV and frequency of 1-100 kHz, which can be named as the microsecond discharge plasma aerodynamic actuation. In recent years, the plasma aerodynamic actuation generated by nanosecond pulsed dielectric barrier discharge has become a hot topic.

Synthetic measurements and analysis of the characteristics of the asymmetric surface dielectric barrier discharge plasma aerodynamic actuation were performed. Fig. 1 shows the schematic diagram of the asymmetric surface dielectric barrier discharge plasma aerodynamic actuator. The dielectric layer used is a RO4350B (Rogers Corporation) plate with a relative permittivity constant of 3.48. The electrodes are made of copper, covered with a thin layer of lead-tin film. The plasma aerodynamic actuator is driven by a high frequency high voltage power supply (CTP-2000K, Suman Electronics). The output voltage range and frequency range of the power supply are 0~40 kV and 6~40 kHz respectively.

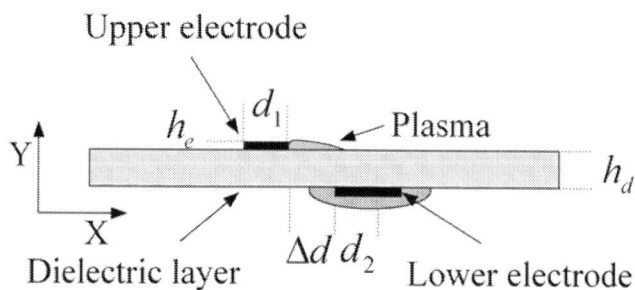

Figure 1. A schematic of the asymmetric surface dielectric barrier discharge plasma aerodynamic actuator

The experimental arrangement is shown in Fig. 2. Synthetic measurements are applied to measure both the plasma and the induced flow characteristics of the plasma aerodynamic actuation. The applied voltage and the total discharge current are measured by a high voltage probe (P6015A, Tektronix Inc.) and a current probe (TCP312+TCPA300, Tektronix Inc.). Signals are recorded on an oscilloscope (DPO4104, Tektronix Inc.). Optical emission spectroscopy is measured by a 0.5m monochrometer (TRIAX550, Jobin Yvon Inc.) through an optical fiber collector, 1 cm over the surface of the dielectric layer. The detector of the monochrometer is a set of photon counting system (Model 76915, Oriel

Inc.). The slit width and the calibrated resolution in this case are 20μm and 0.05nm. The emission intensity is averaged temporally and spatially. Body force induced by the plasma aerodynamic actuation is measured by a electronic balance (ABS 204-S, Mettler Toledo Inc.). Velocity and vorticity induced by the plasma aerodynamic actuation is measured by Particle Image Velocimetry (Lavision). The air is seeded by vaporization of mineral oil with a mean size of about 0.3 μm.

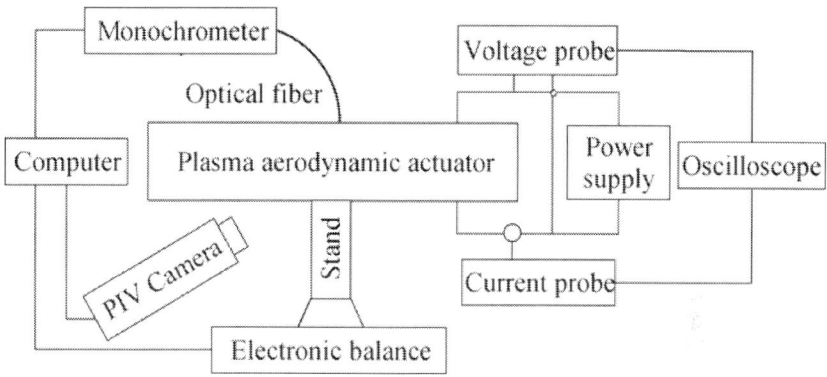

Figure 2. Schematic diagram of the experimental arrangement

2.1.1. Characteristics of the Plasma

Optical emission spectroscopy(OES), as a simple and non-intrusive diagnostic method, is playing a very important role in providing qualitative and quantitative information of plasma properties atmospheric pressure air plasmas. Several plasma parameters, including gas temperature, vibrational temperature, electron temperature and density, have been successfully characterized by OES method in this study. The major spectra come from the second positive system (SPS) of $N_2(C^3\Pi_u \rightarrow B^3\Pi_g)$, and the first negative system (FNS) of $N_2^+(B^2\Sigma_u^+ \rightarrow X^2\Sigma_g^+)$. The relative concentration of $N_2(C^3\Pi_u)$ is much greater than that of $N_2^+(B^2\Sigma_u^+)$. The emitting species of $N_2(C^3\Pi_u)$ and $N_2^+(B^2\Sigma_u^+)$ mainly come from the following excitation processes from the ground state of N_2.

$$e + N_2(X) \rightarrow e + N_2(C) \quad (E_{th} = 11 \text{ eV})$$

$$e + N_2(X) \rightarrow e + e + N_2^+(B) \quad (E_{th} = 19 \text{ eV})$$

Other main dynamics/kinetics processes in this surface discharge include[24-28]:

$$e + N_2(X, v) \rightarrow e + N_2(X, \omega)$$

$$N_2(X, v) + N_2(X, \omega) \rightarrow N_2(X, v - 1) + N_2(X, \omega + 1)$$

$$N_2(X, v) + N_2 \rightarrow N_2(X, v - 1) + N_2$$

$$N_2(C) + N_2, O_2 \rightarrow N_2, O_2 + product$$

$$N_2(C) \rightarrow N_2(B) + h\nu$$

$$N_2^+(B) + N_2, O_2 \rightarrow N_2, O_2 + product$$

$$N_2^+(B) \rightarrow N_2^+(X) + h\nu$$

In atmospheric discharge condition, the gas temperature can be estimated by the rotational temperature of a molecular because the rotational energy levels are closely spaced(10^{-3} eV) and allow for rapid energy transfer between the two energy modes. The population distribution in the rotational energy level of N_2 is also believed to fit Boltzmann distribution due to the dense collisions between molecules. Here, we obtained the rotational temperature of N_2 through fitting the N_2 second positive system band from 378 nm to 381 nm for the 380.5 nm (v'=0, v''=2) N_2 line. By assuming a rotational temperature and considering the dipole radiation probability and the response function of the monochrometer, one can calculate the profile of a certain emission band. The actual rotational temperature(T_r) can be determined through comparing the experimental measurement and theoretical calculation.

In this surface discharge, the vibrational excitation of nitrogen seems to be the most important. $N_2(C^3\Pi_u)$, which is not a metastable state, is generated from the ground-state electron impact excitation and the cascading effect is not important for the $N_2(C^3\Pi_u)$ state population. Therefore, the vibrational temperature can be determined according to the ratio of two lines in the $N_2(C)$ second positive system. The spectra lines at 371.1 nm and 380.5 nm are selected to calculate the vibrational temperature (T_v), as follows:

$$\frac{I_{371.1nm}}{I_{380.5nm}} = 1.1384 \cdot \exp(-0.4952 / T_v)$$

When the applied voltage amplitude and driving frequency are 10kV (peak to peak) and 23kHz respectively, the rotational and vibrational temperatures of $N_2(C^3\Pi_u)$ are 500 K and 0.22 eV respectively. The rotational temperature is insensitive to the applied voltage and the driving frequency. The vibrational temperature also shows minor dependence on the applied voltage and the driving frequency, as shown in Fig. 3.

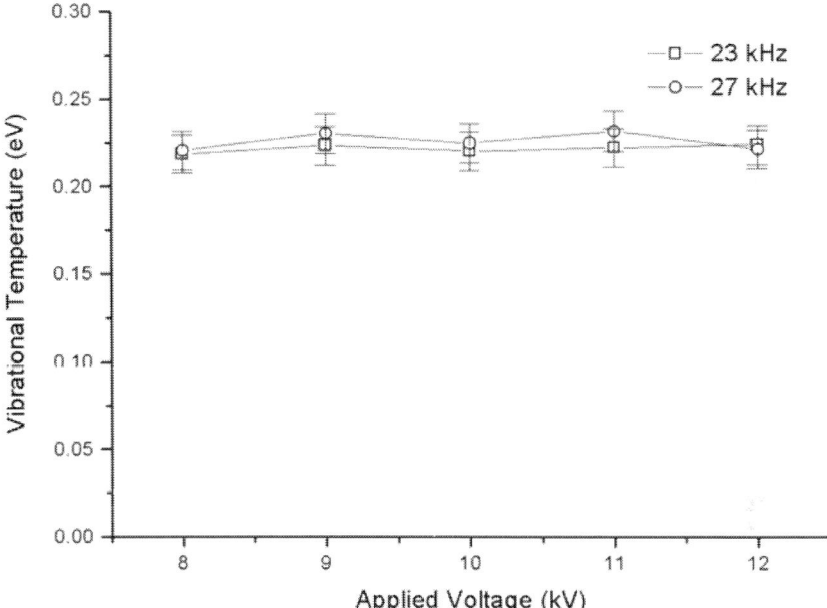

Figure 3. Vibrational temperature versus the applied voltage

The 0D rate balance equation for the concentration, n_v, of molecules in the vth vibrational level of the ground state is

$$\frac{dn_v}{dt} = n_e \sum_{w \neq v} n_w C_w^v - n_e n_v \sum_{w \neq v} C_v^w + n_{v-1} \sum_w n_{w+1} Q_{v,v-1}^{w+1,w}$$

$$+ n_{v+1} \sum_w n_w Q_{v+1,v}^{w,w+1} - n_v \left(\sum_w n_{w+1} Q_{v,v+1}^{w+1,w} + \sum_w n_w Q_{v,v-1}^{w,w+1} \right)$$

$$+ n_{v+1} \sum_w n_w Q_{v+1}^v - n_v \sum_w n_w Q_v^{v-1}$$

Here Q and C are rate coefficients of collision.

The rate balance equation for each excited state is, for state $N_2(C)$,

$$\frac{dn_C}{dt} = n_e n_{N_2} Q_C - A_C n_C - n_C n_{N_2} Q_{N_2} - n_C n_{O_2} Q_{O_2}$$

for state $N_2^+(B)$,

$$\frac{dn_{B+}}{dt} = n_e n_{N_2} Q_{B+} - A_{B+} n_{B+} - n_{B+} n_{N_2} Q_{N_2} - n_{B+} n_{O_2} Q_{O_2}$$

Here A is the Einstein coefficient.

The intensity ratio of optical emission line from FNS to SPS is strongly dependent on the electron temperature or the energy distribution function for the electrons whose energy is more than 11 eV, because the excited thresholds of the two processes have a difference of about 8 eV. According toequation (12) and (13), electron temperature(T_e) can be calculated using the intensity ratio of 391.4 nm and 380.5 nm:

$$\frac{I_{391.4nm}}{I_{380.5nm}} = K_0 \cdot (T_e)^{C_0} \cdot \exp(-\frac{E_0}{T_e})$$

As shown in equation (12) and (13), the relative intensity of different vibrational levels of $N_2(X)$ is mainly determined by the electron density. According to Frank-Condon principle, the vibrational energy level distribution of $N_2(C)$ can be revealed using the changes in vibrational energy level distribution of $N_2(X)$. Therefore, the ionization rate(n) can be calculated using the intensity ratio of 371.1 nm and 380.5 nm:

$$\frac{I_{371.1nm}}{I_{380.5nm}} = C_0 + C_1 \log_{10}(n) + C_2 (\log_{10} n)^2$$

The electron temperature and electron density are 1.63 eV and 1.1×10^{11}cm^{-3} respectively with the applied voltage of 10 kV. The driving frequency is 23 kHz. At one atmospheric pressure, electron temperature is usually 1~2 eV because of the frequent collisions between electrons and molecules. The variations of average electron temperature and density with the applied voltage are shown in Fig. 4. It can be concluded that the average electron density and temperature have minor dependence on the applied voltage and its frequency. The electron temperature of a dielectric barrier discharge is strongly affected by the gas pressure because the collisionless free path of the electrons mainly determines the energy obtained by the electron from the electric field. The frequent collisions between electrons and molecules govern the discharge process at one atmosphere.

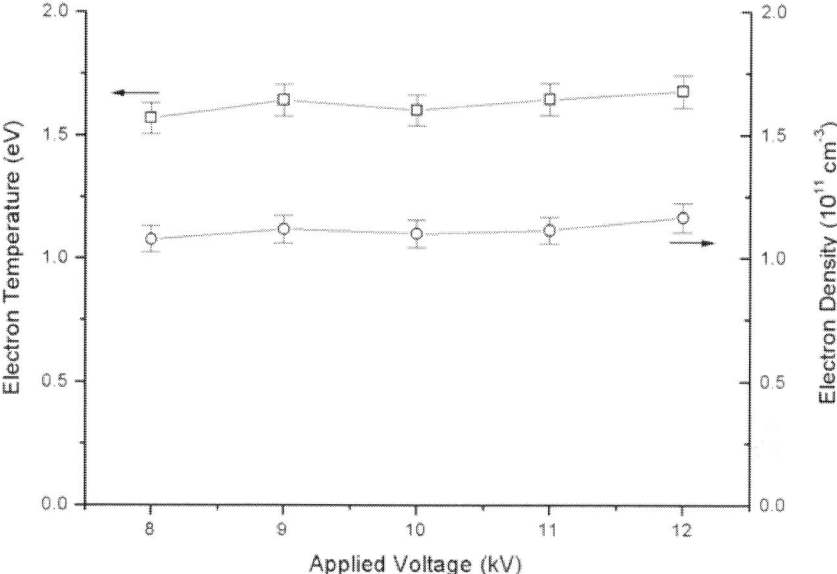

Figure 4. Electron temperature and density versus the applied voltage

2.1.2 Characteristics of The Induced Flow

The induced body force in X direction versus the applied voltage is shown in Fig. 5. The driving frequency is 23 kHz. When the applied voltage increases from 8 kV to 12 kV, the body force increases from 11 mN to 65 mN.

Starting vortex and directed wall jet induced by the steady plasma aerodynamic actuation in static air are shown in Fig. 6. The starting vortex exists for about 1s and then directed wall jet is formed, about 70mm downstream of the upper electrode. The induced flow velocity is 3 m/s when the applied voltage and the driving frequency are 10 kV and 23 kHz respectively.

When the unsteady plasma aerodynamic actuation is on, the starting vortex is stronger and exists for a longer time, which is about 4m/s. Then the directed wall jet along with many small vortexes is also formed, about 50mm downstream of the upper electrode, as shown in Fig. 7. The duty cycle and excitation frequency are 70% and 190Hz respectively. The unsteady plasma aerodynamic actuation induces much more vortex into the flow field.

When the voltage waveform is a nanosecond pulse, not the sinusoidal pulse, the induced flow direction changes remarkably. The induced flow direction by nanosecond discharge plasma aerodynamic actuation is not parallel, but vertical to the dielectric layer surface (see Fig. 8). The voltage pulse has a full wave at half maximum (FWHM) of 190 ns and a rise time of 450 ns. The peak voltage and frequency are 10 kV and 1 kHz, respectively.

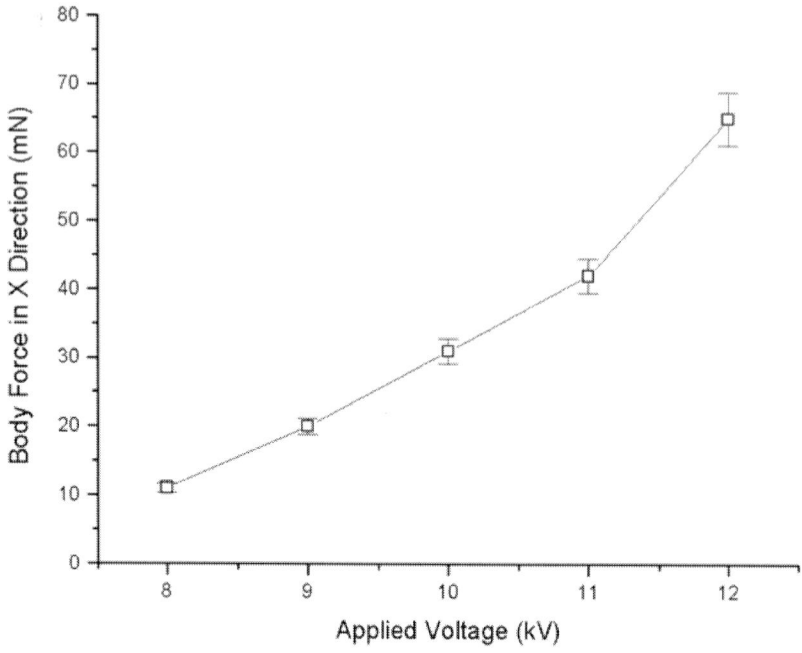

Figure 5. Body force induced by the plasma aerodynamic actuation

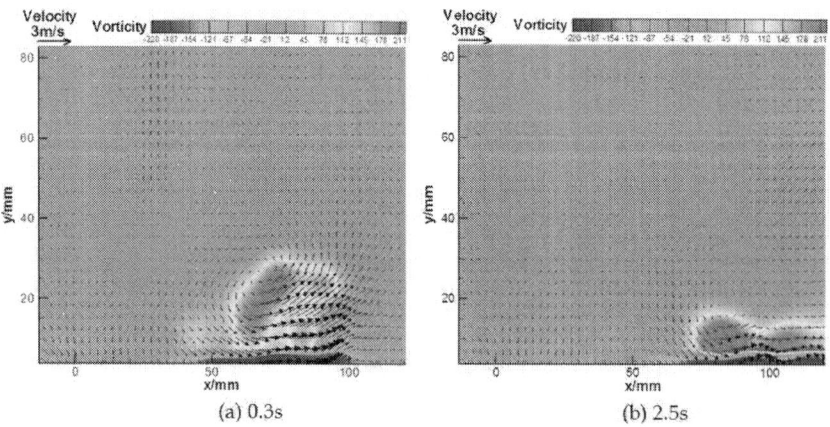

(a) 0.3s   (b) 2.5s

Figure 6. Velocity and vorticity of starting vortex induced by steady plasma aerodynamic actuation

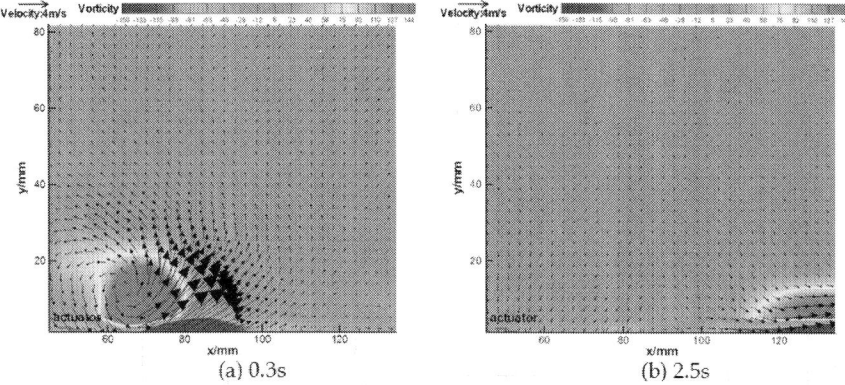

Figure 7. Velocity and vorticity of starting vortex induced by unsteady plasma aerodynamic actuation

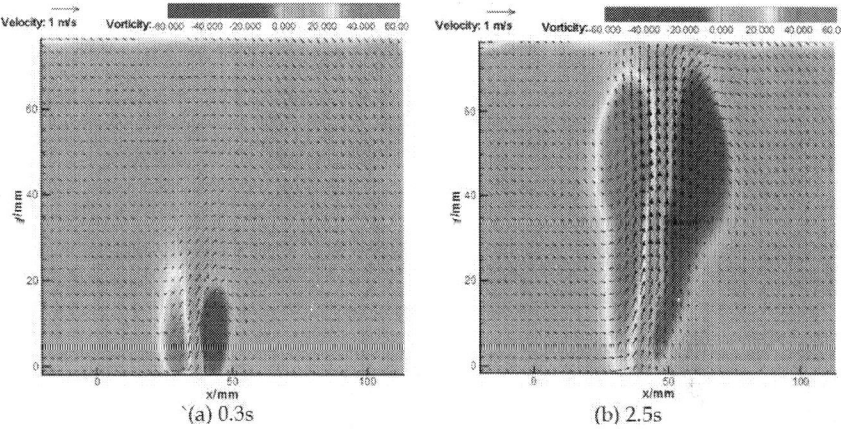

Figure 8. Velocity and vorticity of starting vortex induced by nanosecond discharge plasma aerodynamic actuation

2.2. Arc Discharge Plasma Aerodynamic Actuation

A schematic of the arc discharge plasma aerodynamic actuator is shown in Fig. 9. A dc surface arc discharge is generated between the anode and the cathode. In order to generate a nonuniform electric field that can reduce gas breakdown voltage, a cylindrical structure is designed for the electrodes. The electrodes are oriented in the spanwise direction, which means that the flow direction is perpendicular to the discharge current direction.

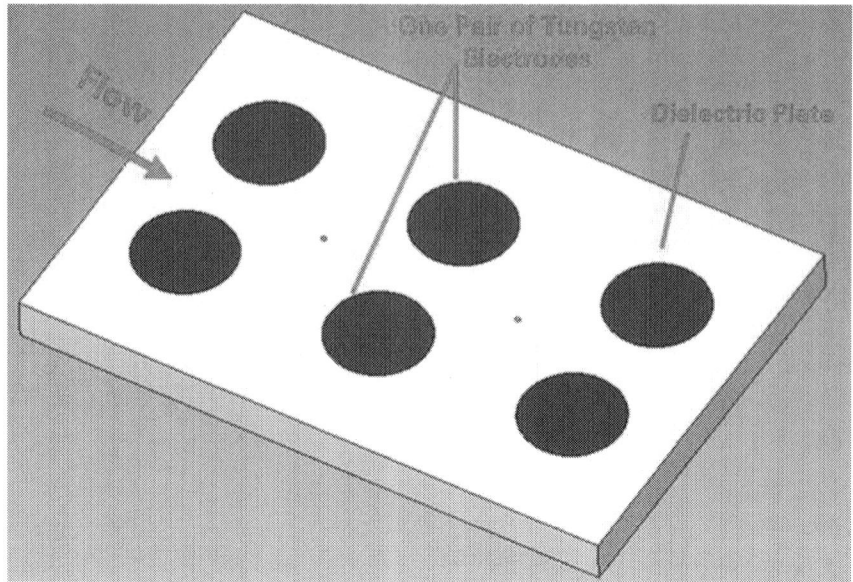

Figure 9. A schematic of the arc discharge plasma aerodynamic actuator

The power supply consists of a high-voltage pulse circuit and a high-voltage dc circuit. The output voltage of the pulse circuit can reach 90 kV, which is used for electrical breakdown of the gas. The dc circuit is the 3 kV-4 kW power source, which is used to ignite the arc discharge. The plasma aerodynamic actuator consists of graphite electrodes and boron-nitride (BN) ceramic dielectric material. Three pairs of graphite electrodes are designed with the cathode-anode interval of 5mm and the individual electrode is designed as a cylindrical structure which is embedded in the BN ceramic. The arc discharge voltage and current are monitored by a voltage probe (P6015A, Tektronix Inc.) and a current probe with a signal amplifier (TCP312+TCPA300, Tektronix Inc.), respectively. The two signals are measured by a four-channel digital oscilloscope (DPO4104, Tektronix Inc.).

The voltage, current and power measurements are shown in Fig. 10. The gas breakdown voltage between the graphite electrodes is about 2 kV and when the input voltage provided by the power supply exceeds this value, arc discharge happens. At the instant of gas breakdown, voltage decreases from 2 kV to about 300 V and current increases to about 1 A. The discharge power is calculated as 300 W. Then the voltage holds at 300 V, but the current decreases gradually. After about 0.5 s, the current sustains at about 440 mA and the discharge power holds at about 130 W. The arc discharge can be separated into two phases, which correspond to the strong pulsed breakdown process and the steady discharge process, respectively.

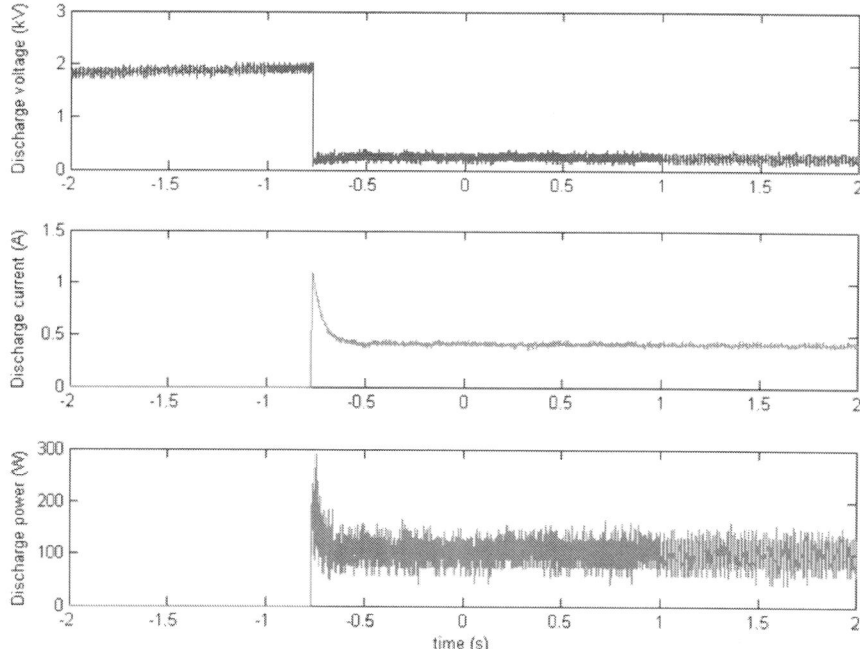

Figure 10. Electrical characteristics of arc discharge

3. SUBSONIC PLASMA FLOW CONTROL

Surface dielectric barrier discharge was proved effective in subsonic plasma flow control. A great number of papers devoted to subsonic plasma flow control have appeared in the past ten years. The use of dielectric barrier discharge for flow control has been demonstrated in many applications. Examples include boundary layer acceleration, transition delay, lift augmentation on wings, separation control for low-pressure turbine blades, jet mixing enhancement, plasma flaps and slats, leading-edge separation control on wing sections, phased plasma arrays for unsteady flow control, and control of the dynamic stall vortex on oscillating airfoils.

3.1. Airfoil Flow Separation Control

More than 70% lift force of aircraft is produced by wings. The lift-to-drag ratio and stall characteristic of the wing is of vital importance to the takeoff distance and climbing speed and the flight quality of the aircrafts. In order to enhance the manoeuvrability and flexibility of the aircrafts, large angle of attack is used frequently. New technology should be employed into the development of aircrafts of the next generation. Active flow control technologies are considered to be the most promising technology in the 21th century.

3.1.1. Flow Separation Control Using Microsecond and Nanosecond Discharge

Flow separation control by microsecond and nanosecond discharge plasma aerodynamic actuation was presented. The control effects influenced by various actuation parameters were investigated.

The airfoil used was a NACA 0015. This shape was chosen because it exhibits well-known and documented steady characteristics as well as leading-edge separation at large angles of attack. The airfoil had a 12 cm chord and a 20 cm span. The airfoil was made of Plexiglas. Twelve pressure ports were used to obtain the pressure distribution along the model surface. Fig. 11 shows location of the pressure ports on the model's surface. Three pairs of plasma aerodynamic actuators were mounted on the suction side of the airfoil. The actuators were positioned 2% and 20% and 45% cord length of the airfoil. The plasma aerodynamic actuators were made from two 0.018mm thick copper electrodes separated by 1mm thick Kapton film layer. The electrodes were 4mm in width and 120mm in length. They were arranged just in the asymmetric arrangement. A 1mm recess was molded into the model to secure the actuator flush to the surface. The pressure distribution along the airfoil surface was obtained by a Scanivalve with 96 channels having a range of ±11 kPa. A pitot static probe was mounted on the traversing mechanism. This was located at different positions downstream of the airfoil, on its spanwise centerline. Discrete points were sampled across the wake to determine the mean-velocity profile. The uncertainty of the measurement was calculated to be less than 1.5%.

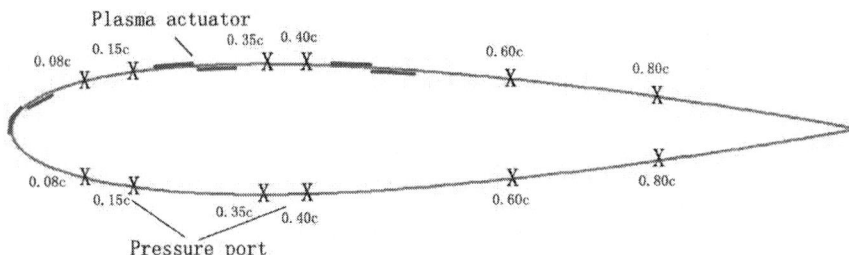

Figure 11. A schematic of NACA 0015 airfoil with dielectric barrier discharge plasma aerodynamic actuator

The power supply used for microsecond discharge is 0-40 kV and 6-40 kHz, respectively. The output voltage and the frequency range of the power supply used for nanosecond discharge are 5-80 kV and 0.1-2 kHz, respectively. The rise time and full width half maximum (FWHM) are 190ns and 450ns, respectively.

The plasma aerodynamic actuation strength, which is related to the discharge voltage, is an important parameter in plasma flow control experiments. The flow control effects influenced by discharge voltage were investigated. Flow separates at the leading edge of the airfoil without discharge. The pressure distribution has a plateau from leading edge to trailing edge which corresponds to global separation from the leading edge. When the microsecond discharge voltage is 13 kV and 14 kV, the flow separation can not be suppressed. As the microsecond discharge voltage increases to 15 kV, the actuation intensity increases and the flow separation is suppressed. There is a 34.0% lift force increase and a 25.3% drag force decrease when the discharge voltage is 15 kV. When the millisecond discharge voltage increases to 16 kV, there is a 35.1% lift force increase and a 25.5% drag force decrease. The control effects for discharge voltage of 15 kV and 16 kV are approximately the same. Thus, a threshold voltage exists for plasma aerodynamic actuation of different time scale. The flow separation can't be suppressed if the discharge voltage is less than the threshold voltage. When the flow separation is suppressed, the lift and drag almost unchanged when the discharge voltage increases. The initial actuation strength is of vital importance in plasma flow control. Once the flow separation is suppressed with a initial discharge voltage higher than the threshold voltage, the flow reattachment can be sustained even the discharge voltage was reduced to a value less than the threshold voltage, that is to say, the voltage to sustain the flow reattachment is lower than the voltage to suppress the flow separation in the same conditions. We can make use of the results by managing the discharge voltage properly. A higher discharge voltage can be used to suppress the separation in the beginning, and then we can use a much lower discharge voltage to sustain the flow reattachment later. Not only the power consumption can be reduced obviously, but also the life-span and the reliability of the actuator can be increased greatly.

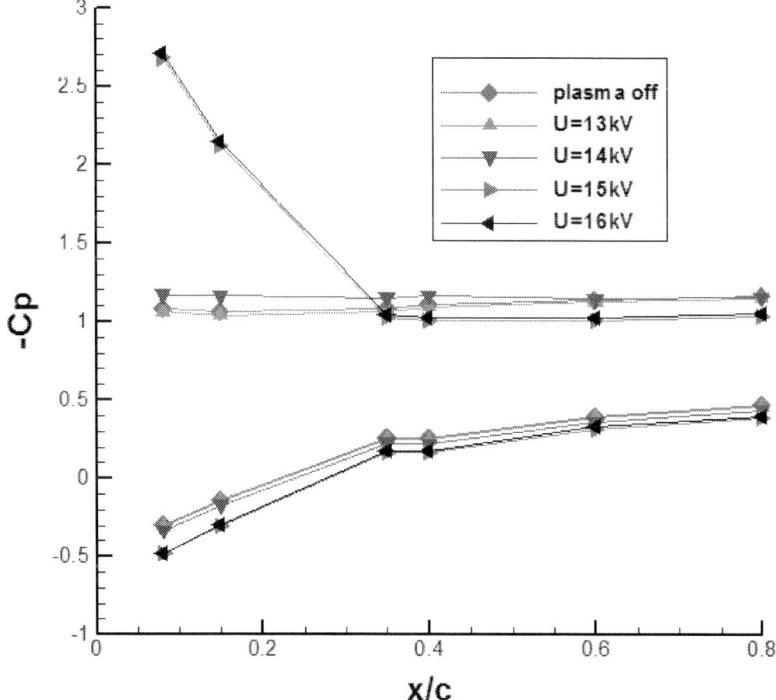

Figure 12. Pressure distribution for microsecond discharge of different voltage ($\alpha=20°$, $V\infty=72$ m/s, $Re=5.8\times10^5$)

The frequency of nanosecond discharge is believed to be optimum when the Strouhal number $S_{tr} = fc_{sep}/v_\infty$ is near unity. The separation region length and inflow velocity are 100% chord length and 100m/s respectively. The Strouhal number is 1 when the pulse frequency is 830 Hz. Experiments of different pulse frequency were made to determine if such an optimum frequency exists for the unsteady actuation used in controlling the airfoil flow separation.

The experimental results are shown in Fig. 13. It is found that there's an optimum pulse frequency in controlling the airfoil flow separation. The inflow velocity and the angle of attack are 100 m/s and 25° respectively. The duty cycle is fixed at 50%. All three electrodes are switched on. The threshold voltage for different discharge frequency was shown Fig. 14. When the pulse frequency is 830 Hz, the threshold voltage to suppress the flow separation is only 10 kV which is the lowest. When the pulse frequency is 200 Hz and 1500 Hz, the threshold voltage is 13 kV and 12 kV respectively.

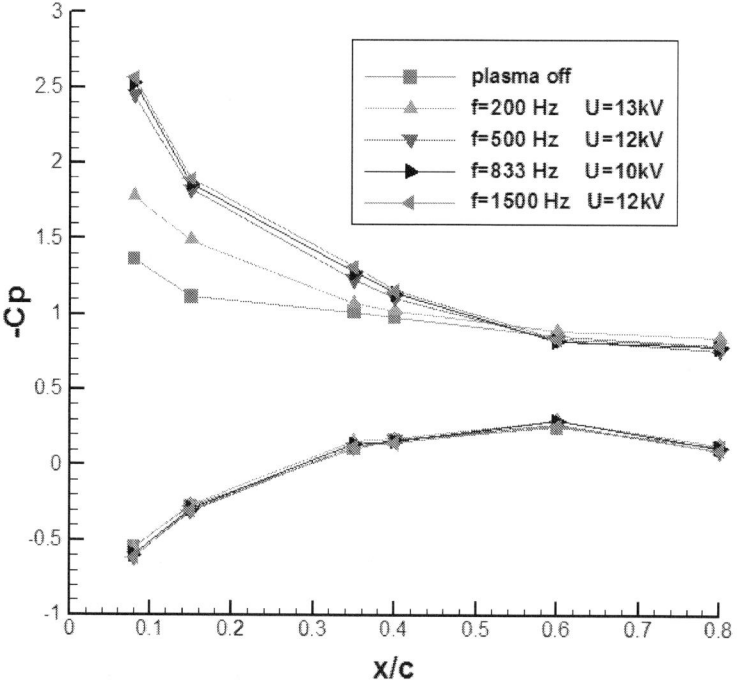

Figure 13. Pressure distribution for nanosecond discharge of different frequency (α=20°, $V\infty$=100 m/s, Re=8.1×10⁵)

Plasma aerodynamic actuation of different time scales was used for flow separation control. The flow control ability for microsecond discharge and nanosecond discharge were analyzed. The pressure distribution along airfoil surface obtained in experiments for inflow velocity of 150 m/s (Re=12.2×10⁵) are presented in Fig. 15. The angle of attack is 25°, which is approximately 5° past the critical angle of attack at the inflow velocity of 150m/s (Re=12.2×10⁵).

The discharge frequency is fixed at 1600 Hz. The discharge voltage for microsecond and nanosecond discharge is 17 kV and 12 kV respectively. When the nanosecond discharge is on, the flow is fully attached at the leading edge. The lift force increases by 22.1% and the drag force decreases by 17.4% with the actuation on. But the microsecond discharge can not suppress the flow separation. The flow still separates at the leading edge with microsecond plasma aerodynamic actuation. It indicates that the flow control ability for nanosecond discharge is stronger than that of the microsecond discharge. The nanosecond discharge is much more effective in leading edge separation control than microsecond discharge.

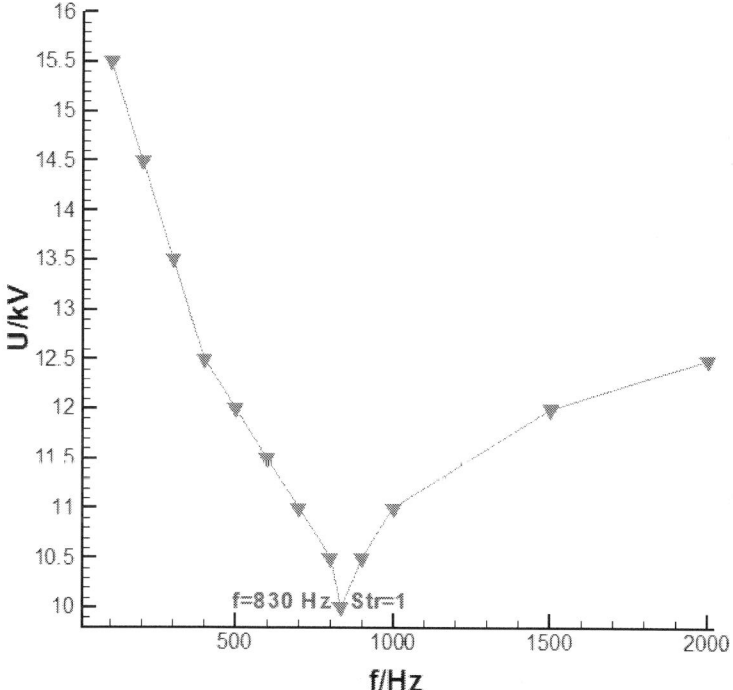

Figure 14. The threshold voltage at different frequencies for nanosecond discharge ($V_\infty=100$ m/s, $\alpha=22°$, $Re=8.1\times10^5$)

3.1.2. Flow Separation Control by Spanwise Nanosecond Discharge

The model used in this study was a NACA 0015 airfoil. Fig. 16 shows the geometry of the airfoil and the actuators. The actuator was made from two 0.018mm thick copper electrodes separated by 1mm thick Kapton film layer. The electrodes were 4mm in width and 60mm in length. They were arranged just in the asymmetric arrangement.

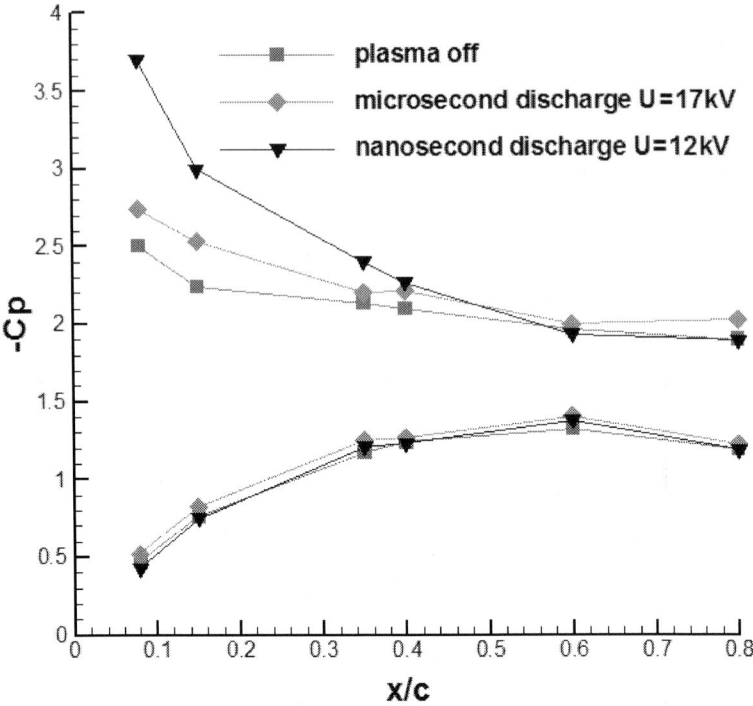

Figure 15. Experimental results for microsecond and nanosecond discharge ($V\infty=150$ m/s, $\alpha=25°$, $Re=12.2\times10^5$)

Experimental results for different angle of attacks (α) at the inflow velocity of 72 m/s ($Re=5.8\times10^5$) are shown in Fig. 17. The discharge voltage and frequency of the nanosecond power supply were fixed at 13 kV and 1000 Hz respectively. Experimental results show that spanwise nanosecond discharge aerodynamic actuation can suppress the flow separation effectively. The lift and drag coefficient are nearly unchanged with actuation when the angle of attack is less than 18° or more than 24°. When the angle of attack is less than the critical value, there is nearly no flow separation on the airfoil surface. The effect of spanwise nanosecond discharge aerodynamic actuation can is not obvious. When the angle of attack is more than 24°, the flow separation on the airfoil surface is so aggressive that spanwise nanosecond discharge aerodynamic actuation can not suppress the flow separation on the suction side of the airfoil. So the lift and drag coefficients nearly the same. There is an obvious lift augmentation and drag reduction after actuation when the angle of attack is between 18° and 24°. The lift coefficient is increased from 0.814 to 1.099 and the drag coefficient is decreased from 0.460 to 0.328 after actuation at the angle of attack 24°. The critical stall angle of attack for NACA 0015 airfoil increased from 18° to 24°. When the angle of attack is 24°, there is a lift force augmentation of 30.2% and a drag force reduction of 22.1% after actuation.

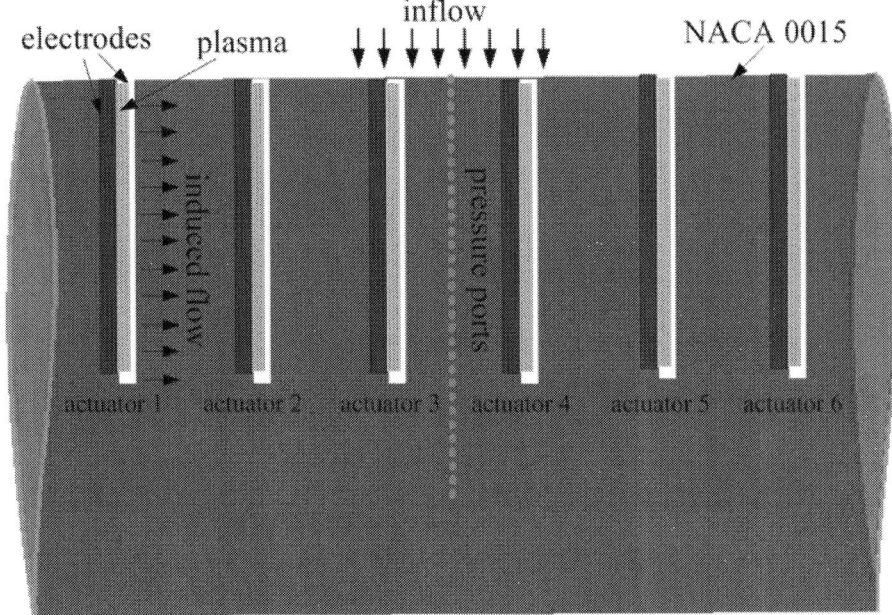

Figure 16. Schematic drawing of the actuators on the airfoil

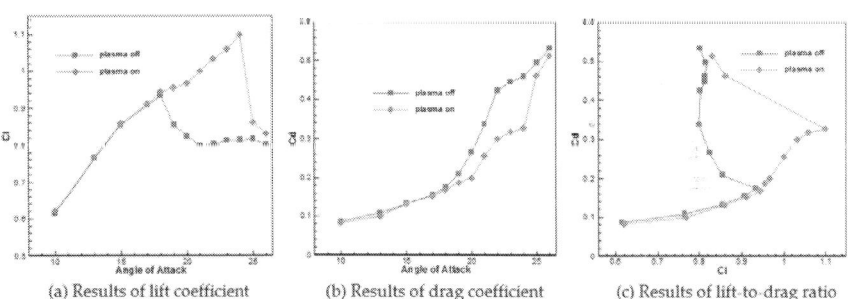

(a) Results of lift coefficient   (b) Results of drag coefficient   (c) Results of lift-to-drag ratio

Figure 17. Experimental results at different angles of attack (V ∞ =72 m/s, Re=5.8×105)

The discharge frequency for microsecond discharge is in the orders of kilo hertz. Spanwise plasma aerodynamic actuation of different time scales was used for flow separation control. The flow control ability for microsecond discharge and nanosecond discharge were analyzed. The pressure distribution along airfoil surface obtained in experiments for inflow velocity of 66 m/s ($Re=5.3\times10^5$) and 100 m/s ($Re=8.1\times10^5$) are presented in Fig. 18 and Fig. 19. At the angle of attack 22° and inflow velocity of 66 m/s (Fig. 18), there is initial separated flow on the suction surface of the airfoil without discharge. The discharge voltage for microsecond and nanosecond discharge is 7 kV and 12 kV respectively. The discharge frequency is 1000 Hz. The flow separation on the suction surface can be suppressed by both microsecond and nanosecond discharge actuation. The control effects are nearly the same for microsecond and nanosecond discharge. The spanwise plasma aerodynamic actuations result in a lift augmentation of 23.6% and a drag reduction of 25.6%.

In Fig. 19, the angle of attack is 24°, which is approximately 4° past the critical angle of attack at the inflow velocity of 100m/s ($Re=5.8\times10^5$). The discharge frequency is fixed at 1000 Hz. The discharge voltage for microsecond and nanosecond discharge is 8.5 kV and 12 kV respectively. When the nanosecond discharge is on, the flow is fully attached at the leading edge. The lift force increases by 25.3% and the drag force decreases by 20.1% with the actuation on. But the microsecond discharge can not suppress the flow separation. The flow still separates at the leading edge with microsecond plasma aerodynamic actuation. It indicates that the flow control ability for nanosecond discharge is stronger than that of the microsecond discharge. The nanosecond discharge actuation is much more effective in leading edge separation control than microsecond discharge actuation.

The dielectric layer will be destroyed when the discharge voltage is strong enough. Kapton is used as the dielectric in our experiments. The threshold voltage to destroy the Kapton layer is 8.5kV for microsecond discharge in our experiments. The actuators will be destroyed when the discharge voltage is more than 8.5kV for microsecond discharge. The threshold voltage to destroy the Kapton layer is 17 kV for nanosecond discharge in our experiments.The instantaneous actuation intensity for nanosecond discharge is much stronger than microsecond discharge. So nanosecond discharge is more effective in flow control than microsecond discharge.

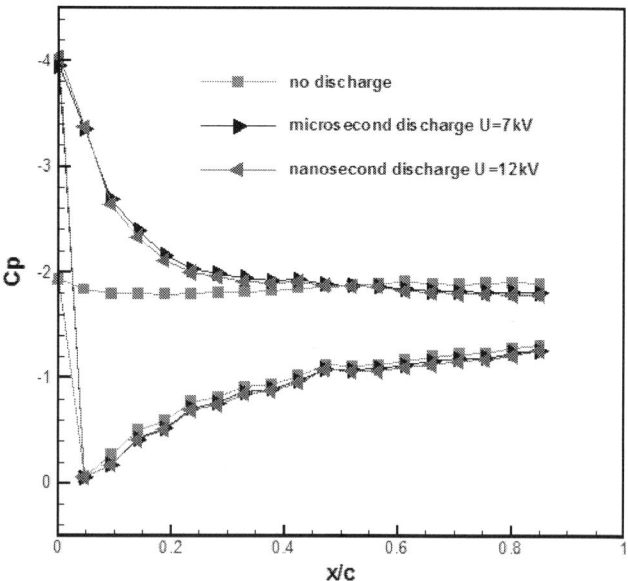

Figure 18. Experimental results for microsecond and nanosecond discharge (V_∞=66 m/s and α=22° $Re=5.3\times10^5$)

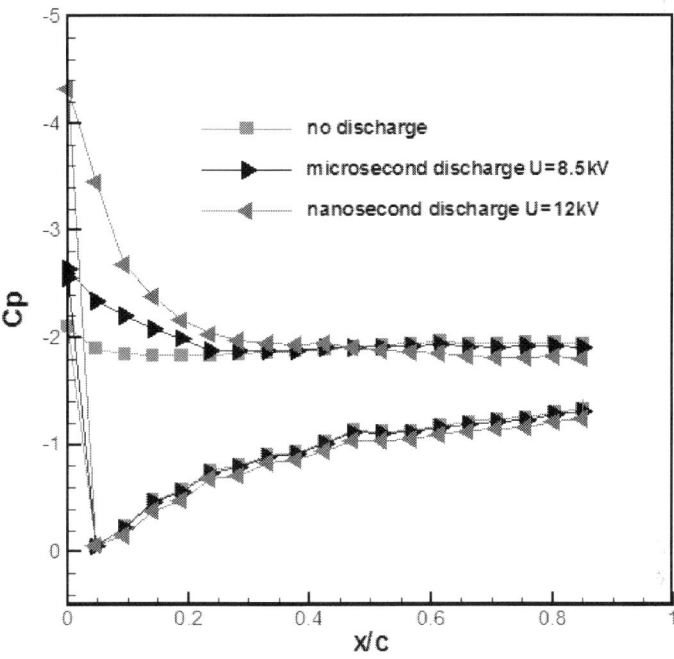

Figure 19. Experimental results for microsecond and nanosecond discharge (V_∞=100 m/s, α=24°, Re=8.1×10⁵)

3.1.3. The Mechanism of Plasma Shock Flow Control

Based on our works, the principle of "plasma-shock-based flow control" was proposed. Energy should be released in extremely short time to intensify the instantaneous actuation strength, such as nanosecond discharge. Nanosecond discharge yields strong turbulence even shock waves which are act on the boundary layer. Shock wave produces stronger turbulent mixing of the flow, which can enhance momentum and energy exchange between the boundary layer and inflow greatly. High momentum fluid was brought into the boundary layer intermittently, enabling the flow to withstand the adverse pressure gradient without flow separation.The spirits of "plasma-shock-based flow control" lay in three aspects. Firstly, "Shock Actuation", nanosecond discharge should be used to increase the instantaneous discharge power. Nanosecond discharge induces strong local pressure or temperature rise in the boundary. Pressure or temperature rise result in strong pulse disturbance or shock waves in the boundary. Secondly, "Vortex control", shock wave disturbance induces vortex in the process of propagation. Vortex enhances energy and momentum mixing between boundary layer and inflow. The velocity of the boundary layer increase and the flow separation is suppressed. Thirdly, "Frequency Coupling", adjust the discharge frequency to the optimal response frequency in flow control. The optimal response frequency is the one which makes the Strouhal number equal to 1. The plasma aerodynamic actuation work best at the optimal response frequency. Nanosecond discharge can increase the capability of plasma flow control effectively while its energy consumption can be reduced greatly.

For microsecond plasma aerodynamic actuation, the momentum effect may be the dominant mechanism. Microsecond plasma aerodynamic actuation induces near-surface boundary layer acceleration. Energy and momentum is added into the boundary layer, which enhances the ability to resist flow separation caused by adverse pressure gradient for boundary. But the maximum induced velocity for microsecond discharge is less than 10m/s. The actuators will be destroyed if the discharge voltage is too high. The momentum added into the boundary layer by microsecond discharge is quite limited. The microsecond plasma aerodynamic actuation can only work effectively when the inflow velocity is several tens of meters per second.

The main mechanism for nanosecond discharge plasma flow control may be not momentum effect, since the induced velocity is less than 1m/s. The velocity and vorticity measurements by the Particle Image Velocimetry show that, the flow direction is vertical, not parallel to the dielectric layer surface. The induce flow is likely to be formed by temperature and pressure gradient caused by nanosecond discharge other than energy exchange between charged and neutral particles. Thus, the main flow control mechanism for nanosecond plasma aerodynamic actuation is local fast heating due to high reduced electric field, which then induces shock wave and vortex near the electrode.

Experimental results indicate that nanosecond discharge is more effective in flow control than microsecond discharge. The latest study showed that nanosecond discharges have demonstrated an extremely high efficiency of operation for aerodynamic plasma actuators over a very wide velocity range (Ma= 0.03-0.75). So shock effect is more important than momentum effect in plasma flow control.

3.2. Corner Separation Control in a Compressor Cascade

Control of the corner separation is one of the important ways of improving axial compressor stability and efficiency. Our approach to control the corner separation is based on the use of plasma aerodynamic actuation. Experiments were carried out on a low speed compressor cascade facility. Main cascade parameters are shown in Fig. 20. Only the middle blade was laid with the plasma aerodynamic actuator.

Total pressure distributions at 10mm, which is 15% of the chord length, downstream of the blade trailing edge along the pitch direction at 50%, 60% and 70% blade spans were measured with and without the plasma aerodynamic actuation. A three-hole probe calibrated for pitch and yaw was used to measure the total pressure at the cascade exit. Two parameters, total pressure recovery coefficient σ and the relative reduction of the total pressure loss coefficient $\delta(\omega)$, were used to quantify the performance improvement due to the plasma aerodynamic actuation.

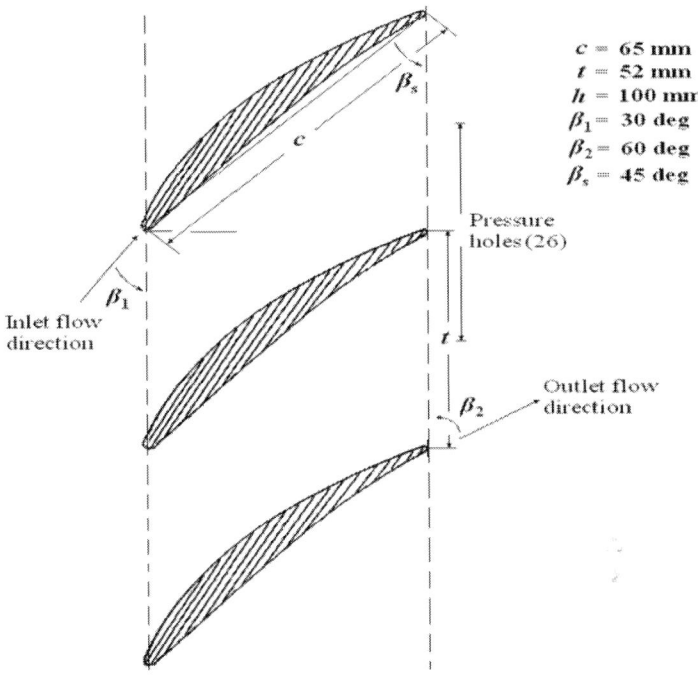

$$c = 65 \text{ mm}$$
$$t = 52 \text{ mm}$$
$$h = 100 \text{ mm}$$
$$\beta_1 = 30 \text{ deg}$$
$$\beta_2 = 60 \text{ deg}$$
$$\beta_s = 45 \text{ deg}$$

Figure 20. Compressor cascade parameters

The plasma aerodynamic actuator used in the present experiments consists of four electrode pairs, located at 5%, 25%, 50% and 75% of the chord length, respectively. The electrode pair at 5% of chord length is named as the 1st electrode pair. A sketch of a blade with the actuator on the surface is shown in Fig. 21. The electrode thickness is not to scale in the figure.

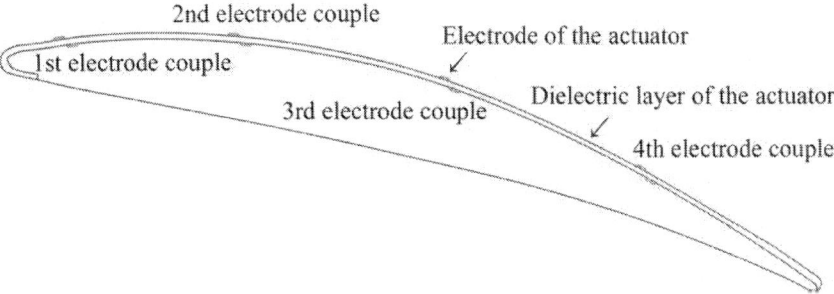

Figure 21. A sketch of a blade with plasma aerodynamic actuator

The plasma aerodynamic actuator is driven by a high frequency high voltage power supply (CTP-2000M+, Suman Electronics). The output waveform is sine wave. The output ranges of the peak-to-peak voltage and the driving frequency of the power supply are $V_{p-p} = 0\sim40$ kV and $F = 6\sim40$ kHz, respectively. The driving frequency is fixed at 23 kHz in the experiments.

The plasma aerodynamic actuator works at steady or unsteady mode in the experiments. In the steady mode, the actuator is operated at the ac frequency. In the unsteady mode of operation, the ac voltage is cycled off and on. Fig. 22 shows a typical signal sent to the plasma aerodynamic actuator during the unsteady actuation. Two important parameters of the unsteady plasma aerodynamic actuation are the excitation frequency f, and the duty cycle α, respectively.

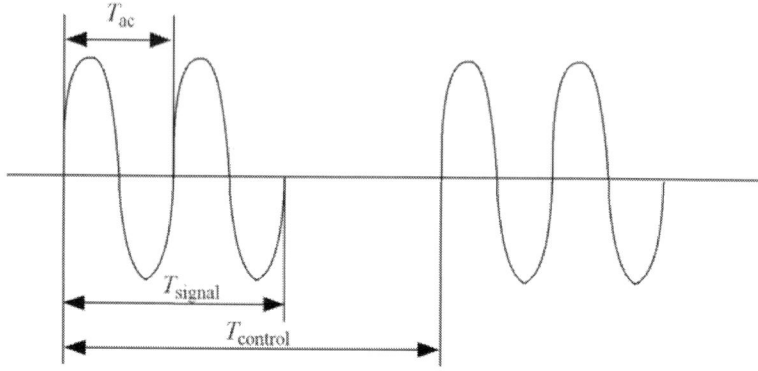

Figure 22. The signal sent to the plasma aerodynamic actuator during unsteady excitation

3.2.1. Steady Plasma Flow Control Experiment Results

The mechanism of steady plasma aerodynamic actuation to control the corner separation may be that the actuation induces a time-averaged body force on the flow due to that the flow can't respond to such high frequency (23 kHz in the experiments) disturbances. A wall jet, which is oriented in the mean flow direction, is produced to add momentum to the near-wall boundary layer near the flow separation location. The energized flow is able to withstand the adverse pressure gradient without separation. The directed wall jet governs the flow control effect of steady plasma aerodynamic actuation. When the electrode length is enlarged, the consumed power increases nonlinearly.

The location of the plasma aerodynamic actuation is a key parameter in plasma flow control experiments. Total pressure recovery coefficients with steady actuation at different locations are shown in Fig. 23.

The applied peak-to-peak voltage and driving frequency are $V_{p\text{-}p} = 10$ kV and $F = 23$ kHz, respectively. $\delta(\omega)_{max}$ is 5.5%, 10.3%, 2.4% and 0.07% when the 1st, 2nd, 3rd and 4th electrode pair is switched on, respectively. The 2nd electrode pair at 25% chord length is most effective and the control effect is as the same as that obtained by all four electrode pairs. The power dissipated by the 2nd electrode pair is just 18.4W, about half of the power dissipated by all four electrode pairs. Therefore, the actuation location is vital to the control effect in corner separation control. In corner separation control by tailored boundary layer suction, the optimum slot should be long enough to be sure to remove the limiting streamline and the suction upstream of the corner separation location at the suction surface is most important for the control effect. Therefore, it can be inferred that the location of the 2nd electrode pair is just upstream of the corner separation.

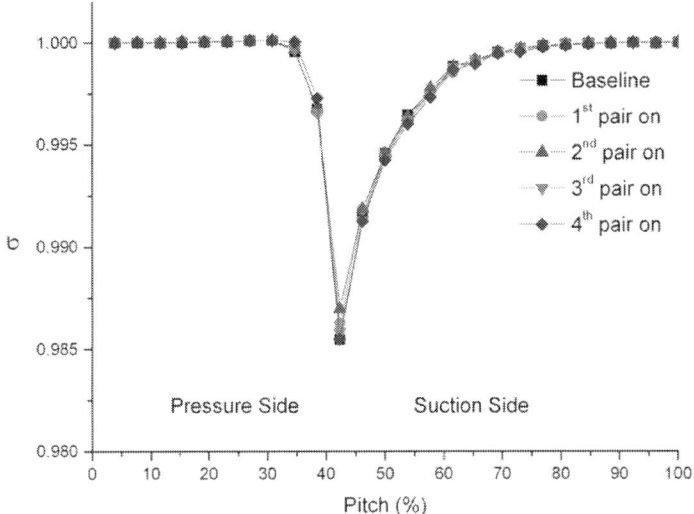

Figure 23. Total pressure recovery coefficients with steady actuation at different locations ($v_\infty = 50$ m/s, $i = 0$ deg, $V_{p-p} = 10$ kV, $F = 23$ kHz, 70% Span)

The plasma aerodynamic actuation strength is another important parameter in plasma flow control experiments. The body force increases with the voltage amplitude in proportion to the volume of plasma (ionized air) and the strength of the electric field gradient. As the applied peak to peak voltage increases from 8 kV to 12 kV, $\delta(\omega)_{max}$ increases from to 2.7% to 11.1%, as shown in Fig. 24. The 2nd electrode pair at 25% chord length is switched on and the driving frequency is 23 kHz. The power dissipation increases from 8.4 W to 23.5 W when the applied peak to peak voltage increases from 8 kV to 12 kV. When the applied voltage is less than 9 kV, the control effect is very tiny. When the applied voltage is higher than 10 kV, the control effect saturates and further increases in the voltage amplitude shows no evident benefit. Furthermore, higher voltage may lead to earlier destruction of the dielectric material, which is not desirable in the experiments.

3.2.2. Unsteady Plasma Flow Control Experiment Results

Optimization of the excitation mode based on coupling between the plasma aerodynamic actuation and the separated flow is one of the important ways of improving plasma flow control effect. It has been shown in the literature that the introduction of unsteady disturbances near the separation location can cause the generation of large coherent vortical structures that could prevent or delay the onset of flow separation. These structures are thought to intermittently bring high momentum fluid to the surface, enabling the flow to withstand the adverse pressure gradient without separation.

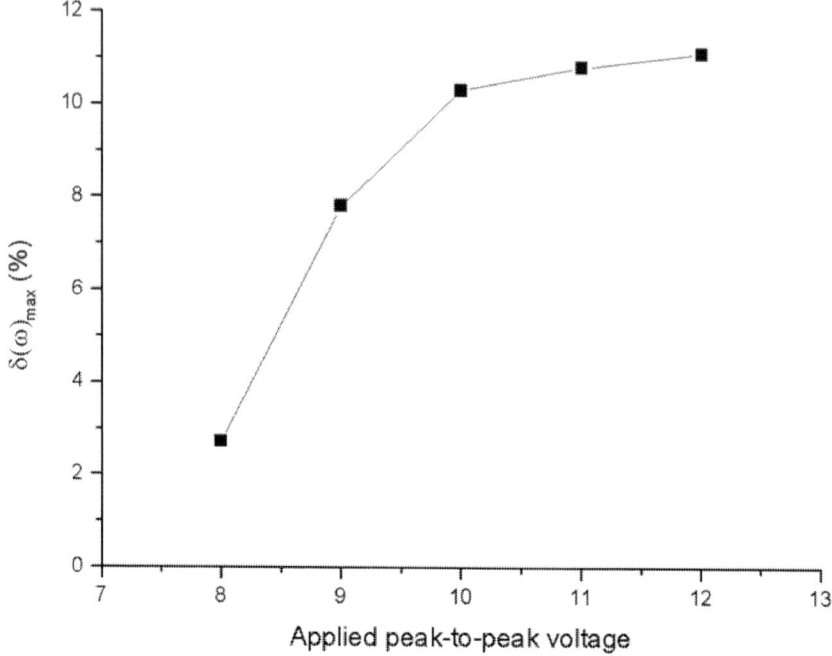

Figure 24. Control effect with steady actuation of different applied voltages ($v_\infty = 50$ m/s, $i = 0$ deg, $F = 23$ kHz, 70% Span)

A sensitive study is performed to determine if such an optimum frequency exists for the unsteady actuation used in controlling the corner separation. Fig. 25 documents the relative reductions of maximum total pressure loss coefficient at 70% blade span for a range of excitation frequencies from 100 Hz to 1000 Hz when the duty cycle is fixed at 60%. All four electrode pairs are switched on. The applied peak-to-peak voltage and driving frequency are $V_{p-p} = 10$ kV and $F = 23$ kHz, respectively.

When the excitation is 100 Hz, $\delta(\omega)_{max}$ is just 11.2%, which is almost as same as the steady control effect that is 10.7%. Along with the excitation frequency increasing, the control effect increases. When the excitation frequency is 400 Hz, $\delta(\omega)_{max}$ increases to 28%. Thus, compared with the steady actuation, the unsteady actuation is much more effective and requires less power. When the excitation frequency is higher than 400 Hz, the control effect saturates and further increases in the excitation frequency shows no evident benefit. The difference between steady and unsteady plasma aerodynamic actuation may be that, the unsteady pulsed operation allows the continuously generation of vortical structures, while the steady operation can't. Vortical structures in the flow field promote momentum transfer in the boundary layer in order to withstand separation. Under different duty cycles and excitation frequencies, the coupling between actuation and flow field leads to different flow control effects.

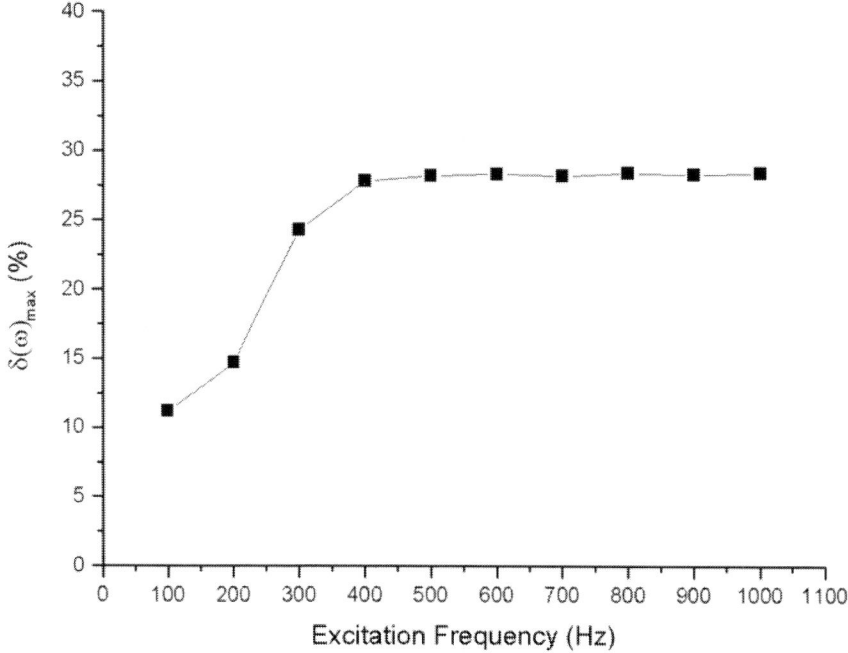

Figure 25. Maximum relative reductions of total pressure loss coefficient with unsteady actuation of different duty cycles ($v \infty = 50$ m/s, i = 0 deg, V p-p = 10 kV, F = 23 kHz, 70% Span)

Each electrode pair is switched on to study the effect of the actuation location. The control effect of all four electrode pairs is almost as same as that obtained by the 2nd electrode pair. The saturation frequency is also 400 Hz. For the 2nd electrode pair, the characteristic length is the remaining chord length downstream of the actuator, which is 75% chord length. Thus, the Strouhal number $Sr = f \times C / v \infty$ is 0.4 when the frequency and freestream velocity are $f = 400$ Hz and $v_\infty = 50$ m/s, respectively. When the Strouhal number exceeds 0.4, the control effect saturates in the unsteady plasma flow control experiments. In the separation control above a NACA 0015 airfoil with unsteady plasma aerodynamic actuation (Benard et al. 2009), the most effective actuation was performed with a Strouhal number of Sr ranging from 0.2 to 1. The optimum excitation frequency depends much on the flow separation state. Under different flow conditions, the optimum excitation frequency is also different.

Fig. 26 documents the maximum relative reductions of total pressure loss coefficient for a range of unsteady duty cycles from 5% to 100% when the excitation frequency is fixed at 400 Hz. All four electrode pairs are switched on. The applied peak-to-peak voltage and driving frequency are $V_{p-p} = 10$ kV and $F = 23$ kHz, respectively.

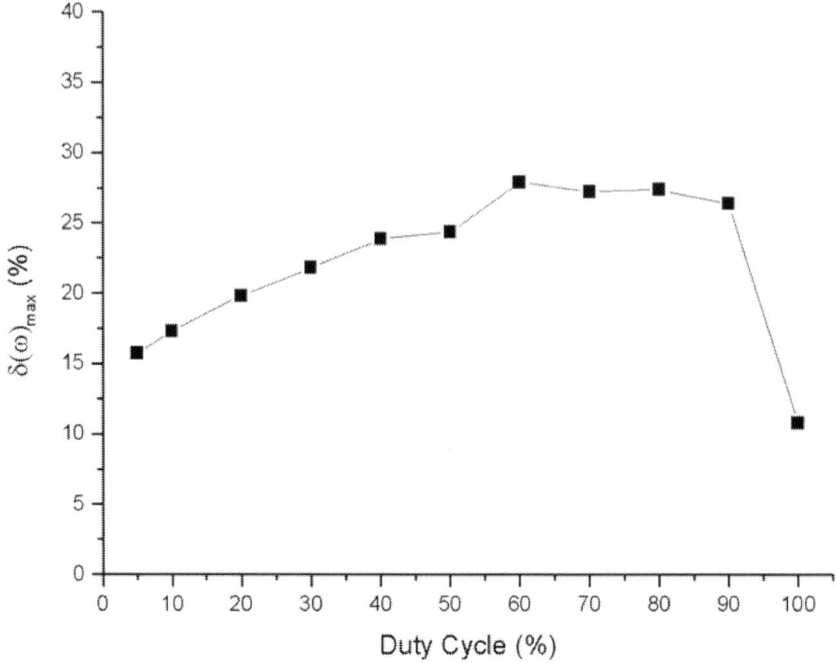

Figure 26. Maximum relative reductions of total pressure loss coefficient with unsteady actuation of different duty cycles ($v_\infty = 50$ m/s, $i = 0$ deg, $V_{p\text{-}p} = 10$ kV, $F = 23$ kHz, 70% Span)

It is found that there's also a duty cycle threshold in controlling the corner separation. When the duty cycle is less than 60%, the control effect increases along with the duty cycle increasing. $\delta(\omega)_{max}$ is 28% at the duty cycle of 60%. Even when the duty cycle is 5%, $\delta(\omega)_{max}$ is 15.7%, much more effective than the steady actuation. When the duty cycle is higher than 60%, the control effect saturates along with the duty cycle increasing. Thus it can be inferred that when the duty cycle is less than 60%, the injected energy is not sufficient to control the corner separation. In the separation control above a NACA 0015 airfoil with unsteady plasma aerodynamic actuation, the most effective duty cycle values range from 10% to 60%. In the separation control of low-pressure turbine blades with unsteady plasma aerodynamic actuation, the lowest plasma duty cycle (10%) was as effective as the highest plasma duty cycle (50%) at the same excitation frequency. Thus, the optimum duty cycle also depends much on the flow separation state.

3.3. Low Speed Axial Compressor Stability Extension

This series of tests were carried out using a low speed axial compressor test rig at Institute of Engineering Thermophysics, Chinese Academy of Sciences. The tested compressor rotor was isolated from the stator to avoid interaction effects generated by the presence of a downstream stator blade row. The isolated compressor rotor selected for this investigation is actually the rotor of the first stage of a low-speed three-stage axial compressor test rig, which has been used for a number of research programs for the flow instability in compression system. The blading is typical of high-pressure ratio compressor design. Previous work indicates that the isolated rotor is prone to tip stall behavior, which is suitable for flow control methods in the end wall flow regions.

The overall compressor performance in terms of pressure rise coefficient Ψ and mass flow coefficient Φ was measured with eight static pressure taps on casing around the annulus in both the inlet and the outlet of the compressor. The measurement uncertainties were: static pressure, $\pm 60 \text{N/m}^2$. Errors in calculated Ψ and Φ were estimated at $\pm 0.2\%$ maximum, as far as relative comparison between the results for a certain condition is concerned.

The basic principle of using plasma actuation reated caseing(PATC) to improve compressor stability range is shown in Fig. 27. When the PATC is energized, plasma forms and induces airflow along the direction of compressor inflow in the end wall flow region.

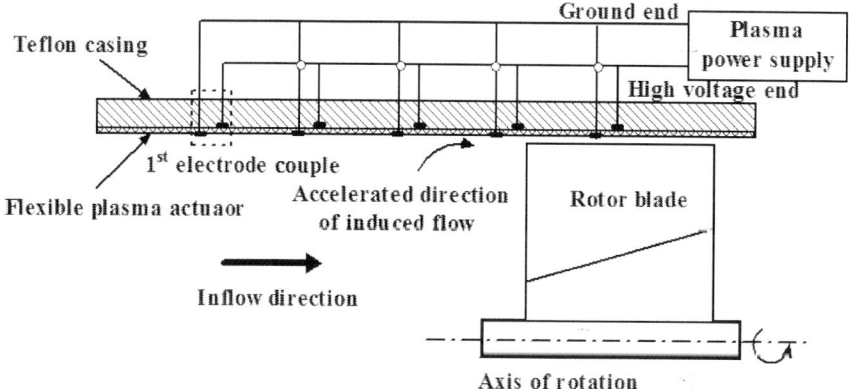

Figure 27. Sketch map of using PATC to improve compressor stability range

The basic mechanism for plasma actuation to extend compressor's stability can be classified to three effects. The first is that plasma actuation induces air acceleration along with the inflow direction in the blade tip end wall region. Energy is added to the low-energy flow in the end wall region, which can increase mass flux at blade leading edge, inhibit development of blade tip secondary flow and leakage flow, and enhance circulating ability in the end wall region. Thus the accumulation of flow build up is minished. The second is that due to the end wall flow acceleration induced by plasma actuation, velocity in flow direction at blade tip channel is enhanced and inflow attack angle is reduced. Thus flow separation at blade suction surface is inhibited. The last effect is that plasma actuation is non-stationary and non-linear actuation, which can enhance mixture among flow with different momentum in the end wall region. Thus flow separation due to low energy is inhibited and compressor stability is extended. Since plasma actuation can minish flow build up extent in the blade tip end wall region, inhibit secondary flow and leakage flow, and enhance circulating ability, compressor pressure rise ability is improved.

PATC consists of a flexible plasma actuator and a casing. The plasma actuator, layout of which is asymmetrical, consists of 5 electrode couples. The 4th electrode couple is located at 3mm away from the blade leading edge, while the 5th couple is located at the 40% blade tip chord. The thickness of teflon layer, h, is 0.5mm. The electrode is 0.035×2mm copper layer. The horizontal displacement between upper and lower electrode for each couple, Δd, is 1 mm. The distance between adjacent electrode couples, D, is 10 mm. The casing is also made of teflon. Fig. 28 and Fig. 29 show the PATC and the low speed axial compressor with PATC, respectively. For the PATC and tested rotor, tip clearance is 0.6 mm, which is 1.65% of the blade chord length.

Figure 28. Plasma actuation treated casing

Plasma actuation casing is energized by a high voltage power supply. The output of the power supply is sine wave. The amplitude and frequency range is 0-30 kV and 6-40 kHz, respectively, which can be adjusted continuously.

The compressor throttling was throttled by the exit rotary cone valve mounted on the shaft and regulated manually when stall was approached. Wall static pressure was collected to calculate pressure rise coefficient Ψ and flow coefficient Φ, which are adopted as representatives of compressor performance and stability with and without plasma actuation at a constant rotor speed.

Figure 29. Low speed axial compressor test rig with PATC

The effect of plasma actuation on the compressor performance and stability range is displayed in Fig. 30 at the rotor speed of 900 rpm. The 4th electrode couple is actuated and the actuation voltage is 9 kV.

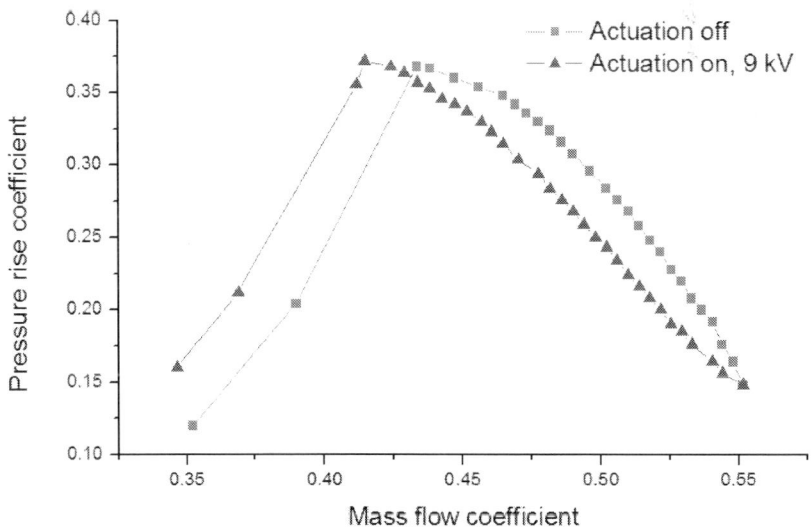

Figure 30. Test results with and without plasma actuation(rotor speed: 900 rpm)

The changes of maximum pressure rise coefficient, Ψ_{max} and mass flow coefficient near stall, Φ_{ns} are summarized in table 1. The Φ_{ns} decreases by 5.2%, while the Ψ_{max} increases by 1.08%.

Table 1. The effect of plasma actuation on compressor performance and stability range

	Ψ_{max}	$\Delta\Psi_{max}/\Psi_{max}$	Φ_{ns}	$\Delta\Phi_{ns}/\Phi_{ns}$
0	0.3721		0.4426	
1	0.3761	1.08%	0.4196	-5.2%

0: PATC off, 1: 4th electrode couple on, 9 kV.

Fig. 31 illustrates the test results with and without plasma actuation at the rotor speed of 1080 rpm. When the 2nd and 3rd electrode couples are switched on, Φ_{ns} decreases by 1.42% and 5.07% when the actuation voltage is 9 kV and 12 kV respectively. Ψ_{max} decreases by 2.21% and 0.74% respectively.

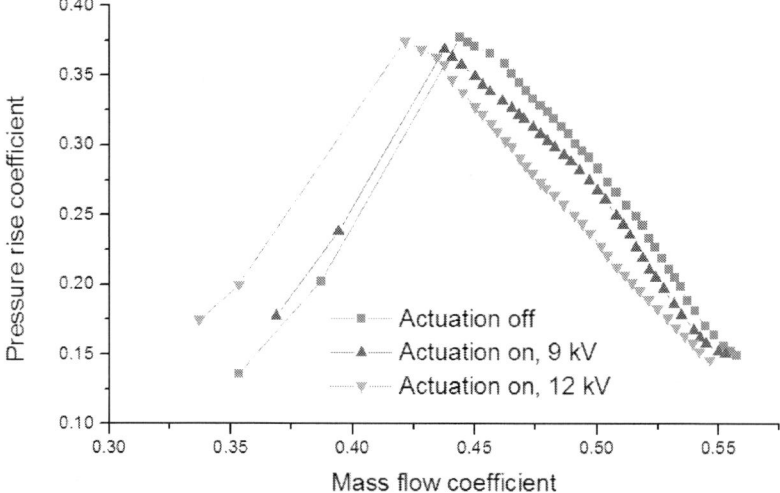

Figure 31. Test results with and without plasma actuation(rotor speed: 1080 rpm)

Along with the actuation voltage ascending, plasma actuation strength increases while plasma actuator's layout form and material remain same. Thus the velocity of induced flow acceleration increases, which can better enhance the circulation ability in end wall region, reduce inflow attack angle and promote flow mixture. As a result, the compressor stability range is much wider. Therefore plasma actuation strength, which can be improved by adjusting actuator layout form or increasing actuation voltage, is one key factor in plasma based stability extension.

Fig. 32 represents the effect of plasma actuation location on the compressor performance and stability range when the rotor speed equals 1080 rpm. When 3rd and 4th electrode couples are switched on at 12 kV, Φ_{ns} decreases by 1.42% and Ψ_{max} decreases by 1.47%. When the 2nd and 3rd electrode couples are switched on, Φ_{ns} and Ψ_{max} decrease by 5.07% and 0.74%, respectively. Therefore, different actuation location results in different stability range extension effect. One possible reason is that the 4^{th}electrode couple is just 3mm(8.3% of axial chord) away from the rotor blade leading edge, where flow build up is very serious and flow separation has well developed in blade tip end wall region at near stall state. Thus plasma actuation at this location can't control the flow field well and the stability extension effect is limited. When the 2nd and 3rd electrode couple is on, because the 3rd electrode couple is 18mm(49.5% of axial chord) away from the rotor blade leading edge, plasma actuation can accelerate the flow boundary layer before flow separation and build up in well development, which can inhibit the end all separation flow, secondary flow and leakage vortex better. Therefore stability extension effect is much better.

The changes of Ψ_{max} and Φ_{ns} are summarized in table 2. Ψ_{max} decreases at every case when plasma actuation is on. The Ψ_{max} decrease is least when the Φ_{ns} decrease is most. So there is no contradiction between stability range extension and pressure rise coefficient improvement. When the ability for plasma actuation to control the blade tip end wall region flow becomes stronger, the stability extension effect is better and the pressure rise ability almost remains same.

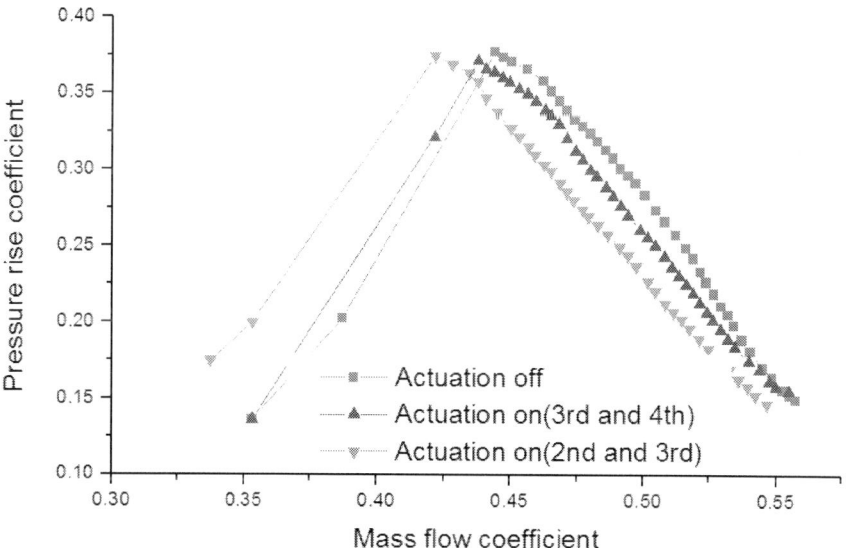

Figure 32. Test results with and without plasma actuation (rotor speed: 1080 rpm)

Table 2. The effect of plasma actuation on compressor performance and stability range

	Ψ_{max}	$\Delta\Psi_{max}/\Psi_{max}$	Φ_{ns}	$\Delta\Phi_{ns}/\Phi_{ns}$
0	0.3772		0.4438	
1	0.3689	-2.21%	0.4375	-1.42%
2	0.3744	-0.74%	0.4213	-5.07%
3	0.3717	-1.47%	0.4375	-1.42%

0: PATC off.
1: 2nd and 3rd electrode couples on, 9kV.
2: 2nd and 3rd electrode couples on, 12kV.
3: 3rd and 4th electrode couples on, 12kV.

4. SUPERSONIC PLASMA FLOW CONTROL

Based on plasma aerodynamic actuation, plasma flow control is a novel active flow control technique and has important applications in the field of supersonic flow control. Shock waves are typical aerodynamic phenomena in supersonic flow. If they are controlled effectively, the aerodynamic performance of both flight vehicles and aeroengines will be greatly enhanced. Conventional mechanical or gasdynamic control methods have disadvantages of complex structure and slow response. Novel plasma flow control method has advantages of simple structure, fast response and wide actuation frequency range. Therefore, plasma flow control method has become a newly-rising focus in the field of shock wave control.

4.1. Experimental Principle and Arrangement

Fig. 33 shows the MHD flow control experimental principle. The high density plasma column which primarily consists of ions and electrons was generated between a pair of electrodes through pulsed DC discharge. There were three pairs of electrodes and an oblique shock wave appeared in front of the ramp in low-temperature supersonic flow. The alphabet "I" and "B" represented the current and magnetic field. The arrows gave their directions.

When magnetic field, normal to the surface, was imposed on the plasma column created in the boundary layer, it affected both the plasma and, through the Lorentz body force (j×B body force), the flow. The direction of Lorentz body force was determined by the directions of current and magnetic field. The alphabet "F" represented the Lorentz body force which could accelerate the flow.

The plasma column was produced by pulsed DC discharge. Therefore the plasma would be influenced by electric field force, magnetic field force and the airflow inertial force. The magnetic field force and the airflow inertial force were dominant. When the direction of magnetic field force was same as that of airflow inertial force and the velocity of plasma was faster than that of the neutral gas molecules, the plasma would strike the neutral gas molecules to transfer momentum and accelerate the flow in the boundary layer. Otherwise, when the direction of magnetic field force was against with that of airflow inertial force, the plasma would strike the neutral gas molecules to transfer momentum and decelerate the flow in the boundary layer.

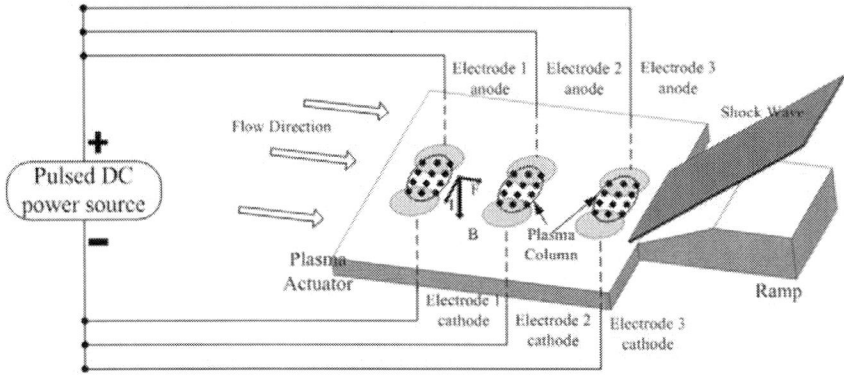

Figure 33. The experimental principle

MHD flow control system consisted of low-temperature supersonic wind tunnel, plasma actuation system, experimental ramp, magnetic field generator, parameter measurement system and schlieren optical system. The inlet total pressure of low-temperature supersonic wind tunnel was about 5-7atm. The stagnation conditions for the tunnel were atmospheric pressure and room temperature. The run time could reach up to 60 seconds dependent on the inlet total pressure. The experimental duct was 115mm(length)×80mm(width) and the designed Mach number was 2.2. The static pressure was 0.5-0.7atm and the static temperature was 152K.

The plasma actuation system included pulsed DC power source, plasma actuator, insulating acrylic base. Pulsed DC power source was the critical equipment which consisted of high voltage pulsed circuit, high voltage DC circuit and feedback circuit. It could provide 0-90kV selected high voltage pulse and 0-3kV selected high voltage direct current. The electrodes were made of plumbago, and were flush-mounted on the top wall of the insulating dielectric. The diameter of the electrode was 10mm. The insulating dielectric was made of BN ceramic. Two kinds of arrangements for the plasma actuator were adopted according to the distance between a pair of electrodes (D=5mm or D=8mm).

The experimental ramp was also constructed of insulating acrylic material. It was installed on the insulating acrylic base. The dimension of the ramp was 34mm(length) ×25mm(width) ×6mm(height). One angle of ramp was A=15°and another was A=20°. As illustrated in Fig. 34, 10 static pressure measurement holes were drilled on the acrylic base, the plasma actuator and the ramp. The holes were numbered with k1-k10 from upstream to downstream. Holes k2-k8 were drilled on the plasma actuator, and holes k9-k10 were drilled on the ramp. The diameter of k1-k10 was 0.5mm. On the plasma actuator the distance between adjacent holes was 10mm except that the distance between k2 and k3 was 7.5mm and the distance between k7 and k8 was also 7.5mm. On the ramp the distance between k9 and plasma actuator edge was 6mm and the distance between k9 and k10 was 6mm.

Nd-Fe-B rare-earth permanent magnets were used as the magnetic field generator which was located normal to experimental duct. The direction of magnetic field was perpendicular to the flow direction and the electric field direction. The magnetic field strength was about 0.3T between two magnets.

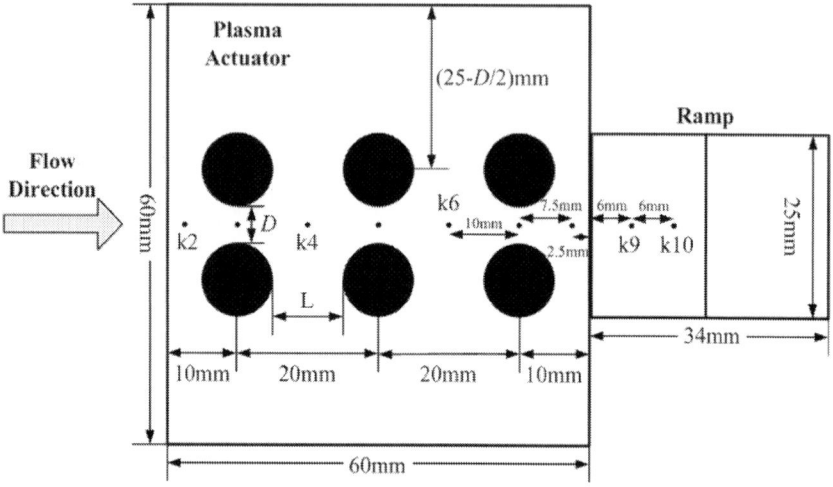

Figure 34. The dimensions of the plasma actuator and the ramp

Parameter measurement system consisted of electric parameter and flow characteristic measurement systems. Electric parameter measurement system included oscillograph (DPO4104, Tektronix Inc.), high voltage probe (P6015A, Tektronix Inc.) and current probe(TCP312+TCPA300, Tektronix Inc.). Flow characteristic measurement system included 10 static pressure sensors and a data collection apparatus. Because the run time of the wind tunnel was above 10 seconds in every experiment, the inlet total pressure decreased slowly during the experimental time. Therefore, in this study the ratio of Pitot pressure after shock-wave to that before shock-wave was adopted to compare the flow characteristic of the airflow around the ramp that was illustrated as P_{k10}/P_{k7}.

Schlieren optical system consisted of a high-speed camera and a storage computer. The high-speed camera was an Optronis® high-speed camera, and the maximum frame frequency was 200k frames per second(FPS). In this study, the schlieren pictures were taken at 8kFPS. The exposure time was 100μs and the run time was 8s.

4.2. Experimental Results

In terms of different magnetic field directions, distances between electrodes, ramp angles and DC Voltage, the results of these four kinds of MHD flow control were compared and analyzed.

Through changing the magnetic field direction, MHD acceleration and MHD deceleration experiments were carried out. Fig. 35 represented two typical flow characteristics.

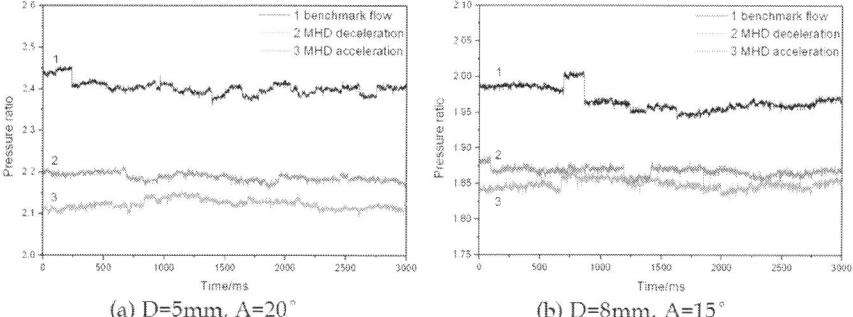

(a) D=5mm, A=20°   (b) D=8mm, A=15°

Figure 35. Two typical flow characteristics

The time-averaged pressure ratio decreased by 6.04% and 5.09% with MHD acceleration and MHD deceleration respectively. Thus, MHD flow control could drastically weaken the oblique shock wave strength and change the flow characteristic of the airflow around the ramp. MHD acceleration was more effective than MHD deceleration.

MHD acceleration experiments were carried out at ramp angle $A=20°$ with $D=8$mm or 5mm. Fig.36 showed the flow characteristics with different distances. At $D=8$mm the time-averaged pressure ratio decreased by 19.66% with MHD acceleration. At $D=5$mm the time-averaged pressure ratio decreased by 11.64% with MHD acceleration. Thus, MHD acceleration was more effective to weaken the shock wave strength when D increased, but there existed a maximum value D *max* exceeding which the airflow could not be ionized restricted by power source.

MHD acceleration experiments were carried out at $D=8$mm with different ramp angles $A=15°$ or 20°.Fig. 37 showed the flow characteristics with different ramp angles. At $A=15°$ the time-averaged pressure ratio decreased by 6.04% with MHD acceleration. At $A=20°$ the time-averaged pressure ratio decreased by 19.66% with MHD acceleration. Thus, MHD acceleration was more effective to weaken the shock wave strength when A increased, but there existed a maximum A *max* exceeding which oblique shock wave strength would be too strong to be changed.

Through changing DC voltage V_{DC}(2kV, 2.5kV, 3kV), MHD acceleration and MHD deceleration experiments were carried out at ramp angle $D=8$mm, $A=15°$. Fig. 38 showed the flow characteristics with different V_{DC}. Fig. 38(a) showed the static pressure ratio varying with MHD acceleration. The time-averaged pressure ratio decreased by 3.95%, 5.19% and 6.04% with $V_{DC}=2$kV, 2.5kV and 3kV respectively. Fig. 38(b) showed the static pressure ratio varying with MHD deceleration. The time-averaged pressure ratio decreased by 3.44%, 4.26% and 5.09% with $V_{DC}=2$kV, 2.5kV and 3kV respectively. Thus, MHD flow control could drastically weaken the oblique shock wave strength and change the flow characteristic of the airflow around the ramp. MHD interaction was more effective when V_{DC} increased.

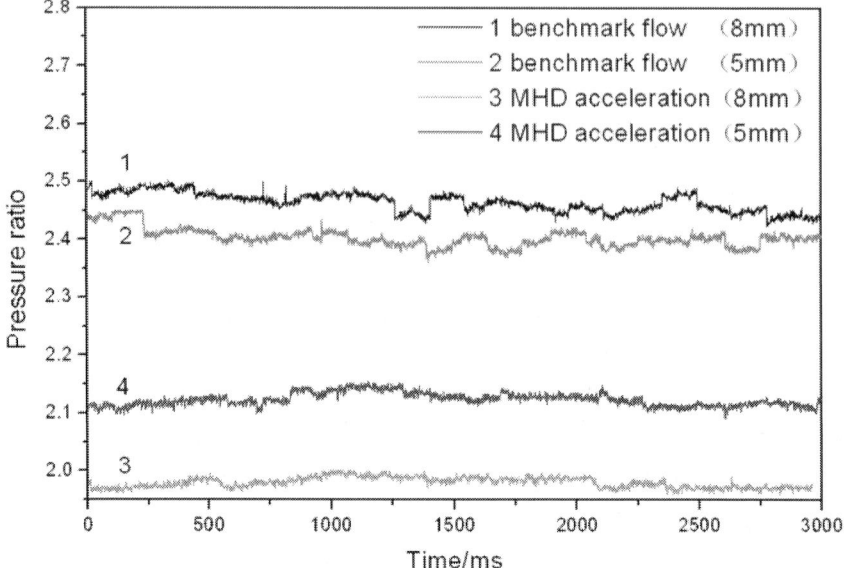

Figure 36. Flow characteristics with different distances

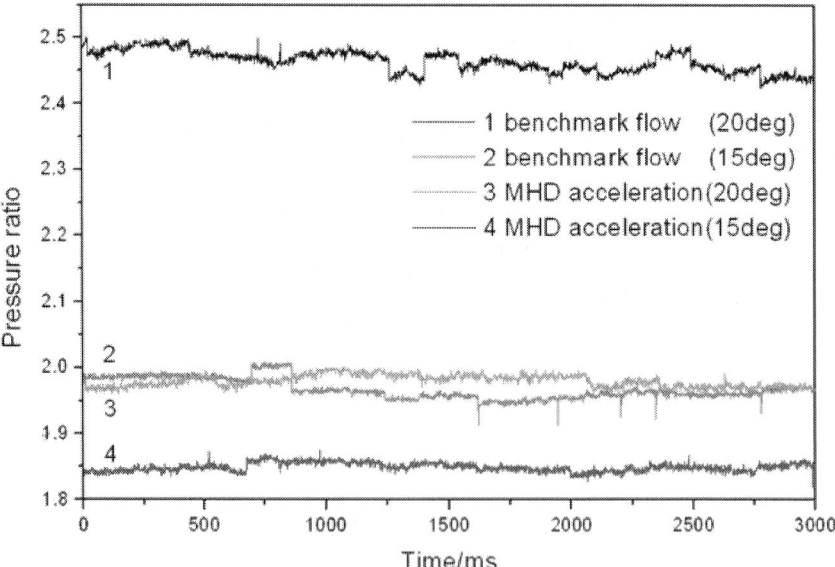

Figure 37. Flow characteristics with different ramp angles

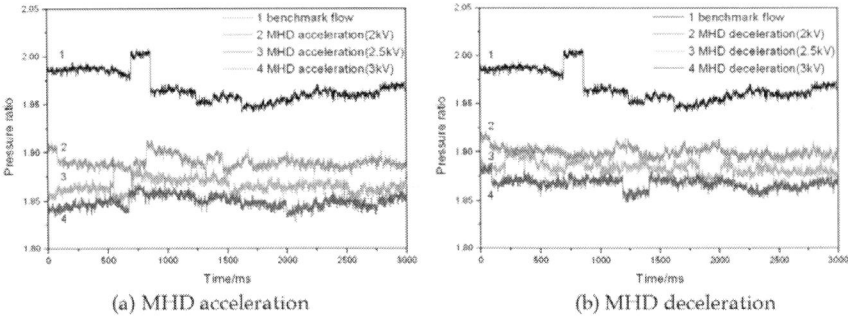

(a) MHD acceleration

(b) MHD deceleration

Figure 38. Flow characteristics with different VDC

The schlieren pictures were captured during the time of flow characteristic measurement at all conditions. The typical schieren pictures were illustrated in Fig. 39 at D=8mm, A=20° with MHD acceleration. Fig. 39(a) showed the benchmark flow with no electric field and no magnetic field whileFig. 39(b) showed the flow with electric field and magnetic field in the case of MHD acceleration. Four shock waves before the ramp were shown in the benchmark picture, three of which were produced by the coarse interface between electrodes and ceramic and the last shock wave was produced by the ramp. Without MHD acceleration the shock wave angle near the ramp was about 37.5°, and the distance between the shock wave location and the ramp was about 7.1mm. With MHD acceleration the shock wave angle near the ramp was reduced to 35°, and the distance was increased to 10mm which meant the shock wave was moved upstream by 2.9mm. Therefore, MHD flow control could change shock wave location, convert a strong shock wave into many weak shock waves, weaken the shock wave strength and change the flow characteristic near the ramp.

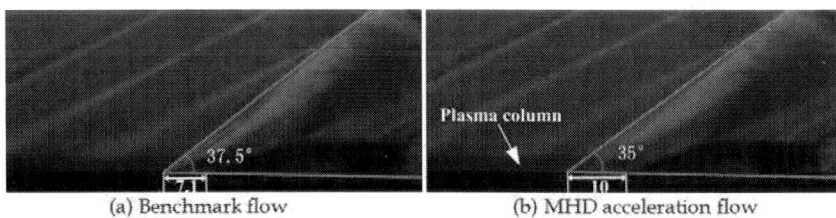

(a) Benchmark flow

(b) MHD acceleration flow

Figure 39. Benchmark and MHD acceleration flow at a typical condition

5. CONCLUSION

The principle of plasma aerodynamic actuation and its application in subsonic and supersonic flow control was summarized. The mechanism for plasma flow control can be summarized as momentum effect, shock effect, and chemical effect. Both the plasma and flow characteristics of the plasma aerodynamic actuation were investigated. Plasma flow control used in airfoil separation control, corner separation control, axial compressor stability extension and shock wave control were studied.

6. ACKNOWLEDGEMENTS

The authors thank Yikang Pu, Shouguo Wang, Junqiang Zhu and Chaoqun Nie for the help in the experiment and analysis. Support for this work was received from the National Natural Science Foundation of China (50906100, 10972236, 51007095) is gratefully acknowledged.

REFERENCES

1. T. Corke, M. Post, D. Orlov, Dielectric Barrier Discharge Plasma Actuators for Flow Control, Progress in Aerospace Sciences, 43January 2010 505 529 0376-0421

2. E. Moreau, Airflow Control by Non-thermal Plasma Actuators, Journal of Physics D: Applied Physics, 40 3January 2007 605 636 0022-3727

3. J. Roth, Aerodynamic Flow Acceleration Using Paraelectric and Peristaltic Electrohydrodynamic Effects of a One Atmosphere Uniform Glow Discharge Plasma, Physics of Plasmas, 10 5May 2003 2117 2126 0107-0664X

4. P. Bletzinger, B. Ganguly, D. Van Wie, 2005Plasmas in High Speed Aerodynamics, Journal of Physics D: Applied Physics, 38 4April 2005), R33 R57 0022-3727

5. Y. Wu, Y. Li, M. Jia, H. Song, Z. G. Guo, X. Zhu, Y. Pu, Influence of Operating Pressure on Surface Dielectric Barrier Discharge Plasma Aerodynamic Actuation Characteristics, Applied Physics Letters, 93 3July 2008 031503p), 0003-6951

6. Y. Li, Y. Wu, M. Zhou, C. Su, X. Zhang, J. Zhu, Control of the Corner Separation in a Compressor Cascade by Steady and Unsteady Plasma Aerodynamic Actuation, Experiments in Fluids, 48 6June 2010 1015 1023 0723-4864

7. Y. Li, Y. Wu, M. Jia, Z. Zhou, Z. Guo, Y. Pu, Optical Emission Spectroscopy Investigation of a Surface Dielectric Barrier Discharge Plasma Aerodynamic Actuator, Chinese Physics Letters, 25 11November 2008 4068 4071 0025-6307X

8. Y. Wu, Y. Li, M. Jia, H. Song, C. Su, Y. Pu, Experimental Investigation into Characteristics of Plasma Aerodynamic Actuation Generated by Dielectric Barrier Discharge, Chinese Journal of Aeronautics, 23 1January 2010 39 45 1000-9361

9. Y. Li, J. Wang, C. Wang, Z. An, S. Hou, F. Xing, Properties of Surface Arc Discharge in a Supersonic Airflow, Plasma Sources Science and Technology, 19 2March 2010 025016p), 0963-0252

10. J. Wang, Y. Li, F. Xing, Investigation on Oblique Shock Wave Control by Arc Discharge Plasma in Supersonic Airflow, Journal of Applied Physics, 106 7October 2009 073307p), 0021-8979

11. J. Wang, Y. Li, B. Cheng, C. Su, H. Song, Y. Wu, Effects of Plasma Aerodynamic Actuation on Oblique Shock Wave in a Cold Supersonic Flow, Journal of Physics D: Applied Physics, 42 16June 2009 165503p), 0022-3727

12. C. Su, Y. Li, B. Cheng, J. Wang, J. Cao, Y. Li, Experimental Investigation of MHD Flow Control for the Oblique Shock Wave Around the Ramp in Low-temperature Supersonic Flow, Chinese Journal of Aeronautics, 23 1January 2010 22 32 1000-9361

13. Y. Wu, Y. Li, J. Zhu, C. Su, H. Liang, G. Li, 2007Experimental Investigation of a Subsonic Compressor with Plasma Actuation Treated Casing, 37th AIAA Fluid Dynamics Conference and Exhibit, 1 8Miami, FL, USA, June 25-28, 2007

14. Y. Li, Y. Wu, H. Liang, H. Song, M. Jia, 2010. The mechanism of plasma shock flow control for enhancing flow separation control capability, Chinese Sci Bull (Chinese Ver), 55 31November 2010), 3060 3068 0002-3074X

15. Y. Li, H. Liang, Q. , Y. Wu, H. Song, W. Wu, 2008Experimental investigation on airfoil suction side flow separation suppression by pulse plasma aerodynamic actuation, Acta Aeronautica et Astronautica Sinica (Chinese Ver), 29 6November 2008), 1429 1435 1000-6893

CHAPTER 3

New Sensors and Techniques for the Structural Health Monitoring of Propulsion Systems

Mark Woike,[1] Ali Abdul-Aziz,[1] Nikunj Oza,[2] and Bryan Matthews[2]

[1]*National Aeronautics and Space Administration, Glenn Research Center, Cleveland, OH 44135, USA*

[2]*National Aeronautics and Space Administration, Ames Research Center, Moffett Field, CA 94035, USA*

ABSTRACT

The ability to monitor the structural health of the rotating components, especially in the hot sections of turbine engines, is of major interest to aero community in improving engine safety and reliability. The use of instrumentation for these applications remains very challenging. It requires sensors and techniques that are highly accurate, are able to operate in a high temperature environment, and can detect minute changes and hidden flaws before catastrophic events occur. The National Aeronautics and Space Administration (NASA), through the Aviation Safety Program (AVSP), has taken a lead role in the development of new sensor technologies and techniques for the in situ structural health monitoring of gas turbine engines. This paper presents a summary of key results and findings obtained from three different structural health monitoring approaches that have been investigated. This includes evaluating the performance of a novel microwave blade tip clearance sensor; a vibration based crack detection technique using an externally mounted capacitive blade tip clearance sensor; and lastly the results of using data driven anomaly detection algorithms for detecting cracks in a rotating disk.

1. INTRODUCTION

The development of in situ measurement technologies and fault-detection techniques for the structural health monitoring of gas turbine engines is of major interest to NASA's Aviation Safety Program and the aeronautical community. The rotating components of modern gas turbine engines operate in severe environmental conditions and are exposed to high thermal and mechanical loads. The cumulative effects of these loads often lead to high stresses, structural deformities, cracks, and eventual component failure. Current structural health monitoring practices involve periodic inspections and schedule-based maintenance of engine components to ensure their integrity over the lifetime of the engine. However, these methods have their limitations, and failures are experienced leading to unscheduled maintenance and unplanned engine shutdowns. To prevent these failures and enhance aviation safety, the NASA Glenn Research Center has investigated new sensor technologies and techniques for the in situ structural health monitoring and detection of defects in gas turbine engines.

Much of the research effort to date has dealt with the development of low technology readiness level (TRL) structural health concepts with the intent of validating these concepts and transitioning them to higher TRLs for usage on actual aero engine hardware. In this study, microwave blade tip clearance and blade tip timing sensor technology is being investigated as a means of making high temperature noncontact structural health measurements in the hot sections of gas turbine engines. It is specifically being targeted for use in the high pressure turbine (HPT) and high pressure compressor (HPC) sections to directly monitor the structural health of the rotating components. The capability to make in situ health measurements is a need that has been identified by the aero engine community as blade damage in the HPT and HPC sections account for 12 percent of the inflight engine shutdown events and 32 percent of the damage events that caused engine removal for unscheduled maintenance [1].

Currently there are no off-the-shelf blade tip clearance sensors that are used in commercial turbine engines for in-situ structural health monitoring, as there are challenges with the sensors being able to operate in and survive the harsh high temperature environment. Microwave sensor technology is appealing in that it is accurate, it has the ability to operate at extremely high temperatures, and is unaffected by contaminants that are present in turbine engines. In addition to its use for structural health monitoring, this type of sensor also has parallel usage in improving engine performance through its use in active turbine tip clearance control schemes. As a means of better understanding the issues associated with the microwave sensors, a series of evaluation experiments were conducted to evaluate their performance on aero turbine engine like hardware.

Along with the development and evaluation of new types of blade tip clearance sensors, efforts have been placed into developing techniques that utilize these sensors for the structural health monitoring of engine components. The vibration based crack detection experiment that will be discussed involved introducing a notch to simulate a crack on a subscale turbine engine disk and monitoring its vibration response as the disk was rotated at speeds up to 12 000 rpm. The vibration response was characterized by using externally mounted capacitive blade tip clearance sensors to measure the combined disk-rotor system's whirl amplitude and phase during operation. Testing was performed on a clean undamaged baseline disk and a disk with a 50.8 mm long notch machined into the disk to simulate a crack. The responses were compared and evaluated against the theoretical models to investigate the applicability and success of detecting the notch.

And finally, in parallel with the vibration based crack detection investigation an experiment using data driven anomaly detection algorithms was undertaken as a means of evaluating whether these data mining techniques could be used to detect a fault or an anomaly in a rotating engine disk. In this experiment blade tip clearance data sets were acquired on an undamaged subscale engine rotor disk as it was operated at speeds up to 10 000 rpm. This baseline data was used to train three different data mining algorithms. Data was then acquired from a damaged disk with a known crack and the data mining algorithms were used to detect the presence of the crack.

This paper presents a summary of key results and findings obtained from three different structural health monitoring approaches that have been investigated. This includes evaluating the performance of a novel microwave blade tip clearance sensor; a crack detection technique using externally mounted blade tip clearance sensors; and lastly discussing the results of a data driven anomaly detection technique for sensing cracks in a rotating rotor disk.

2. MICROWAVE BLADE TIP CLEARANCE/BLADE TIP TIMING SENSORS

The ability to monitor the structural health of the rotating components, especially in the hot sections of turbine engines, is of major interest to aero community in improving engine safety and reliability [2]. In addition, the active control and minimization of the gap between the rotating turbine blades and the stationary case of gas turbine engines is being sought as a means of increasing engine efficiency, reducing fuel consumption, reducing emissions, and increasing engine service life [3]. The use of instrumentation for these applications in gas turbine engines requires sensors that are highly accurate and can operate in a high temperature environment. To address this need, microwave sensor technology is being investigated as a means of making high temperature non-contact blade tip clearance and tip timing measurements for use in structural health monitoring and active clearance control applications in turbine engines. This technology is appealing due to its high accuracy and its potential to operate at extremely high temperatures that are present in turbine engines. It is intended to use blade tip clearance to monitor blade growth and wear and blade tip timing to monitor blade vibration and deflection.

2.1. Sensor Background and Theory

NASA has worked with Radatec (now Meggitt Inc.) through the Small Business Innovation Research (SBIR) program for the development of microwave sensor technology for high temperature noncontact blade tip clearance and blade tip timing measurements. The initial development of the technology was accomplished through a phase II SBIR contract awarded in 2002. Further development of the technology was accomplished in 2004 through 2005 as part of NASA's Ultra Efficient Engine Technology (UEET) Program. A prototype first generation 5.8 GHZ system was delivered as part of a phase III SBIR commercialization contract in 2007 and a second generation 24 GHZ system along with upgraded electronics was delivered as part of subsequent follow-on contracts in 2009 and 2010, respectively.

The microwave blade tip clearance sensor operates essentially as a field disturbance device. The tip clearance probe contains both a transmitting and receiving antenna. The sensor emits a continuous microwave signal and measures the signal that is reflected off a rotating blade. The sensor measures the changes in the microwave field due to the blade passing through the field. The motion of the blade phase modulates the reflected signal and this reflected signal is compared to an internal reference. Changes in amplitude and phase directly correspond to the distance to the blade. The time interval of when the blade passes through the field is measured to provide blade tip timing. More detailed information on the sensor's theory of operation can be found in [4–6]. The microwave blade tip clearance probes are made of high temperature material and are designed to operate in temperatures up to 900°C. The first generation probes (Figure 1(a)) operate at 5.8 GHz and can measure clearance distances up to one-half the radiating wavelength which is 25 mm. The second generation probes (Figure 1(b)) operate at 24 GHz and can measure clearance distances up to 6 mm. The 5.8 GHz sensor is targeted for use on large rotating machinery such as land based power turbines or in the fan sections of aero gas turbine engines. The 24 GHz sensor is being targeted for use in smaller rotating machinery applications such as the turbine and compressor sections of aero engines. This technology has an ultimate goal of obtaining clearance measurement accuracies approaching +/−25 μm. A frequency response of up to 5 MHz is typical, with up to 25 MHz being possible with this technology which lends itself for use in structural health monitoring applications for the measurement of blade deflection and vibration.

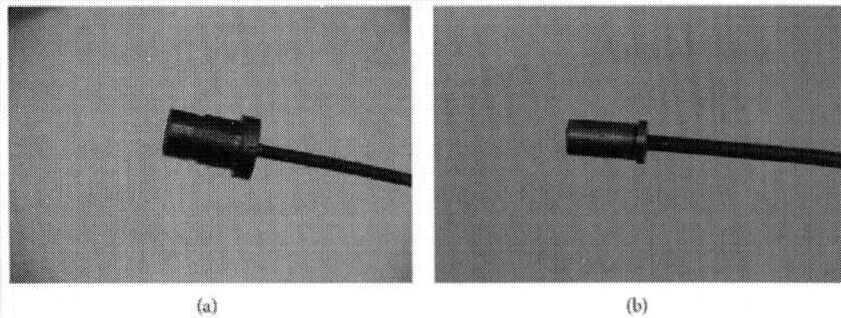

(a)   (b)

Figure 1: Microwave blade tip clearance probe (Meggitt). (a) 5.8 GHz probe. (b) 24 GHz probe.

2.2. Experimental Results

NASA's primary goal is to demonstrate this microwave blade tip clearance sensor technology on an actual gas turbine engine in a relevant high temperature environment. However, the use of microwave sensors for this application is an emerging concept. Techniques on their use and calibration needed to be understood and developed. In addition, the microwave sensor's accuracy and ability to make blade tip clearance and deflection measurements had to be assessed prior to use on actual engine hardware. As a means of better understanding the issues associated with the microwave sensors, a series of experiments were conducted to evaluate the sensor's performance on aero engine type applications. A summary of these experiments and their results are as follows.

The first generation 5.8 GHz microwave sensors were used to a make blade tip clearance measurements on a large axial vane fan and a subscale NASA Turbofan. The purpose of the test on the large Axial Vane Fan (Figure 2) was to develop the infrastructure required for the calibration of the sensors and the techniques for their use in the field to make clearance measurements on large rotating machinery. The motivation behind their use on the NASA Turbofan (Figure 3) was to evaluate the first generation sensor's ability to acquire blade tip clearance data on an aero engine size test article and blades. Blade tip clearance data sets were acquired for several test runs of the NASA Turbofan. Data was acquired at a variable sampling rate that was synchronized to the fan's speed. Each measurement consisted of two revolutions of data with 10 000 samples taken per revolution. Figure 4 shows the individual blade clearances measured for several fan speeds. It is clearly noted from the polar plot that the tip clearances decreased as the fan speed is increased.

Figure 2: Axial vane fan located at the SWT.

Figure 3: NASA Turbofan.

This result is expected and is due to the growth and expansion of the turbofan's composite blades as the fan operates at higher speeds. An average decrease of 0.22 mm was observed as the rig speed was increased to 8 875 rpm. The change in clearance detected in this experiment was within the range predicted for these blades. In addition, the change in tip clearances measured by the microwave sensors was nearly identical to previously recorded values obtained with capacitive clearance sensors. In previous test entries, changes in tip clearances of up to 0.22 mm were also noted when the turbofan was operated over the same speed range. These two initial experiments served as test beds for the development of techniques and infrastructure required for the calibration of the sensors and successfully demonstrated the microwave clearance sensor's ability to make measurements on aero engine size hardware.

The second generation 24 GHz sensors were used to make measurements on a 32 blade subscale turbine engine like disk and on an actual compressor stage from a small aero engine. These experiments were conducted in a laboratory environment using a calibration spin rig, a High Precision Spin Rig, and a High Temperature Spin Rig for the purposes of evaluating the second generation sensor's capability of making both clearance and timing measurements on small aero engine hardware. The 32 blade subscale engine disk (Figure 5) was manufactured to have 6 equally spaced blades that were bent at predefined angles or deflection distances. The purpose of this experiment was to evaluate the sensors ability to generate blade tip clearance and blade tip timing measurements simultaneously. During this experiment it was successfully demonstrated that the sensors were able to reliably make absolute measurements down to a clearance level of 100 µm. This was an improvement over the previously demonstrated minimum measurement of 500 µm made with the first generation sensors. This ability to measure very low clearance ranges is critical for use in closed loop clearance control applications. In addition, the sensor was able to simultaneously detect the minimum deflection of 0.70 mm (2 degrees) that was machined onto the disk. It is planned to further explore this capability by fabricating a disk that has smaller deflections in order to see the minimum deflection that can be reliably measured.

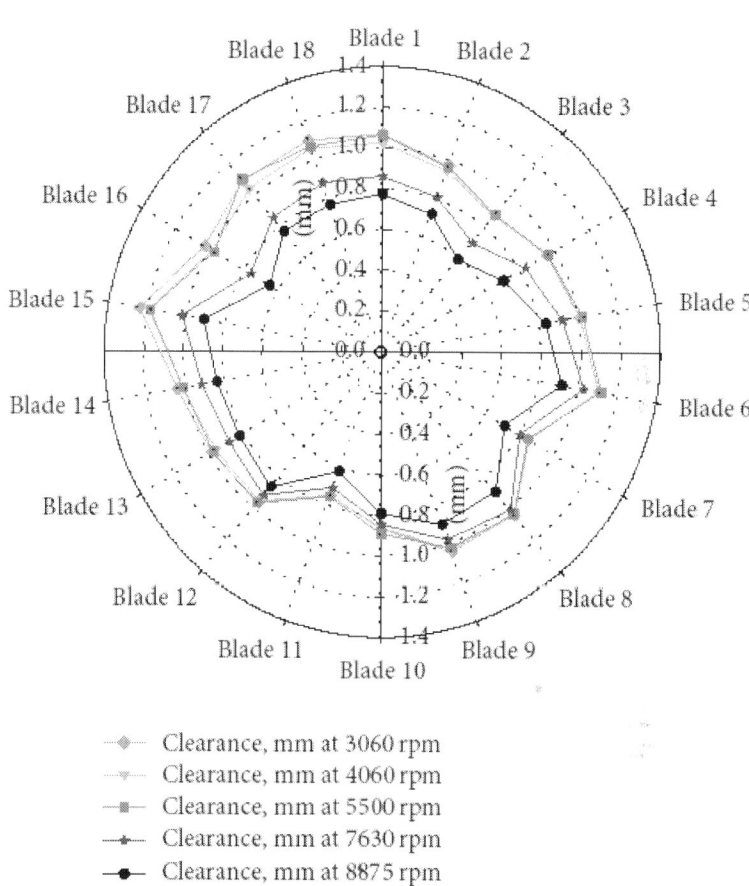

Figure 4: Blade tip clearance polar plot for probe #1, 90° location—NASA Turbofan experiment.

The investigation using a compressor stage from a small aero engine was conducted for the purposes of calibrating the sensors for future use on a small compressor and evaluating their ability to measure blade tip clearances on actual engine hardware over the very low clearance ranges typically associated with aero engines. For this experiment clearances were successfully measured over a range from 0.10 to 1.50 mm with a maximum observed error of +/− 0.021 mm. These results were encouraging in that the sensor was able to make measurements on actual aero engine hardware over a relatively low clearance range within the desired accuracy of +/− 0.025 mm. However, it is acknowledged that these results were achieved in the ideal set-up of a laboratory and that the results may differ in an actual engine due to noise and other environmental effects.

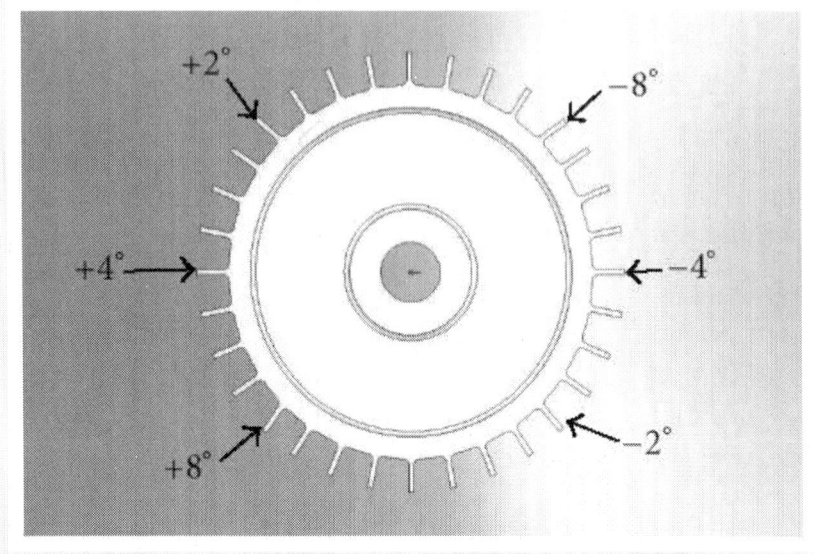

Figure 5: Simulated turbine disk with prebent blades for blade tip deflection testing.

In summary, the evaluation testing that was accomplished on the microwave blade tip clearance technology has shown that sensors are a viable option for propulsion health monitoring and clearance control applications. Techniques for the calibration, integration, and use of the sensors to make measurements on aero engine hardware were successfully developed and both the first and second generation sensors have been successfully demonstrated on rotating machinery and aero engine hardware. A demonstration test of these sensors along with other advanced technologies on an engine ground test is being planned to occur in the 2013-2014 timeframe.

3. VIBRATION BASED CRACK DETECTION TECHNIQUE

This crack-detection methodology involved introducing a notch on a simulated turbine engine disk and monitoring its vibration response as the disk was rotated at speeds up to 12 000 rpm. The vibration response was characterized by monitoring the disk-rotor system's whirl amplitude and phase during operation. The whirl amplitude and phase were derived from the blade tip clearance profile that was measured using externally mounted blade tip clearance sensors. This type of sensor was chosen as it represents the type of sensor that is most likely to be installed on a commercial engine due to its minimal installation impact, low overhead, and parallel benefits that it can bring to improve engine performance through its use in active turbine tip clearance control. The testing was performed on a clean undamaged baseline disk and on a damaged disk with a 50.8 mm long notch machined into the disk to simulate a crack. The experiment was conducted at the NASA Glenn Research Center's High Precision Spin Rig, a rig which is ideally suited for the development and validation of low technology readiness level (TRL) concepts before implementation on more expensive and complicated rotating machinery. A description of the High Precision Spin Rig, the baseline theory behind this technique, the experimental setup, and the results are discussed.

3.1. High Precision Spin Rig Description

Figure 6 shows the Rotordynamics Laboratory's High Precision Spin Rig, which can accommodate simulated engine rotor disks of up to 235 mm in diameter. It has a stainless steel shaft with a length of 781 mm and diameter of 20 mm. The shaft is supported by precision contact ball bearings on each end and has adjustable dampers that were positioned along the length of the shaft for these experiments. An encoder mounted on the end of the shaft was used by the control system to provide closed-loop control of rig speed. A secondary optical tachometer was used to record the speed into the data system and to synchronize the data to the rig's rotation. A 12-hp custom-built, brushless direct-current (dc) motor was used to rotate the spin rig and the subscale engine disks at speeds up to 12 000 rpm.

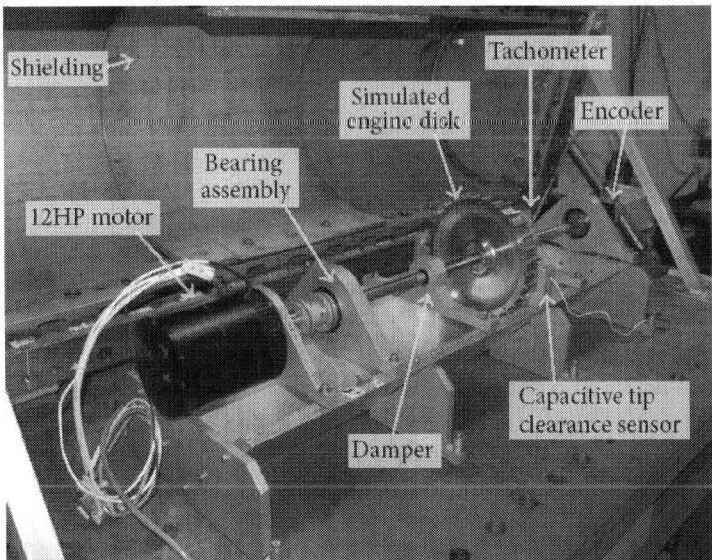

Figure 6: High Precision Spin Rig.

The rig was set up to acquire two channels of radial blade tip clearance data from the simulated engine disk using capacitive displacement sensors (Figure 7). These sensors were developed as part of a NASA Small Business Innovation Research (SBIR) contract and were different from traditional capacitive sensors in that their operation was based on a dc offset technique instead of the typical modulation technique. A National Instruments System was used to acquire data from the capacitive displacement probes at a fixed sampling rate of 1 MHz. This system and its application software were delivered as part of the SBIR contract for the capacitive blade tip clearance sensors. The system used custom data acquisition and processing applications that were tailored for acquiring and processing data from the blade tip clearance sensors.

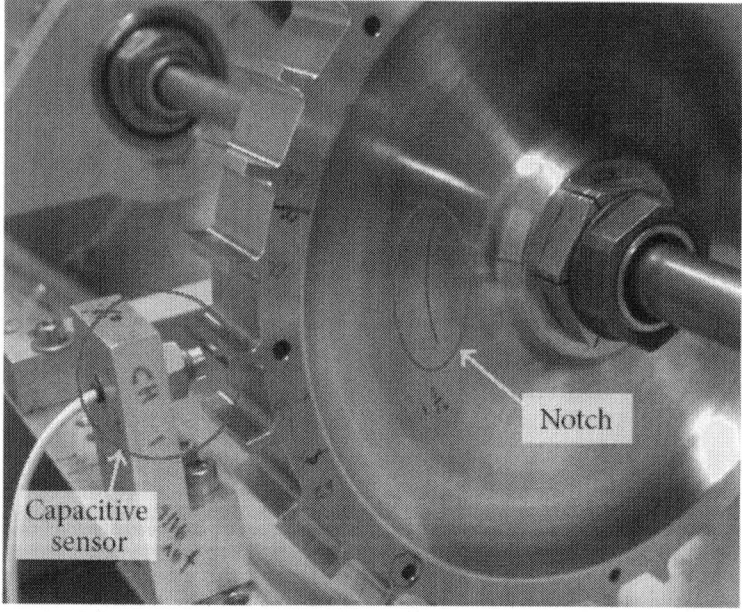

Figure 7: Capacitive blade tip clearance sensor and subscale turbine engine disk with notch.

3.2. Experimental Theory and Setup

The theory behind the crack-detection methodology that was investigated was based on previous theoretical and experimental work performed by Abdul-Aziz et al. [7, 8], Gyekenyesi et al. [9, 10], and Haase and Drumm [11]. The goal of this experiment was to determine if the crack-detection methodology investigated in these earlier studies could be validated by using the spin rig to conduct tests on simulated engine rotor disks with a notch introduced to replicate a crack.

The detection methodology is based on monitoring the vibration response of rotating disks to determine if a crack has developed. The theory implies that a defect, such as a crack, creates minute deformations in the disk as it is rotated. The deformation, in turn, creates a speed-dependent shift in the disk's center of mass. It was theorized that this shift could be detected by analyzing the vibration whirl amplitude and phase as measured by the blade tip clearance profile. The system's behavior was modeled after a 2-degree-of-freedom Jeffcott rotor. The model predicts that the vibration amplitude peaks when a clean, undamaged disk goes through the first critical speed but heads to a lower steady-state value as the speed is increased above the critical speed. Correspondingly, the phase shifts 180° when going through the first critical speed and then stabilizes to a steady-state value as the speed is increased past critical. At this point, the rotor is rotating about the combined system's center of mass and has stabilized. However, as the speed of a cracked disk is increased, centrifugal forces open the crack. This, in turn, deforms the disk and shifts its center of mass. At speeds above critical, the crack-induced shift in the disk's center of mass starts to grow and dominate the overall system's vibration response. The modeling predicts that, instead of heading toward a steady-state value, the vibration amplitude will change as a second-order function of rotational speed as it is increased beyond the first critical speed. This can be detected by analyzing the vibration response of the disk-rotor system, particularly the amplitude and phase of the first harmonic, as the system is operated over a range of speeds.

The testing approach used in this experiment was to spin a subscale engine disk over a speed range from 0 to 12 000 rpm and simultaneously record its vibration response. The vibration response was acquired as previously described using capacitive displacement probes to measure the blade tip clearance profile. Two subscale simulated engine turbine disks were tested. The first disk was undamaged and was used to acquire the baseline vibration response data. The disk had an outside diameter of 235 mm, a bore thickness of 25.4 mm, and an outside rim thickness of 31.75 mm. The thinnest portion of the disk's web was 2.54 mm. Thirty-two teeth to simulate blades were evenly spaced around the circumference of the disk. Each simulated blade had a cross section of 31.75 mm by 3.30 mm and a height of 8.38 mm. The disk was made of a nickel base alloy, Haynes X-750 (Haynes International, Inc.) and had a weight of 4.88 kg.

The damaged disk is shown in Figure 7. It was identical to the baseline disk with the exception that a 50.8 mm long notch had been introduced in its mid-span region to imitate a crack in the disk. The mid-span region was selected since prior finite-element analysis had shown that this area experienced high stress levels during operation. The data acquired from this disk were compared with the data from the undamaged baseline disk to determine if the simulated crack could be detected by analyzing the vibration response. Supportive analytical calculations were made using finite-element analysis to complement the experimental work and determine the expected radial growth of the disk. The finite-element analysis predicted a radial growth on the order of ~0.075 mm at the highest operating speed, 12 000 rpm. This was within the detection limit of the blade-tip-clearance instrumentation. However, it should be noted that the major effect that was to be monitored was how the notch changed the combined center of mass of the disk-rotor system and its vibration amplitude and phase, not the radial blade tip growth of the disk.

3.3. Experimental Results

The experimentation consisted of operating the baseline and notched disks at several speed profiles and recording their vibration responses using the capacitive blade tip clearance sensors. The first harmonic component of the tip-clearance profile for each revolution was then analyzed to determine if the amplitude and phase exhibited the expected characteristics associated with a crack in the disk. Approximately 15 test runs were conducted for each disk. A typical test run or cycle consisted of ramping up the disk's speed from 0 rpm to a predefined maximum speed, remaining on condition for 0 to 60 s, then ramping back down to 0 rpm. Blade tip clearance data were acquired over the entire cycle. Data sets were acquired at ramp rates of 60 and 100 rpm/s for peak operating speeds of 5 000, 10 000, and 12 000 rpm. In addition, complex mission profiles were conducted where the speed was ramped to various power levels in an attempt to simulate the rigorous loading conditions that an engine would experience during a typical mission.

The run profile presented in this paper is shown in Figure 8. It is typical of the profiles that were used in the investigation. For this test case, the speed of the disk was increased from 0 to 10 000 rpm at a ramp rate of 100 rpm/s, cycled between 10 000 and 5 000 rpm for three cycles, then ramped down from 10 000 to 0 rpm. This was done to simulate a mission profile from start-up to full power and to observe the effects of cyclical loading on the disks. This ramping up and down also maximized the amount of data that could be acquired over a wide range of speeds, which was critical for the validation of this concept.

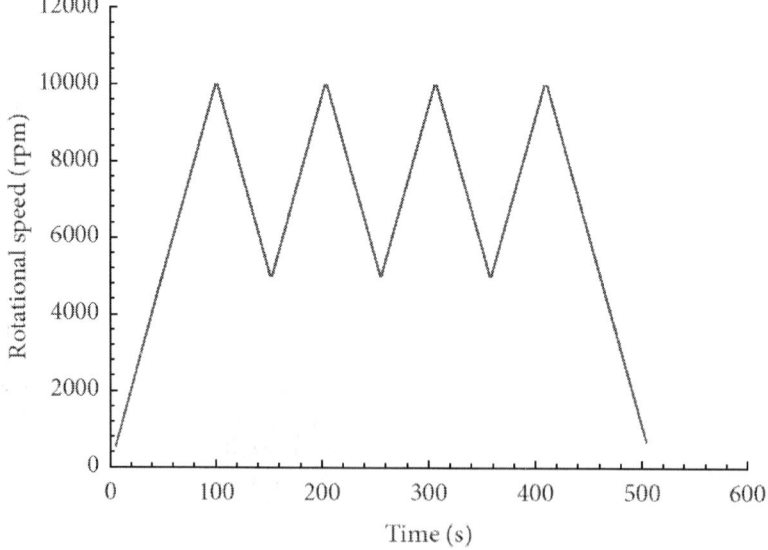

Figure 8: Simulated mission profile.

Figure 9 shows the results from this multiple-cycle test run for both the baseline and notched disks. The figure shows the analysis for the last cycle, starting at 10 000 rpm and ending at 0 rpm. As previously discussed, the presence of a crack is indicated if a second-order speed-dependent variation can be observed in the vibration amplitude as it is operated in the postcritical speed regime. In this test case, a speed-dependent rise was observed in the amplitude data for the notched disk. The area of interest showing the speed-dependent rise is highlighted within the circle of Figure 9(b). A curve-fit analysis, shown in Figure 10, was conducted on the vibration data in this region and it was found to closely follow a second-order polynomial fit, thus following the predicted behavior and potentially indicating the presence of the simulated crack.

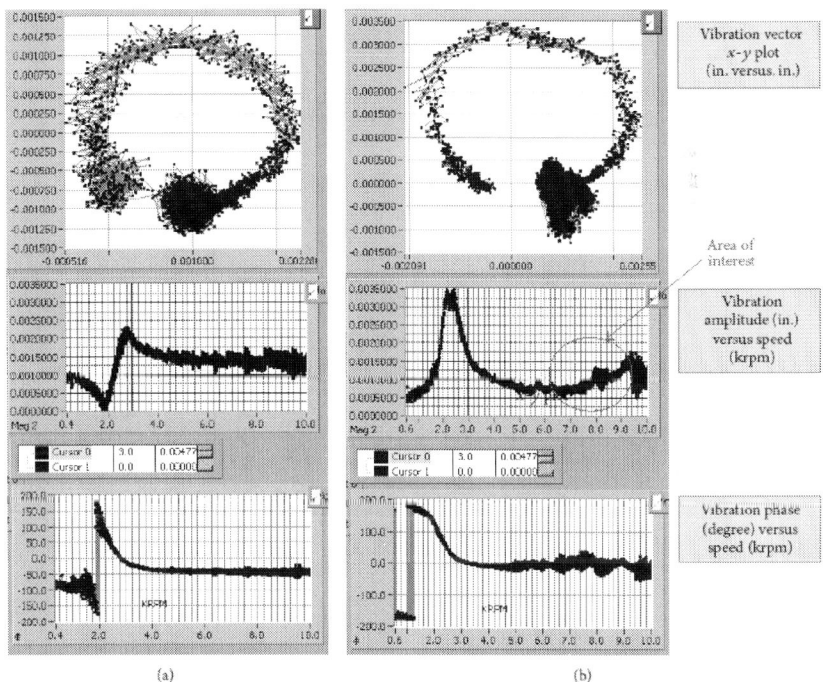

(a)   (b)

Figure 9: 10 000 rpm multiple-cycle run comparison. (a) Baseline disk. (b) Notched disk.

Although the vibration data looked promising in this region, it was also observed that the second-order speed-dependent variation was not consistent over the entire range and that it appeared to reset itself at a speed of ~9500 rpm. Moreover, this relatively large amplitude variation was not consistently observed in other test runs, which cast some doubt on whether the effect was entirely due to the notch or to some other factor that is yet to be determined. However, this case yielded the best data so far and showed positive indications of being able to detect a defect such as a crack by analyzing the disk's vibration response as it is operated over a range of speeds. It is theorized that what made this case different, and more promising than other test runs, is that this test was conducted towards the end of the test series and many cycles had been placed on the disk prior to this test run. In addition, the disk experienced more loading because of the cyclical nature of the test profile. Overall, the results were promising and plans are in place to further investigate and refine this crack-detection technique as part of future studies.

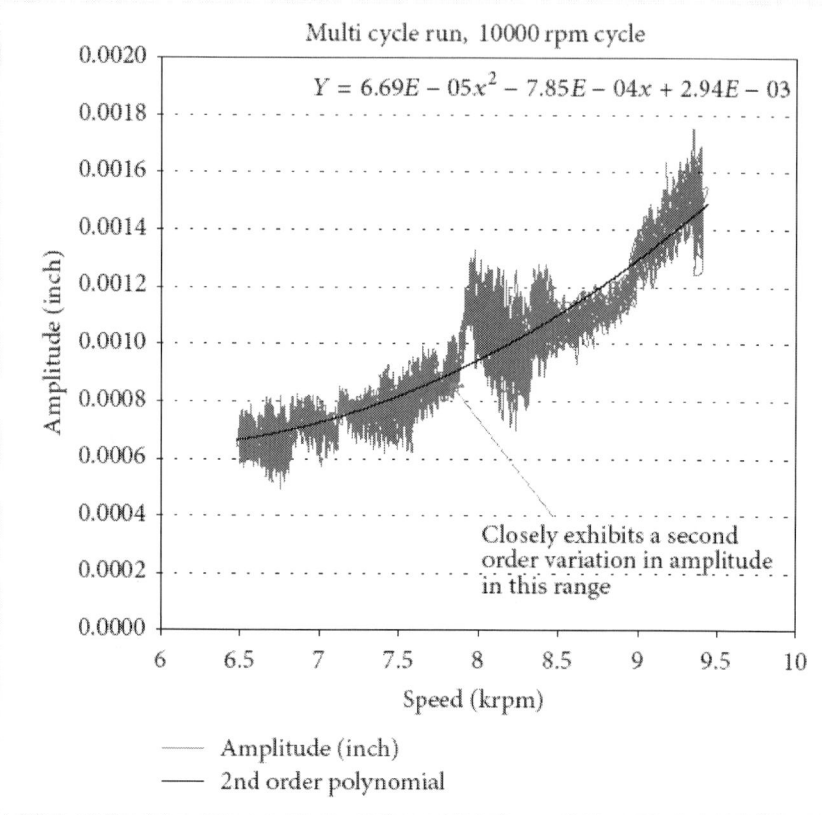

Figure 10: Detailed vibration amplitude plot for 10 000 rpm multiple-cycle run.

4. DATA DRIVEN ANOMALY DETECTION TECHNIQUE

In addition to the experimental work previously described, parallel health monitoring assessments have been conducted using data mining algorithms in order to detect flaws, such as cracks, in a rotating engine disk. In this technique blade tip clearance data were acquired on an undamaged subscale engine disk using the same 10 000 rpm multi-cycle run profile that was used for the previously described vibration based technique. These baseline data were used to train three different data mining algorithms: Orca, Inductive Monitoring System (IMS), and One-Class Support Vector Machine (OCSVM) [12–15]. These algorithms were then used to analyze the blade tip clearance data that was acquired from a damaged disk which had a notch machined into it to simulate a crack in order to evaluate their performance and determine if the algorithms could detect the presence of the notch due to changes in the disks operating characteristics as measured by the blade tip clearance sensors.

As stated previously, three different techniques were investigated: Orca, Inductive Monitoring System (IMS), and One-Class Support Vector Machine (OCSVM). These methodologies provide a technique that can monitor the health of a system with fidelity by training the model to identify normal system parameters from abnormal ones. This is implemented by defining groups of consistent system parameter data using nominal system data. These training data are then used to model the system. Upon learning how the system should behave under nominal operating conditions, these models are then used to identify abnormal behavior, such as a crack or flaw in the disk which causes a change in the known operation of the system.

4.1. Detection Algorithms

4.1.1. Orca

Orca is an outlier detection algorithm which uses a Euclidean distance nearest neighbor based approach to determine outliers. For computational efficiency, it employs a modified pruning technique which allows it to perform in near linear time. For each point in the test data set, where a point is a row in the data set consisting of measurements taken at a single point in time, Orca calculates the nearest neighbor points from the reference data set. The output from Orca is a distance score which represents the average distance to its k-nearest neighbors; the more anomalous the point is the higher the score, since the nearest neighbors are farther away. More information about this algorithm can be found in [13].

4.1.2. Inductive Monitoring System (IMS)

IMS is a cluster based modeling method. The algorithm is given a set of nominal data points and builds a model by agglomerative clustering of the data points. The resulting model is used to generate anomaly scores for new data. For each test data point, IMS finds its distance to the nearest cluster's boundary. The score that is reported is the sum of the squares of the distances from the test data point to the dimensional bounds of the closest cluster. If a data point falls entirely within the cluster bounds, the point is expected to represent normal behavior and it is assigned a score of zero. More information about IMS can be found in [12].

4.1.3. One-Class Support Vector Machine (OCSVM)

OCSVM is a one-class nonlinear kernel based algorithm that maps the training data to a higher-dimensional feature space and then linearly separates nominal data from anomalies in that feature space. The idea is that such a model corresponds to a nonlinear model in the original data space but still maintains the benefit of a linear model in that it is guaranteed to return the model with the lowest error over the training set. The algorithm identifies a subset of the training data, called the "support vectors," which is used to generate a hyperplane model. The anomaly score that is reported is the distance from the test data point to the hyperplane as measured in the feature space. More information on OCSVM can be found in [14, 15].

4.2. Analysis-Results

For the multiple-cycle test run previously shown in Figure 8 blade tip clearance data were recorded for both the undamaged and damaged, notched, disk. For the analysis, the 32 individual blade tip clearance measurements acquired for each revolution were utilized in training and evaluating the three anomaly detection algorithms. The undamaged disk's data was randomly divided in half. The first half was used for training of the algorithm and second half was used for validation. The means and standard deviations for all 32 channels were calculated for the training data and used to normalize both the training and testing data sets.

The plots of Figure 11 are the global anomaly scores for each of the algorithms for both the validation undamaged data set and the damaged data set over time. The anomaly scores are portrayed as positive values and any nominal points are represented as zero. This was done to allow for similar comparison across the algorithms. It is important to note that OCSVM allows some nominal sample points in the training set to be classified as anomalous. This percentage is governed by the Nu parameter, which is set by the user, but for which we set currently at the default value of 10%. Due to this characteristic of the algorithm 10% of all nominal data tested may result in anomalous classification. In analyzing the effectiveness of the three techniques in Table 1 the threshold for correct detection was set to 90% to make the comparisons fair between all three algorithms, since OCSVM, in this case, was optimized for a correct detection of 90%.

Table 1: Multiple 10 000 rpm cycle test results.

	Algorithm	Correct detection rate	False alarm rate	Accuracy	Area under ROC
5% false alarm rate	Orca	100%	5%	97.55%	1.00
	IMS	93%	5%	94.02%	0.93
	OCSVM	100%	5%	97.55%	0.99
90% correct detection	Orca	90%	0.58%	94.62%	1.00
	IMS	90%	0.36%	94.73%	0.93
	OCSVM	90%	1.26%	94.28%	0.99

Figure 11 shows all three algorithms and they appear to have performed quite well in differentiating between the nominal undamaged disk and the damaged, notched disk test runs. Orca and OCSVM seem to show slightly better performance across a few of the metrics. In Table 1 the metrics used for comparison are correct detection, false alarm rate, accuracy, and area under the Receiver Operating Characteristic (ROC) curve. The ROC curve is a plot of false positives versus true positives. The ideal curve is one that has a 90° bend in it shooting straight up with the false positive = 0 and holding the true positive = 1 across resulting in an area of 1.00. When fixing the false alarm rate at 5%,

Orca and OCSVM both have 100% correct detection rate of the notched anomalous data, where IMS's correct detection rate is at 93%. When fixing a threshold so that the correct detection rate is 90%, all three show very good false alarm rates. An additional metric, that is independent of choosing a threshold, is measuring the area under the ROC curve. When comparing this metric across the three algorithms, all three methods are reporting very good areas with Orca and OCSVM doing slightly better than IMS. The results obtained showed that the detection algorithms are capable of predicting anomalies in the rotor disk with very good accuracy. Each detection scheme performed differently under the same experimental conditions and each delivered a different level of precision in terms of detecting a fault in the rotor.

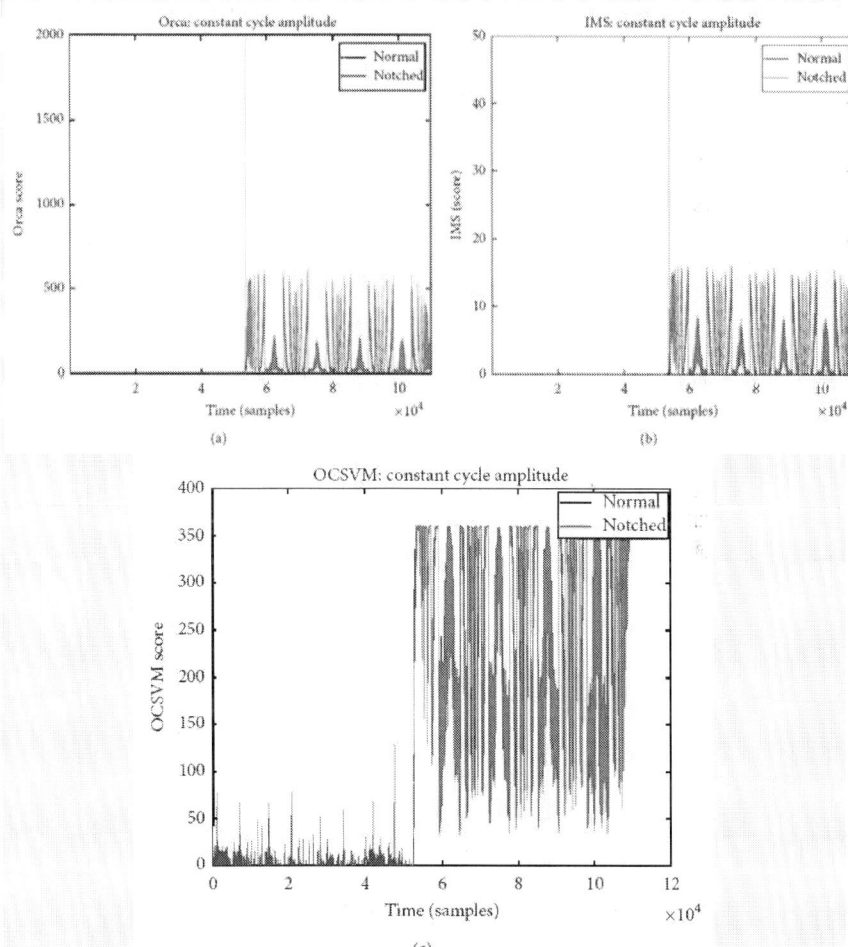

Figure 11: Results of the data driven anomaly detection techniques. (a) Orca. (b) IMS. (c) OCSVM.

5. CONCLUSION AND FUTURE PLANS

The work presented in this paper represents a summary overview of the on-going research activities that are being conducted at the NASA Glenn Research Center for the development of new sensors and fault detection techniques for the structural health monitoring of rotating turbine engine components. The main focus of these efforts is to first develop and validate low technology readiness level (TRL) structural health monitoring concepts and then mature them for further use on more complicated aero engine hardware.

The microwave blade tip clearance sensor technology was found to be a promising option for use in structural health monitoring applications for gas turbine engines. The sensors have been successfully demonstrated on aero engine like hardware. The first generation 5.8 GHz sensors were used on a large axial vane fan and on a NASA Turbofan for the purposes of evaluating the sensor's ability to acquire blade tip clearance data on an aero engine size test article and blades. The second generation 24 GHz sensors were used to make basic blade tip deflection and low range clearance measurements on a simulated engine and on a small aero engine's compressor disk in order to evaluate the sensor's capability of making both clearance and timing measurements on small aero engine hardware. A ground test with these sensors installed on a full scale engine is being planned for the 2013-2014 timeframe.

The vibration based crack detection experimental results showed that this technique has the potential and merit to be used for detecting cracks on a rotating disk. The data acquired from the 10 000 rpm multiple-cycle test run showed a speed-dependent rise in the vibration amplitude, indicating that a crack-induced shift in the disk's center of mass was present. Other test cases equally showed that the system was on the verge of being able to detect the simulated crack using this technique, but not all of the expected trends were consistently observed. Nevertheless, this technique looks very promising and appears to have the sensitivity to detect faults in the rotor disk system. However, additional work remains warranted to further validate and refine this crack detection concept.

Additional health monitoring was performed through combined experimental and data-driven anomaly detection techniques. Three different automatic data-driven detection algorithms, Orca, OCSVM, and IMS, were used. These techniques were utilized to analyze blade tip clearance data acquired from the rotor operating under undamaged and damaged conditions to check the viability of the detection methodology and to evaluate the performance of each approach. The results obtained showed that the detection algorithms are capable of predicting anomalies in the rotor disk with very good accuracy. Each detection scheme performed differently under the same experimental conditions and each delivered a different level of precision in terms of detecting a fault in the rotor. Overall rating showed that both the Orca and OCVSM performed better than the IMS technique, but in the end all of the algorithms showed promise and will be further refined for eventual use in detecting flaws in rotating components of an engine.

REFERENCES

1. U.S. Department of Transportation, Federal Aviation Administration, "Engine Damage-Related Propulsion System Malfunctions," DOT/FAA/AR-08/24, 2008.
2. T. Kurtoglu, K. Leone, M. Revely, and C. Sandifer, "A study on current and emerging technologies and future research requirements for integrated vehicle health management," Internal NASA Report, 2008.
3. S. B. Lattime and B. M. Steinetz, "Turbine engine clearance control systems: current practices and future directions," NASA TM 2002-211794, AIAA-2002-3790, 2002.
4. T. A. Holst, T. R. Kurfess, S. A. Billington, J. L. Geisheimer, and J. L. Littles, "Development of an optical-electromagnetic model of a microwave blade tip sensor," AIAA 2005-4377, 2005.
5. J. L. Geisheimer, S. A. Billington, and D. W. Burgess, "A microwave blade tip clearance sensor for active clearance control applications," AIAA 2004-3720, 2004.
6. T. A. Holst, Analysis of spatial filtering in phase-based microwave measurements of turbine blade tips [M.S. thesis], Georgia Institute of Technology, Atlanta, Ga, USA, 2005.
7. A. Abdul-Aziz, M. R. Woike, J. D. Lekki, and G. Y. Baaklini, "Health monitoring of a rotating disk using a combined analytical experimental approach," NASA/TM 2009-215675, 2009.
8. A. Abdul-Aziz, M. R. Woike, G. Abumeri, J. D. Lekki, and G. Y. Baakilini, "NDE using sensor based approach to propulsion health monitoring of a turbine engine disk," in Health Monitoring of Structural and Biological Systems, Proceedings of SPIE, pp. 9–12, San Diego, Calif, USA, March 2009.
9. A. L. Gyekenyesi, J. T. Sawicki, and G. Y. Baaklini, "Vibration based crack detection in a rotating disk, part 1—an analytical study," NASA/TM 2003-212624, 2003.
10. A. L. Gyekenyesi, J. T. Sawicki, R. E. Martin, W. C. Haase, and G. Y. Baaklini, "Vibration based crack detection in a rotating sisk, part 2—experimental results," NASA/TM 2005-212624/PART2, 2005.
11. W. Haase and M. Drumm, "Detection, discrimination and real-time tracking of cracks in rotating disks," IEEE 0-7803-7321-X/01, 2002.
12. D. L. Iverson, "Inductive system health monitoring," in Proceedings of the International Conference on Artificial Intelligence, Proceedings of the International Conference on Machine Learning, Models, Technologies & Applications (MLMTA '04), vol. 2, June 2004.
13. S. D. Bay and M. Schwabacher, "Mining distance-based outliers in near linear time with randomization and a simple pruning rule," in Proceedings

of the 9th ACM SIGKDD International Conference on Knowledge Discovery and Data Mining (KDD '03), pp. 29–38, Association for Computing Machinery, August 2003.

14. D. M. J. Tax and R. P. W. Duin, "Support vector domain description," Pattern Recognition Letters, vol. 20, no. 11–13, pp. 1191–1199, 1999.

15. B. Schölkopf, J. C. Platt, J. Shawe-Taylor, A. J. Smola, and R. C. Williamson, "Estimating the support of a high-dimensional distribution," Neural Computation, vol. 13, no. 7, pp. 1443–1471, 2001.

CHAPTER 4

A Methodology to Assess the Safety of Aircraft Operations When Aerodrome Obstacle Standards Cannot Be Met[*]

Hartmut Fricke[1], Christoph Thiel[2]

[1]Dresden University of Technology, Chair of Air Transport Technology and Logistics, Dresden, Germany
[2]GfL—Gesellschaft für Luftverkehrsforschung, Dresden, Germany

ABSTRACT

When Aerodrome Obstacle Standards cannot be met as a result of urban or technical development, an aeronautical study can be carried out with the permission of EASA, in conjunction with ICAO, to prove how aircrafts can achieve an equivalent level of safety. However currently, no detailed guidance for this procedure exists. This paper proposes such a safety assessment methodology in order to value obstacle clearance violations around airports. This method has already been applied to a safety case at Frankfurt Airport where a tower elevating 4 km out of threshold 25R severely violates obstacle limitation surfaces. The model data refers to a take-off and landing performance model (TLPM) computing precisely aircraft trajectories for both standard and engine out conditions at ground proximity. The generated tracks are used to estimate collision risk incrementally considering EASA/FAA, EU-OPS & ICAO clearance criteria. Normal operations are assessed with a probabilistic analysis of empirical take-off/landing track data generating the local actual navigation performance (ANP) on site. The ANP shows integration to collision risk for an aircraft with any obstacle. The obstacle is tested for clearance within a "5-step-plan" against all performance requirements for landing climb and take-off climb. The methodology thereby deli- vers a comprehensive risk picture: The presented safety case for Frankfurt Airport showed an equivalent safety level despite the violation of standards. The collision risk during both normal and degraded performance operations was still found to be within ICAO Collision Risk Model

(CRM) limits, requiring only limited risk mitigation measures. The presented work should complement ICAO Doc 9774 Appendix 3.

KEYWORDS:

Aeronautical Study, Aircraft Performance, Obstacle Clearance, Collision Risk, Engine out Operations

1. INTRODUCTION TO OBSTACLE CLEARANCE

Today's obstacle clearance criteria according to ICAO Annex 14 [1] , PANS OPS Doc 8168 [2] , recently came into European law through the EASA CS ADR DSN [3] , which still refers to the 50 years old collision risk model (CRM). Meanwhile, aircraft (navigation) performance has significantly improved [4] [5] , thereby often making today's obstacle clearance requirements too conservative. Estates around airports have become scarce, valuable resources for industry suffering from over-conservative clearance requirements. Large airport projects such as new runways at Frankfurt, Vienna, Berlin or Munich airports demonstrate the potential conflict between urban planning and air traffic operators' interests. From a scientific standpoint, existing regulations for departure/arrival procedure design, obstacle clearance evaluation and collision risk determination do not show congruent requirements with two target levels of safety (TLS) values in place: The CRM TLS [6] is valid for the precision approach segment ending at OCA/H (which is often far above the obstacle hot spots around airports) and the A-SMGCS TLS [7] is valid during ground taxi only. A systematic approach specifically detailing the evaluation of non-compliant obstacles through an aeronautical study as depicted in ICAO Doc 9774 Appendix 3 [8] is therefore crucial for making the best use of land around airports without hampering the safe operation of aircrafts. Also this approach should bring transparency into the various safety margin concepts and be incorporated in the above standards.

This paper recalls existing guiding material for assessing the compliance of obstacles with clearance requirements in Section 2, summarizes the analysis processes for valuing obstacles against these requirements in Section 3, and presents a model built to assess the non-conformant obstacle induced collision risk for departing and landing aircrafts while also bridging the various clearance considerations under normal and critically degraded performance conditions facing mechanical and operational flight aspects in Section 4 (engine out, crosswind etc.). This paper also collects the findings of a safety case in Frankfurt Airport which was successfully investigated with the presented approach in Section 5. Section 6 closes with a vision on how the presented concept may complement ICAO/EASA recommended practices to foster a transparent, formalized assessment of non-com- pliant obstacles in the complex airport environment.

2. OBSTACLE CLEARANCE DETERMINATION

Obstacle clearance is of relevance for runway design according to ICAO Annex 14 [1] , and Doc ADM [9] , for procedure design (ICAO Doc 8168 PANS OPS) [2] , aircraft operation (EU-OPS) [10] , and aircraft certification (EASA CS 25 [11] , FAR Part 25 Large Aeroplanes [12]). All these different requirements lead to given height margins above the ground respective of any obstacle (in fact vertical safety margins). They implicitly dictate a level of safety to avoid the risk of aircraft collision either with a ground plane or other obstacle. However, all three subjects refer to different geometric design standards and aircraft conditions. A safety model of Aircraft Operations when Aerodrome Obstacle Standards cannot be met therefore has to consider all three subjects and decipher any discrepancies and then build either transfer functions or determine the most stringent, limiting case. They are recalled briefly in the next subsections.

2.1. Obstacle Limitation Surfaces at Runway Design

A comprehensive set of standards and recommended practices has been internationally established based on ICAO Annex 14 and Doc 8168. Centered on each take-off/landing Runway, the ANC Obstacle Clearance Panel (OPC) defined limitation surfaces as early as 1976 and is intended to protect aircrafts in flight at ground proximity (see Figure 1).

The design of these horizontal and inclined surfaces obviously follows aircraft performance aspects (e.g. reference angle of climb/descent after take-off/during final approach) by adding safety margins to the design path. An e.g. 5.2% (equals 3.0°) design angle of descent during final approach is enveloped by the approach surface with a slope of 2.0% (1st section) to 2.5% (2nd section) inclination for a precision runway CAT I Code 3, 4 (Annex 14; Table 4-1, ch 4-8 [1]). Similarly, the take-off climb surface for a Code 3, 4 take-off runway inclines with 2.0%; again inducing safety margins as aircrafts typically climb with a gradient of 5% to 8%. To conclude, the

OLS grant a certain, however unspecified maximum collision risk for aircraft providing a given navigation performance (ANP) operating on that runway. The following Figure 2 shows the current obstacle collision risk during a conventional ILS approach based on realistic ANP values [5] . It further highlights the gradients of ICAO's obstacle clearance surfaces OLS and OAS (which apply to precision approaches only).

As we assume regardless of the chosen approach procedure of an equal safety target level, the difference in altitude for both dotted lines can only be considered in addition to the "safety margin" necessary for operations with lower ANP e.g. for non-precision approaches (NPA), which will be the subject of the following Section 2.2.

2.2. Obstacle Clearance Applied during Procedure Design

Closely connected to runway design, procedure design covers the detailed three-dimensional construction of ap- proach and departure procedures to and from an instrument runway. For e.g. Code 3, 4 precision instrument runways, usually a large set of alternative procedures have been developed to fulfill environmental (noise abatement), operational (direct routings), safety aspects (e.g. non precision approaches offered as contingency procedures), as well as to consider geographically dispersed in- and outbound traffic flows.

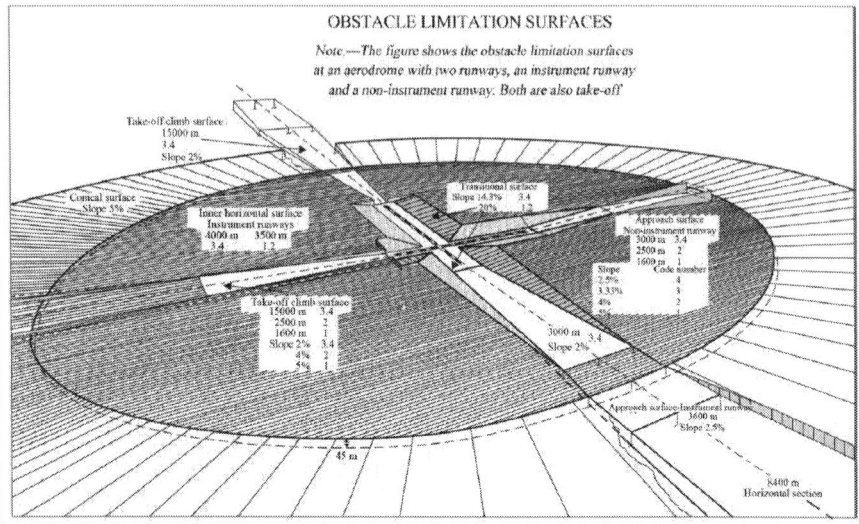

Figure 1. Obstacle limitation surfaces, Annex 14, Att. B; p. 313 [1] .

Comparison of OLS, OAS vertical profiles and flown trajectories (ANP)

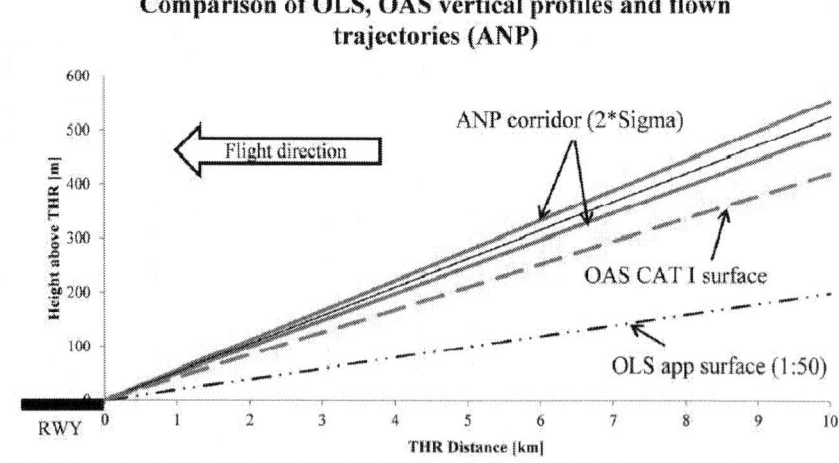

Figure 2. Heterogeneous safety margins resulting from OLS and OAS (during approach).

With regard to precision or APV approaches, ICAO Doc 8168 [2] dictates the consideration of Obstacle Assessment Surfaces (OAS) consisting of inclined surfaces referenced to the threshold of that runway. They look similar to the OLS with a less conservative character (see Figure 2) but extending into the missed approach (not shown): Two alternative safety conclusions can be drawn when comparing OLS versus OAS: Either the OLS allows lower ANP to also methodologically cover NPA (as derived in Section 2.1) with the OAS laying above the OLS, or a different level of obstacle collision risk is implicitly accepted. Consequently, safety assessment will have to include all obstacle related procedure surfaces extending to the visual (approach) segment surface (VSS), the outbound obstacle identification surface (OIS) and the on-following take-off funnel. The VSS prolongs the OAS from the Obstacle Clearance Altitude/Height (OCA/H) down to the ground plane, assuming there is only visual guidance capability for the cockpit crew leading to lower ANP. As the OCA/H itself is also specific for each procedure and depends on the obstacle situation, the whole design process has potential for closed loop configurations (obstacle drives OCA/H and so the length of the VSS). So we can conclude that the OAS, different from the OLS, is not only representing significantly differing CR values but also allowing for two different design concepts: Either it determines those obstacles to be considered in the calculation of the associated OCA/H to the procedure[1] or once set, refuses further adoptions induced by new obstacles by just interpreting OAS as obstacle limitation surface (which de facto is often the case). The ICAO Collision Risk Model (CRM) [6] as introduced in the 70s, however, does not cover the testing of these different design options; it only considers the OAS as an approximated 10^{-7} per operation collision iso-risk line neglecting time and PA procedure specific aircraft ANP figures. The presented assessment will also have to overcome this (second) deficiency.

For departures (see Figure 3), consideration includes, beside the OLS, the OIS extending from the DER with a 2.5% inclination, which is meant to be compared to the Procedure Design Gradient (PDG) for each procedure. To know the local CR and considering all such identified obstacles, ideally we have to refer to the local ANP for each specific procedure during both normal and also "abnormal" conditions. Abnormal conditions assume one engine inoperative (OEI) conditions and so cover rare but safety critical occurrences along the risk estimation.

2.3. Aircraft Certification

The various surfaces (OLS, OAS, VSS, and OIS) analyses all assume that aircraft operate at normal perfor- mance conditions. Certification of aircraft according to FAA Part 25 [12] resp. EASA referring to CS-25 [11] , however, takes additionally degraded, OEI performance test cases during take-off, final approach and landing climb into consideration. The aircraft, therefore, must be accelerated on the ground to the engine failure speed v_{EF}, at which point the critical OEI condition begins, lasting for the rest of the take-off (EASA CS-25.109/ CS25.111 ff. [11]). The take-off section ends, the climb phase begins at 35 ft. (jet aircraft, 50 ft. prop aircraft) above the take-off surface at the end of the take-off distance (15 ft. for wet runways). From there on, the aircraft must demonstrate its climb performance as a so-called gross flight path which will be diminished by aircraft type specific climb gradient reductions, which generate yet another safety margin: e.g. 0.8% for two-engine aircrafts along the take-off flight path. So far there is no evidence for the correlation of the safety impact to that of the regulatory requirement. A comprehensive safety assessment will also have to tackle this (third) aspect.

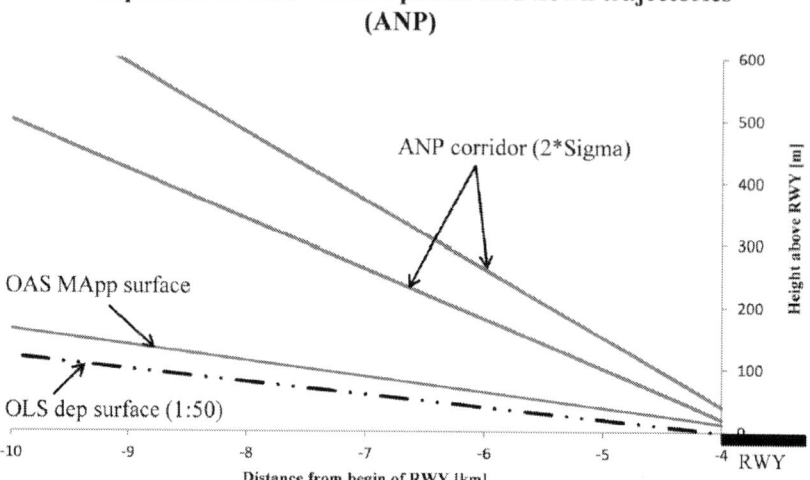

Comparison of OLS vertical profile and flown trajectories (ANP)

Figure 3. Heterogeneous safety margins resulting from OLS and OAS (departure).

To conclude, all certified aircrafts must demonstrate their certified required take-off, landing distances and climb out performance. A second branch of test cases for obstacle clearance will have to be considered aside those tackled in the first and second bunch of clearance considerations like degraded aircraft performance combined with unfavorable ambient conditions like crosswind. These effects can best be handled by defining hazard scenarios in which the aircraft must prove obstacle clearance in terms of e.g. demonstrating minimum climb performance.

The following Sections 3 and 4 present the concept for a Safety Assessment (SA) aiming at satisfying all of the above requirements; thusly bridging the different requirements coming from airport planning, procedure planning and the certification of aircrafts. In so doing, we hope to generate a transparent and homogeneous risk picture.

3. ANALYSIS OF NON-COMPLIANT OBSTACLES

When conducting the SA, we start with systematically spotting obstacles in the vicinity of the runway system as candidates for violating one of the clearance criteria as presented in Section 2. This process is not executed obstacle by obstacle but by limitation criteria starting from the most stringent to the least with reference to Figure 2 and Figure 3. Respectively, the following process forms the hazard assessment phase of the SA.

3.1. Approach and Take-Off OLS Investigations

Independent from specific aircraft performance criteria, EASA generally requires OLS considerations as follows: "OLS means a series of surfaces that define the limits to which objects may project into the airspace around aerodrome to be ideally maintained free from obstacles" [3] . Of course, we can convert these surfaces into performance criteria and thereby determine the implicit climb/descent requirement:

$$\tan\gamma = 1/50 = 0.02 \ \left(\text{OLS dep/arr surface gradient}\right)$$

$$\gamma_{OLS} = 1.145°$$

Regardless of any obstacle clearance considerations, we can calculate the minimum climb gradient and minimum rate of climb (and descent) ROC/ROD as follows for the given low altitude and small climb angles:

$$\text{ROC/ROD}_{minOLS} = \text{TAS}\sin\gamma_{minOLS} \approx \text{CAS}\gamma_{minOLS}$$

For a typical final approach $^{(v_{REF})}$ resp. climb out safety calibrated airspeed $^{(v_2)}$ of 150 kt, this leads to a $\text{ROC/ROD}_{minOLS} = 300 \text{ ft./min}$, proving a very conservative safety character of the OLS surfaces, equaling very low safety requirements as is shown in Section 4.

3.2. Approach and Missed Approach OAS Investigations

As shown in Figure 2 for a given ANP distribution along the flight track, the OAS are less conservative than the OLS; however, they do not equally consider specific aircraft performance categories but solely the PA category, localizer location and further system related parameters. For keeping those surfaces free of obstacles, the following vertical flight path requirements for a typical ILS, CAT I, 3 glideslope OAS result:

$$\gamma_{\text{App OAS}} = -1.62°$$

$$\gamma_{\text{MissedApp OAS}} = 1.43°$$

Which leads with (3) to:

$$\text{ROD}_{\text{App}} = 430 \text{ ft./min}$$

$$\text{ROC}_{\text{MissedApp}} = 380 \text{ ft./min}$$

The higher gradient values in Equations (4) and (5) show the usefulness of the chosen process. It should, however, be noticed that in few cases for very close-in obstacles the OAS may formally be more restrictive than the OLS. Still in those few cases the OLS prevails (PANS-OPS Volume II Part III Attachment B [2]).

3.3. The Role of the Obstacle Clearance Altitude

The final instrument approach ends and the missed approach segment begins at the OCA/H. The higher the value, the earlier the pilot needs visual contact to the runway to continue landing. Non-precision approaches with OCH $\geq$ 600 ft. suffer limited usability due to insufficient visibility and ceiling for the sake of safety. With regard to CR, we may further notice that the OCA/H induces far more conservative safety margins than for PA with up to 250 ft. clearance for close-in obstacles in the primary area compared to 35 ft. for take-off (see above). The OCA/H as third test case gains further complexity as it must be calculated for each (stall speed related) aircraft category [13] . Finally, we may further take into consideration when obstacles are typically closer to the runway then where the OCA/H is being reached; this test case is sequenced at third position.

3.4. Net Flight Take-Off Path Analysis

During type and aircraft specific certification, the aircraft now has to demonstrate landing climb, take-off and climb out performance under both normal and degraded conditions. This demonstrated performance will potentially limit its operational capabilities in terms of maximum allowable masses, payload, and range. E.g. if a given climb gradient or a declared distance cannot be met, take-off mass (as fuel or traffic load) must be reduced and will limit profitability of the flight. This process is driven by the parameter weight (W), required thrust (T), power (P) and selected climb out safety speed (v_2) , as those trigger climb performance as shown in Equation (3). T and v_2 are coupled at constant speed through:

$$T = \frac{\mathrm{d}m_{\text{Air}} + \mathrm{d}m_{\text{Fuel}}}{\mathrm{d}t} v_{\text{EG}} - \frac{\mathrm{d}m_{\text{Air}}}{\mathrm{d}t} v_2 - K \quad \text{and} \quad \sin\gamma = \frac{T-D}{W}$$

$$P = T * v_2 \quad \text{and} \quad v_2 * \sin\gamma = \text{ROC} = v_2 \left(\frac{T-D}{W} \right)$$

EG = exhaust gas, K= pressure effect forces, D=drag

Equation (8) shows the linear correlation of climb gradient (CG) and weight, Equation (9) the respective linearity between operating speed and ROC/ROD. Along the take-off flight path, the "2nd segment" implies the highest CG. Obstacles must be overflown with 35 ft. clearance. As climb-out profiles are more inclined, obstacle limitations likely become more limiting in the approach and landing phase. Independently, we again can determine a minimum clearance requirement resulting from a procedure design similar to the process in Section 3.1. With the procedure design gradient (PDG) [2] at 3.3% (OIS gradient of 2.5% plus 0.8% safety margin) we equally can calculate a rate of climb (ROC) for a given climb out speed v_2:

$$\gamma_{Dep} = 1.89°$$

with $v_2 = 150$ kt we come to $ROC_{Dep} = 500$ ft./min

The slightly higher ROC values in (11) compared to (7) imply to rank this test to this subsequent position. Figure 4 depicts the additional safety margin versus Figure 2 and Figure 3 resulting from the 35 ft. clearance requirement which reduces the CR provided in the same ANP value and distribution (shown in Figure 4 as sketch).

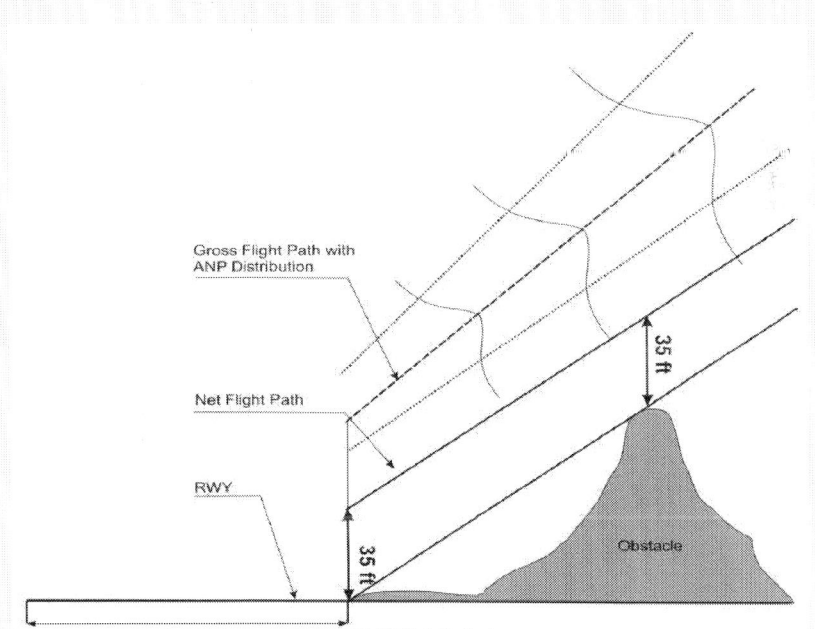

Figure 4. Additional obstacle clearance and required net flight path corrections resulting in lower CR as for OLS and OAS.

With today's economic pressure in the ATM System, limitations for take-off mass are hardly acceptable especially at large (Code 3, 4) airports. As such, the safety test case will usually be performed with maximum structural take-off and respective landing mass considerations.

It shall be noticed that this fourth step also demonstrates some complexity; however, it does not consider— compared to step three—detailed stall speed dependent aircraft categories but a simplified threefold concept only relying on the number of engines (2, 3 or 4) installed. An obstacle therefore only becomes relevant if the respective CG requirements are not met leaving a 35 ft. clearance above the object. The clearance surface can so be represented by the inclined (net flight path) surface when reduced by 35 ft.

Section 4 incorporates this analysis concept and adds systematic algorithms to the presented analysis to form the SA.

4. SAFETY ASSESSMENT METHODOLOGY

4.1. Legitimation

A general philosophy of handling exceptions from the presented standards has been adopted by ICAO[2] in Doc 9774 [8] and has recently been significantly relaxed by EASA in CS-ADR DSN [3] . Accordingly, a so-called Aeronautical Study can be applied as Acceptable Means of Compliance for certain violations such as obstacles conflicting with the large horizontal obstacle limitation surface. Explicitly, ICAO states "An aeronautical study is a study of an aeronautical problem to identify possible solutions and select a solution that is acceptable without degrading safety" [8] . As such, supervisory authorities in ICAO member states generally agree with ICAO non-compliant constellations or procedures as long as such study proves an equivalent safety level. However, there is no specific guideline given by the ICAO about how to conduct these safety assessments.

Therefore, the main goal of the presented methodology is to design a uniform concept for obstacle approval verification clearly having in mind the implicitly deferring safety margins as presented in Section 2.

4.2. Architecture

In line with Eurocontrol's SAM [14] , an Aeronautical Study shall comprise the following main sections: Hazard Assessment and Safety Risk Analysis, including potential risk mitigation. The main hazard to be assessed here is very simple: The collision with any formally non-compliant obstacle according to Section 3.

For the Safety Assessment Analysis, the chart in following Figure 5 has been built to comprehensively consider the (main) hazard with the findings of Section 2.

As shown, the model comprises of two main components: The analysis of normal operations and of quite rare but safety critical, degraded operational performance (emergency related operations). It so covers all potential flight scenarios related to the (main) hazard. For each of the next steps, the probability to divert from the desired flight path towards the obstacle is being determined and valued against the safety target in place according to Section 2.

Normal operations are flight procedures compliant with the standard operational procedures (SOP) as published for the specific aircraft and following the ATC clearances as filed and expected. As such, also a take-off with one engine inoperative is basically a normal operation, whereas the same take-off e.g. leaving the cleared departure route (SID) for unspecified reasons is considered an emergency operation. For safety considerations, emergency operations always assume a degraded aircraft performance behavior when assessing the effects to the declared hazard.

4.2.1. Normal Operations

Normal operations statistically represent more than 99% of all operations at large (Code 3, 4) airports. The Normal Operations model (NOM) processes historic aircraft track data from departures resp. arrivals at the investigated airport for a significant time period (typically six months) gathered from radar or multi-lateration sensors. Similar to the ICAO CRM methodology [6] , the real track data will be compared to the defined track data and modeled as offset probability density functions (PDF). The PDF concept has also been adopted by ICAO with their RNAV/RNP or PBN documents [15] .

We proved in [5] that for all studied cases, aircraft navigation performance behaves normally. Nonetheless, we allow the NOM to begin again, first only assuming two dimensional (y as cross track reference, z as vertical reference) Gauss PDFs, $f(y)$, $f(z)$ with a 2σ value as procedure level of quality identifier (e.g. RNP 0.3 for a non-precision approach procedure equals a square root variance of 0.3 NM left and right to the desired flight path for 95% of the flying time) [15] . This is analytically described at each location along the flight track as shown in Equation (3) for the cross track plane y:

$$f(y) = \frac{1}{\sigma_y \sqrt{2\pi}} e^{-\frac{(y-\mu_y)^2}{\sigma_y^2}}$$

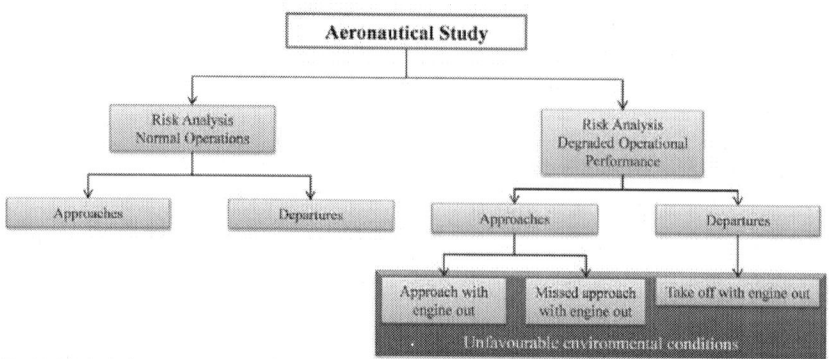

Figure 5. Model architecture of safety assessment methodology.

So y and z represent the track coordinates as sampled, the mean with μ and its variance with σ^2. It shall be noticed that track considerations are unnecessary in the given context since CR is considered as time independent.

After verifying the assumption of normality of the track data by means of chi-squared tests along the flight track, a Grubbs' test is applied to detect potential outliers. Grubbs' test detects one outlier at a time. Each potential outlier is expunged from the dataset and the test is iterated until no (more) outliers are detected. Along the defined take-off or approach path, vertical cuts through the cross track plane provide insight into the shape characteristics. This methodology has already been effectively applied in [5] . Following the normal distribution, mean and standard deviation reflect the graphical center and height of each "bell" curve (see Figure 6).

The NOM systematically integrates the local PDF to a dedicated obstacle distance both vertical and lateral at the abeam location along the flight path to calculate the local collision risk (CR) for that object. The CR per movement can then be compared to a pre-set TLS such as the ICAO CRM TLS, as shown in the following Figure 7 for the lateral plane with different iso-risk contours (1×10^{-3}, 1×10^{-5}, 1×10^{-7}, and 1×10^{-10}).

By extending the integration boundaries accordingly, the CR can also be calculated against the ground plane or any height above ground to estimate a ground/height projected risk contour laterally enveloping the flight path. It may, however, be noticed that close to the ground, track data typically is remarkable unreliability due to radar beam refraction and multi path phenomena. This should be overcome by applying multi-sensor strategies such as additionally using multi-lateration track data. In conjunction with these sensor aspects, the NOM allows safety comparisons of any procedural obstacle clearance standard.

Accordingly, the following Table 1 collects the computed risk levels using ANP from [5] at a preselected height above ground for the OAS and OAS (CAT I) approach surfaces.

Table 1 clearly shows the different implicitly (ICAO) accepted CR if we start from identical ANP values. The target CR varies between 2.2×10^{-32} and 2.2×10^{-23} at a 3000 m distance from the runway location. This difference rises even higher for larger threshold (THR) distances. The NOM, therefore, proves to be crucial for generating a comprehensive obstacle clearance risk picture.

Of course, the ANP distribution may (slightly) vary depending on the (time dependent) environmental condi- tions or differing procedures, e.g. a conventional straight-in precision approach (PA) vs. a more innovative segmented non precision approach (NPA). However, such differing ANP distributions do not alter the general conclusions drawn here, as we prove in [16].

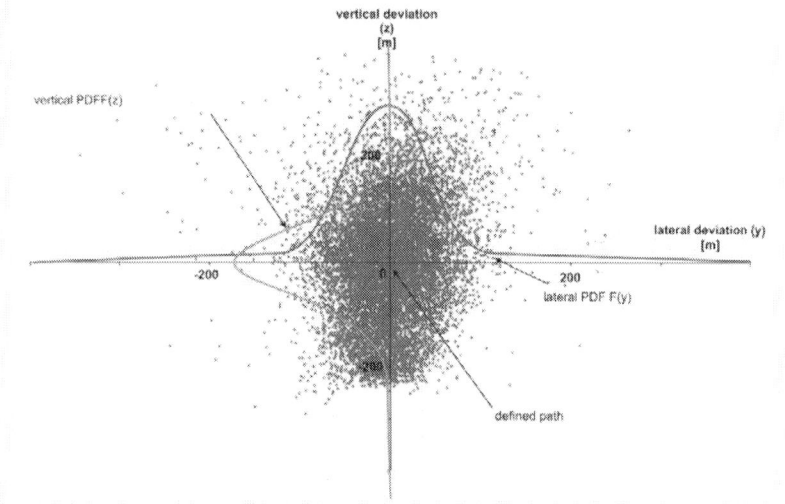

Figure 6. Normally distributed track offsets around the flight path.

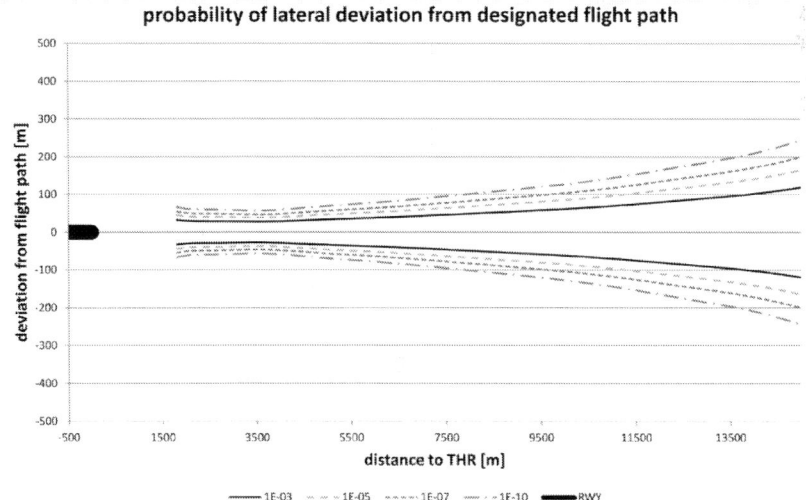

Figure 7. Probability of lateral deviation from designated flight path or risk to hit an obstacle with CR(z) = 1 in lateral plane.

4.2.2. Degraded Operational Performance

The degraded operational performance model (DOM) considers those effects of the remaining less than 1% traffic operations, those emergency or unspecified conditions which do not follow normal procedures according to SOP. These rare instances show a significantly increased collision risk as a result of flying un-cleared procedures combined with engine failure-what we assume here to be in reference to our safety case for a degraded performance of the aircraft during take-off or landing climb. From a statistical perspective, the DOM looks specifically at the (far) tails of the normal distributions used in the NOM. These tails cover only a small percentage of all track data (<0.3% for a 3 Sigma threshold), their functional approximation being remarkably unreliable. To overcome this weakness, the DOM reverts to a scenario technique representing these seldom cases on a functional level, allowing to deterministically analyze a collision potential ("will pass the obstacle or not").

Table 1. Collision risks for different approach surfaces, related to 3° glide path and ANP.

Distance to threshold [m]	Height of the 3° GP	Calc. sigma vertical	Probability to undercut a defined surface		
			Surface	Height of the surface	Calc. collision risk
1000	67.6 m	7.4 m	OLS	18.1 m	1.8E-11
			OAS (CATI)	20.5 m	8.3E-11
2000	120.6 m	8.5 m	OLS	38.8 m	6.6E-22
			OAS (CATI)	49.0 m	3.3E-17
3000	172.5 m	9.6 m	OLS	58.8 m	2.2E-32
			OAS (CATI)	77.5 m	3.4E-23

Regardless of the technical, meteorological or human factors related to the causes leading to these scenarios, the DOM computes the effect of degraded performance while also assuming unfavorable operational conditions such as a strong cross wind coming windward with respect to the remaining operative engine(s). This constellation increases at max the required wind correction angle (WCA, see Figure 8), thus degrading the lateral/ver- tical aircraft navigation performance.

This model is applied to all three constellations while assuming the most critical engine out case a few seconds ahead of passing the non-compliant obstacle:

1. Approach with two different settings (flaps extended, gear up) and landing configuration (flaps high/full, gear down).
2. Missed Approach at FAF as default (see Figure 9).
3. Take-off (at v_{EF} as default) with two different flap/slat settings to weigh-in corresponding CG and ROC effects on obstacle clearance. It minimizes these effects according to the following generalized aerodynamic effects (see Figure 10).

The Take-off and Landing Performance Model (TLPM) which evolves from the EJPM [17] calculates the resulting 3D tracks for all DOM scenarios. Validation was also performed with licensed flight planning software [18] .

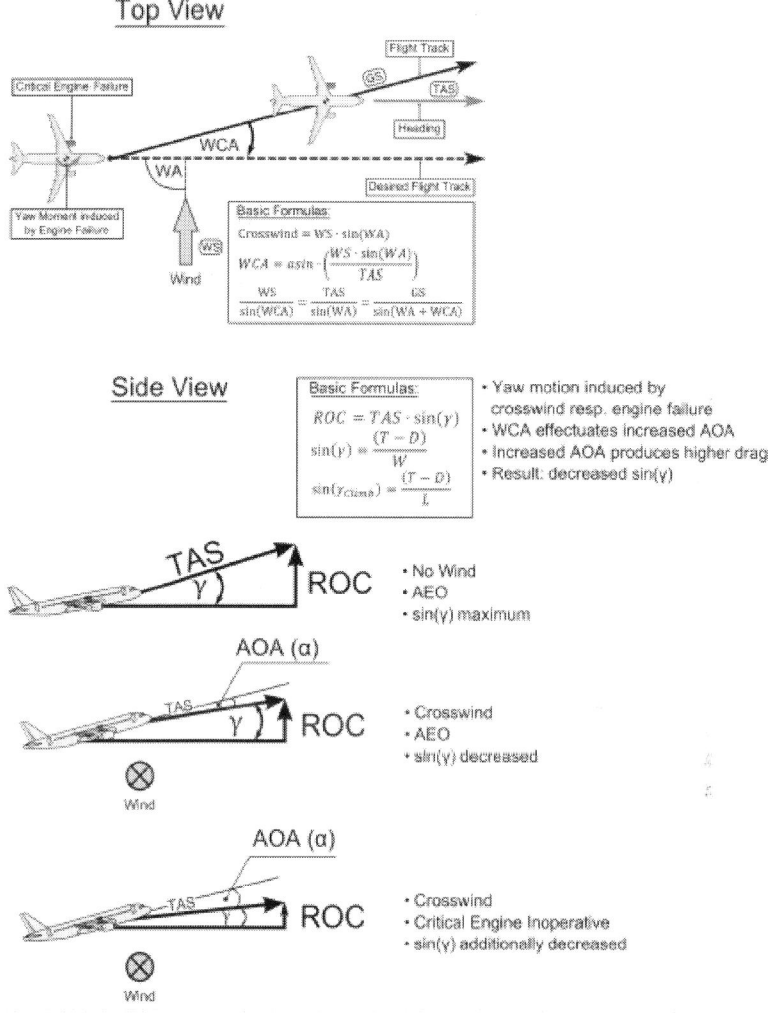

Figure 8. Underlying flight performance in the DOM. AOA = angle of attack, WA/WS = wind angle/speed, AEO = all engines operative.

The presented threefold flight performance analysis was embedded into a five-step evaluation scheme as follows:

DOM Step 1: Vertical Performance Analysis
Step 1 determines the vertical obstacle clearance by applying minimal vertical performance requirements according to EASA CS-25 for each scenario which contains the formulation of a "direct to" trajectory with the obstacle. It answers the following question.

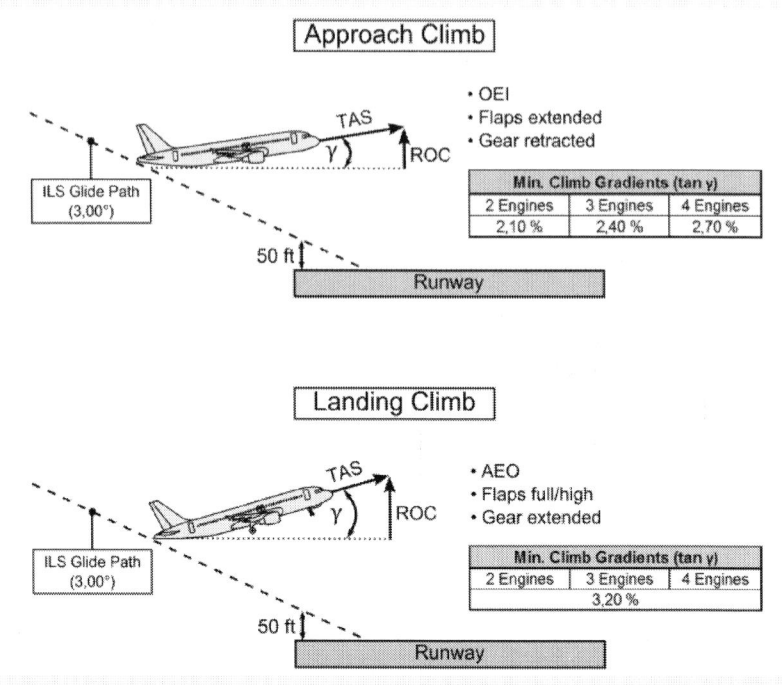

Figure 9. Worst case analysis for approach and landing climb obstacle clearance: Approach/landing configuration impact on ANP and rate of climb.

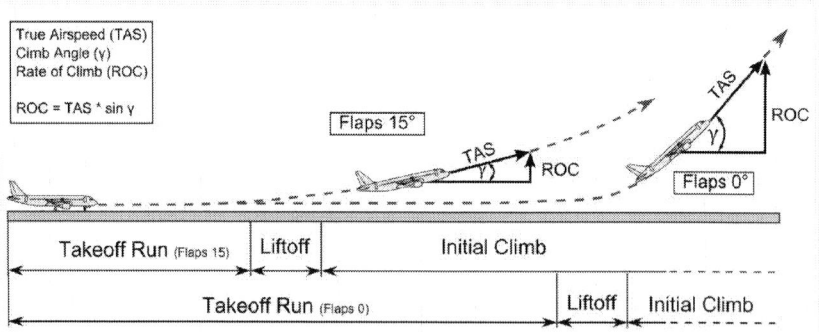

Figure 10. Lowest pass altitude analysis for take-off obstacle clearance.

"Based on purely vertical certification requirements: Do all aircrafts overfly the obstacle safely even if a direct trajectory towards the obstacle is assumed (regardless of the aircraft's capability to fly this horizontal trajectory)?"

For a positive result of all scenarios (the answer being yes), considering all aircraft performance types relevant to the airport, the DOM analysis is completed at step 1. Otherwise, we continue as follows (this is consecutively true for all on-following steps):

DOM Step 2: Procedure Design Analysis

Step 2 additionally contains lateral consideration of the obstacle with respect to all existing approach/depar- ture procedures and their respective clearance requirements according to EU-OPS 1 [10] and ICAO PANS-OPS [2] . Therefore the question of step 2 is:

"Does the obstacle violate any departure or approach procedure clearance requirements?"

DOM Step 3: Lateral Performance Analysis

Additionally, step 3 considers, based on the trajectory resulting from step 1 (vertical) and step 2 (lateral), both additional lateral and vertical divergences from the intended route during approach or departure resulting from OEI and adverse wind assumption. Step 3 uses the TLPM to compute these critical trajectories to answer the question of step 3:

"Is the obstacle's location critical even if we assume OEI and adverse wind conditions?"

DOM Step 4: Lateral Flyability of the Critical Trajectory

Step 4 investigates the capability to fly that trajectory found to be most critical in step 3 through consideration of lateral performance aspects; e.g. maximum bank angle with flaps extended leading to limited turn radii. Step 4 answers the question:

"Is the developed, most critical trajectory flyable at all in terms of flight performance and flight mechanics?"

DOM Step 5: Vertical Performance Analysis of the Critical Trajectory

Step 5 investigates the aircraft's vertical flight profile as fixed in step 4 considering OEI yaw momentum induced drag. This one leads to generally degraded CG values (see also Figure 8). The question to be answered in this step can be concluded as follows:

"Are aircrafts flying the most critical trajectory possible in order to ensure sufficient flight altitude to cross the obstacle safely?"

Often step 5 concentrates on identifying aircraft classes with relatively poor climb performance (such as e.g. light twin engine prop aircraft or heavy aircraft such as A340).

With these five steps, the DOM thereby covers all effects resulting from adverse conditions resulting in a collision threat that is not yet covered by the NOM. This SA methodology was already applied for several safety cases in Germany. Section 5 reveals such application for demonstration and verification purposes.

5. SAFETY CASE CLOSE-IN OBSTACLE AT FRANKFURT AIRPORT

5.1. Environmental Settings

As stated in Section 4.2.1, the hazards, here the obstacle subject to investigation, need to be located relative to the runway system of Frankfurt Airport (FRA). We focus on an 80 m tower building located 600 m ahead of threshold 25C and 1100 m across its extended centerline, close-by the extended centerline of Frankfurt's new landing runway 25R (see Figure 11).

The obstacle clearance analysis with regard to the OLS of all relevant RWY directions proved a violation of the horizontal surface of the center RWY 25C/07C and of the south RWY 25L/07R. None of the OAS, however, was violated, as shown in following Figure 12.

5.2. NOM Application

Flight track data for a six month period at FRA was used for the ANP analysis preceding the CR calculations [19].

5.2.1. Outbound Traffic Analysis

All departure routes passing nearby the considered obstacle operating from runway 07C were analyzed (07R routes are farer away, 07L/25R allows landings only). As shown in Figure 13, flight tracks of the northbound departure route (SID) BIBTI 3E were identified as the most relevant ones.

The relevant cross section (the normal plane to the route through the obstacle) was found almost in the straight out segment of BIBTI 3E, at 600 m beyond DER. The route specific flight track's distribution are shown in Figure 14 at this cross section (BIBTI 3E in red, all other departures in blue).

Figure 14 clearly shows a tendency for aircrafts on BIBTI 3E to climb faster than other traffic following a required PDG of 6.3%. They also initiate the tendency of aircrafts to veer north (more located to the left in the figure) compared to the other traffic.

The SID specific ANP analyses were then performed along the statistical methods as explained in Section 3. Figure 15 shows the determined ANP values and resulting CR iso-risk lines for departing aircraft on BIBTI 3E and all other routes.

The poorest ANP values (XTT and VTT) per class of aircraft and per route are identified in Table 2. It also shows the probability density function's (PDF) parameter shape before/after the critical 600 m cross section.

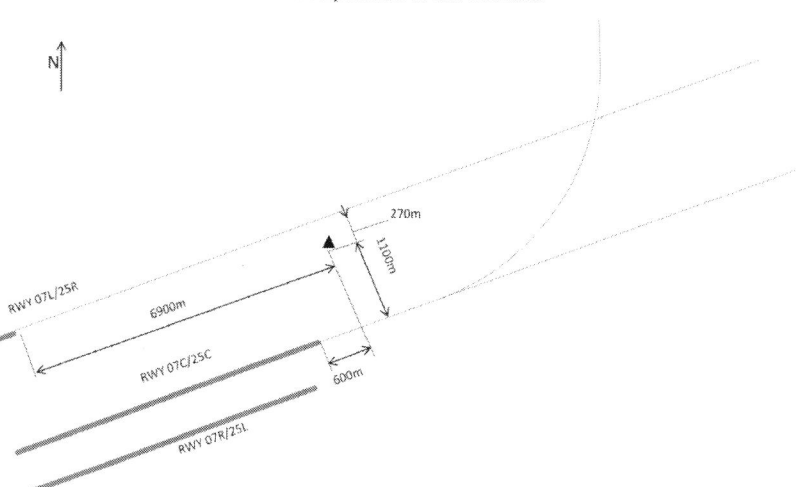

Figure 11. Location of the critical obstacle-the safety case EDDF.

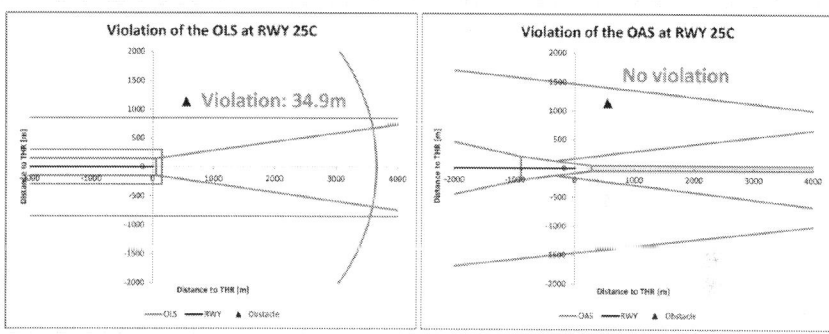

Figure 12. Excerpt of the obstacle clearance analysis for the runway 07/25C, OLS—left figure, OAS CAT I—right figure.

The double integration of the PDF both vertically and laterally to the obstacle equals the specific CR per take-off, as shown in following Table 3.

Due to the position of the obstacle at 600 m from DER, the CR VTT value is relatively high with 1.97×10^{-2} per departure; whereas the XTT related CR is—at a lateral offset of 1100 m for the obstacle from the route —negligible with values below 1×10^{-100} per departure, result in an overall collision risk of 4.13×10^{-117} per departure.

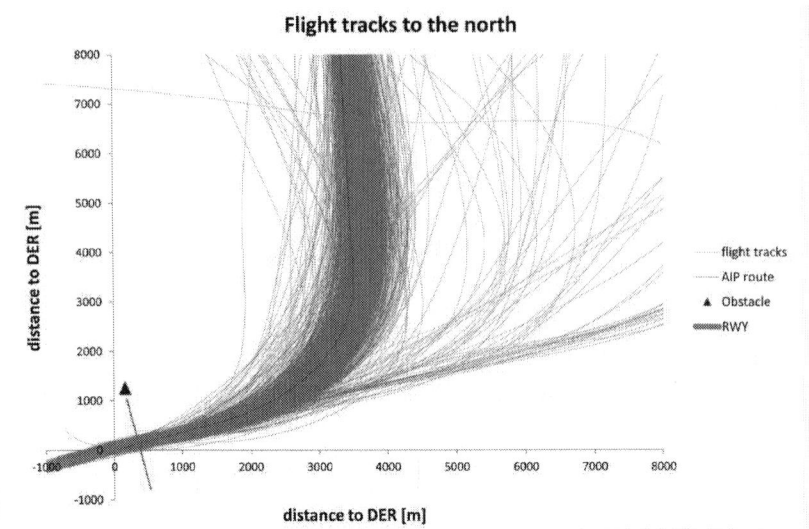

Figure 13. FRA Flight tracks for runways in use 07, outbound flights, FANOMOS Data, with obstacle critical cross section (red line).

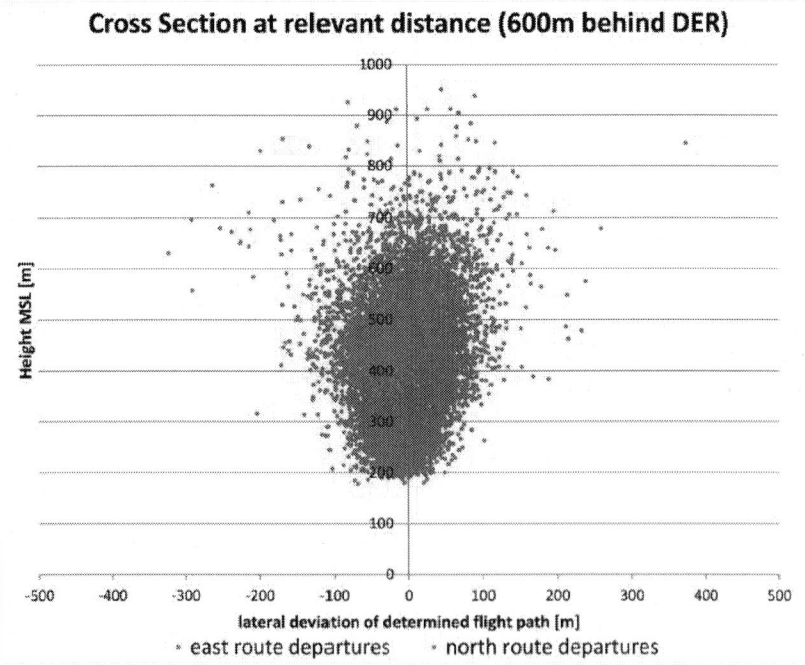

Figure 14. Relevant cross section passing through the critical obstacle at 600 m DER distance, departures from runway 07C.

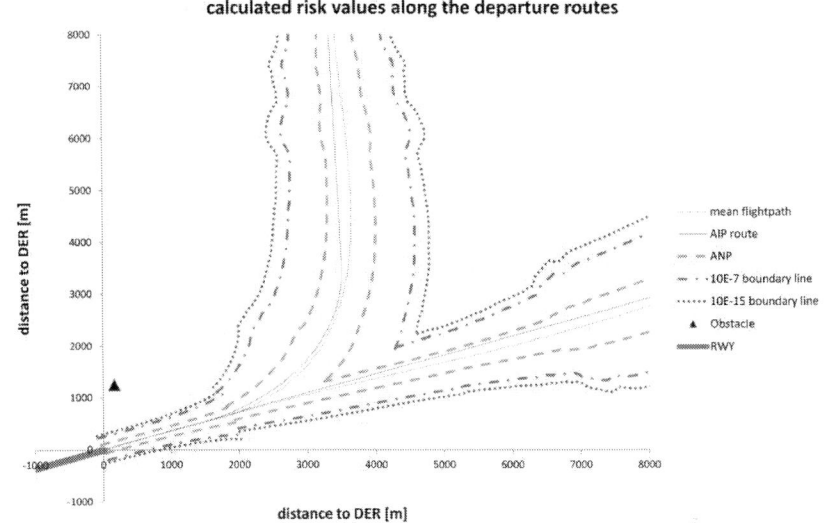

Figure 15. Iso-risk lines and ANP values, outbound flights on runway 07C.

Table 2. Statistical parameter along the flight track (critical cross section highlighted in gray).

Distance from DER [m]	VERTICAL		LATERAL	
	Sigma [m]	VTT [NM]	Sigma [m]	XTT [NM]
400	85.0	0.092	43.5	0.047
500	85.9	0.093	44.8	0.048
600	86.6	0.094	45.7	0.049
700	87.2	0.094	46.7	0.050
800	87.9	0.095	47.7	0.051

Table 3. Calculated collision risk for departures.

Collision risk		
Lateral (XTT)	Vertical (VTT)	Per departure
2.10E–115	1.97E–02	4.13E–117

5.2.2. Inbound Traffic Analysis

For approaches, both north and center runway are relevant. Applying the same investigation steps as for the departures, we computed, compared to the outbound case, clearly higher ANP values as expected with 99% of all aircraft performing ILS approaches. This leads to respectively smaller CR figures below 1×10^{-117}. The calculated CR via ICAO's CRM [6] showed as well a negligible CR outside the calculation limits of the CRM program ($<1 \times 10^{-15}$ per approach).

We can thereby conclude that for the normal operations at Frankfurt Airport a collision risk for take-off and landing is below ICAO's TLS at 1×10^{-7} per operation. The calculated CR below 1×10^{-100} shows both the necessity for scenario based risk analyses, as provided with the DOM, and a further validation of ICAO's CRM.

5.3. DOM Safety Case Application

5.3.1. Scenario Set-Up

Based on the obstacle's location as shown in Section 5.1 the following DOM hazard scenarios were identified:

1. Approach RWY 25R
2. Missed Approach RWY 07L
3. Missed Approach RWY 07C
4. Take-off RWY 07C

The following Figure 16 depicts all four scenarios.

All operations on runway 25L/07R and runway 18 were excluded from the investigation as explained in Section 5.2.

5.3.2. DOM Step 1: Vertical Performance Analysis

Applying the minimum climb requirements according to CS-25, we can prove compliance for scenarios 1, 2, and 3. Scenario 4 "Take-off RWY 07C" does not pass the evaluation. Table 4 shows the obstacle conflicting profiles for multi-engine aircraft (negative values indicate a flight path below the highest point of the obstacle):

Consequently, we declare all scenarios but this one as non-critical. Scenario 4 is subject to further investigation along with step 2 to 5.

5.3.3. DOM Step 2: Procedure Design Analysis

In this step three separate phases will be passed:

Phase 1: Analyze the most critical departure route[3] along EU-OPS 1.495 standards.

Results: The examination revealed that the obstacle does not penetrate the "take-off funnel", and so at least does not violate the departure clearance requirements. As explained in Section 4, this is not yet proof of compliance but a pass indicator. Further investigation leads us to:

Phase 2: Examination of the Obstacle Identification Surface (OIS) and PDG according to ICAO PANS-OPS Vol. II.

Results: This sub-step shows that the obstacle violates the OIS both laterally and vertically. As such, a PDG update is necessary to prove a further pass indicator (see Table 5).

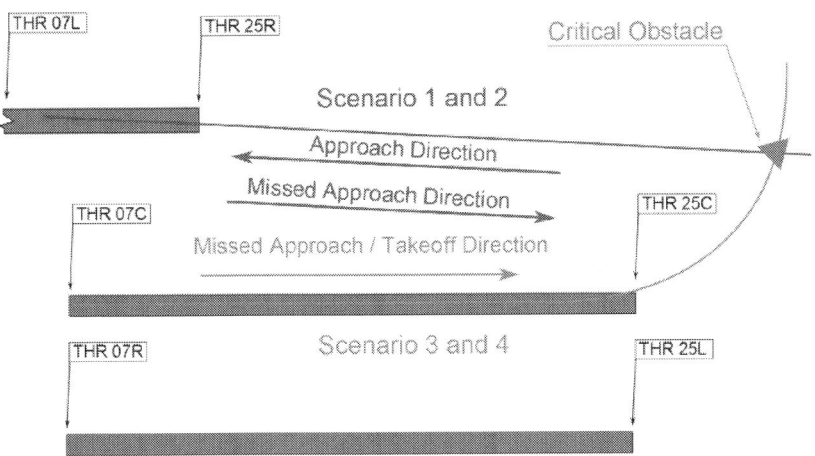

Figure 16. Identified DOM hazard scenarios—safety case FRA.

Table 4. Violation altitudes scenario 4.

Scenario 4—obstacle clearance [m]		
2 engines	3 engines	4 engines
−46.10	−40.00	−33.90

Table 5. Comparison of minimum procedure design gradients.

Minimum PDG [%]	
As published[4]	Required
6.30	5.90

Consequently, despite the identified OIS violation, the obstacle will not vertically impact the published take- off procedure, the pass indicator is positive, leaving us with:

Phase 3: Examination of the protection area for turns according to ICAO PANS-OPS.

Results: We derive that the obstacle is located inside the protection area of BIBTI 3E. However, a calculation of the maximum allowable object height at the given location shows remaining clearance so that this pass indicator is also true.

In total, we find all three pass indicators are true without granting combined lateral and vertical compliance, so we will have to continue to step 3:

5.3.4. DOM Step 3: Lateral Performance Analysis

By additionally considering the uncertainties resulting from adverse conditions and engine failure for all relevant aircraft types (e.g. A321, A340, B777F) in the

TLPM, Figure 17 depicts the exemplary results of the missed approach performance analysis (OEI right prior passing the obstacle):

Results: The calculations prove that yaw motion induced by engine failure and crosswind (up to 20 kt) can be fully compensated by aircraft flight control. Consequently, no relevant lateral nor vertical deviations from the intended trajectory were identified for scenario 1 to 3 (partly shown in Figure 17). Scenario 4, however, showed vertical violations requiring the execution of step 4.

5.3.5. DOM Step 4: Lateral Flyability of the Critical Trajectory

All aircraft types able to complete the turning departure as set out with the critical trajectory are being identified. So we calculate the required climb out speed v_2 to reach the pre-set turn radius   linked as follows:

$$v = \sqrt{r \cdot g \cdot \tan(\Phi)} \geq v_2$$

This delivers the following figures in Table 6.

Comparison of design speed (v) with v_2 for all relevant aircraft models revealed that the required speed can only be achieved by small multi-engine jet or turboprop aircraft (e.g. Cessna C525A CJ2 or Beechcraft King Air B200GT). Even though this aircraft category is rather rare at FRA (close to pass indicator), we will have to identify the remaining risk, which is assessed in the final step 5.

5.3.6. DOM Step 5: Vertical Performance Analysis of the Critical Trajectory

We finally determine CG and lift-off points under the prescribed unfavorable OEI conditions to calculate the pass altitudes above the obstacle for this aircraft category, again using TLPM.

Table 7 depicts the results for the exemplary members of this aircraft category.

As a result, we prove in step 5 that the critical aircraft category can also safely overfly the critical obstacle with significant clearance according to PANS OPS (e.g. 35 ft./10m). So the SA for this safety case closes with a positive result.

If, however, even step 5 fails (or any preceding step beforehand), the SA methodology allows the investigation to set out mitigation measures, such as cancelling a published route, setting stricter prerequisites to allow flying that route (increased PDG or turn radius requirements to aircraft) or just generating appropriate awareness through hot spot advisories in the Aeronautical Information Publications (AIP).

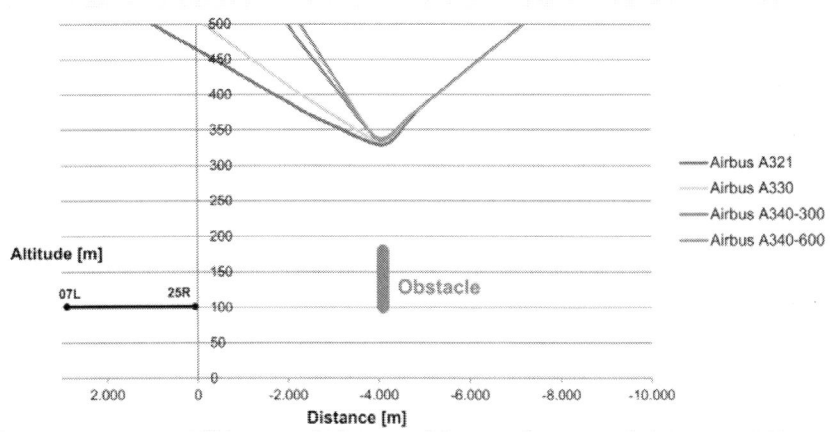

Figure 17. Missed app degraded performance analysis (scenario 2 & 3).

Table 6. Required operational parameter configuration for the critical trajectory.

Value	Unit	Numerical value
Design speed (v)	[m/s]	≈54 (105 kt)
Turn radius (r)	[m]	1100
Maximum bank angle (Φ)[5]	[°]	15
Take-off safety speed v₂	[m/s]	≤54 (105 kt)

Table 7. Values for example calculation of crossing altitude.

Aircraft	Climb gradient [%]	Lift-off point [m]	Crossing altitude [m]
B200GT	42129	840	160.80
C525A	42158	1550	52.24

6. CONCLUSIONS—INTEGRATION CONCEPT INTO ICAO DOC 9774

Aeronautical studies are supportive means to assess the safety and regularity of operations around airports with ICAO non-compliant obstacles in place (e.g. one that violates the OLS). Doc 9774 [8] , however, does not provide any guidance on how to perform such a study.

This paper presents a methodology which may contribute to standardizing the process. Dealing with both statistically representative hazard scenarios and infrequent events systematically investigated through scenario techniques, the presented model considers all potential flight situations by handling both normal and degraded performance triggered operations. In that sense, it generates a complete risk picture which has already proven useful in several certification processes with the German Ministry of Transport and the German ANS

Supervisory Authority BAF. The statistical part relies on dedicated ANP value calculations, stochastic functional approximation leading to validated, procedure-specific probability density functions along any flight track allowing calculation of obstacle CR through double integration. In rare cases (often called as "PDF tails") we developed a scenario configuration technique assuming worst case environment and aircraft performance related conditions. The resulting 3D trajectories generated through the author's take-off and landing performance model (TLPM) allows a deterministic (yes/no) collision potential determination by calculating minimum horizontal and vertical performance under unfavorable conditions for all aircraft categories operating at the investigated airport. Beyond this application case, we recall that the model reviews systematically all safety related design criteria for departure and arrival procedures. As such, its application is universe. On-going investigations run in compliance e.g. for Zurich Airport.

As such, we see strong potential for the presented methodology to become a potential candidate for an ICAO DOC 9774, Appendix 3 supplement in order to give specialists a guideline for how to adequately judge formally non-compliant obstacles for safe and regular operations at the airport. It also reveals the need for updating and extending the current ICAO's collision risk model with correct procedure and flight phase specific ANP values.

ACKNOWLEDGEMENTS

The authors thank Deutsche Flugsicherung DFS and Condor Flugdienst GmbH CFG for the provision of data for verification (NOM) and validation purposes (DOM). We also thank Fraport for the kind provision of detailed geometric data and our colleagues Christian Seiß and Martin Schlosser for their tremendous help with the extensive data analysis.

REFERENCES

1. ICAO (2009) Aerodrome Design and Operations. Annex 14, Volume 1, 5th Edition, ICAO, Montreal.
2. ICAO (2006) Procedures for Air Navigation Services—Aircraft Operations. Doc 8168, Volume II, 5th Edition, ICAO, Montreal.
3. EASA, Notice of Proposed Amendment (NPA) 2011-20, CS ADR DSN, Cologne, November 2011.
4. Frauenkorn, M. (2001) FLIP—Flight Performance Using Frankfurt ILS, DFS, Langen, Germany.
5. Thiel, C. and Fricke, H. (2010) Collision Risk on Final Approach—A Radar-Data Based Evaluation Method to Assess Safety. Proceedings of the 4th International Conference on Research in Air Transportation (ICRAT), Budapest, 1-4 June 2010, 473-480.
6. ICAO (1980) Manual on the Use of the Collision Risk Model (CRM) for ILS Operations. Doc 9274-AN/904, ICAO, Montreal.

7. ICAO (2004) Advanced Surface Movement Guidance and Control Systems (ASMGCS) Manual. Doc 9830, Montreal.
8. ICAO (2001) Manual on Certification of Aerodromes. Doc 9774, Montreal.
9. ICAO (2006) Aerodrome Design Manual (ADM), Part I Runways. Doc 9157, 3rd Edition, Montreal.
10. EU-OPS (2008) Council Regulation (EEC) No 3922/91 on the Harmonization of Technical Requirements and Administrative Procedures in the Field of Civil Aviation. EU-OPS, Brussels.
11. EASA (2011) Certification Specifications and Acceptable Means of Compliance for Large Aeroplanes—CS-25. Amendment 11, Cologne.
12. FAA: Federal Aviation Regulations (FAR) Part 25—Airworthiness Standards: Transport Category Airplanes, USA.
13. ICAO (2007) PANS ATM (Air Traffic Management). Doc 4444, 5th Edition, Montreal.
14. EUROCONTROL (2006) Air Navigation System Safety Assessment Methodology (SAM), SAF.ET1.ST03.1000- MAN-01, Edition 2.1.
15. ICAO (2008) Performance-Based Navigation (PBN) Manual. Doc 9613-AN/937, 3rd Edition, Montreal.
16. Thiel, C., Seiß, C., Vogel, M. and Fricke, H. (2012) Safety Monitoring of New Implemented Approach Procedures by Means of Radar Data Analysis. Proceedings of the 5th International Conference on Research in Air Transportation (ICRAT), Berkeley, 22-25 May 2012.
17. Kaiser, M., Schultz, M. and Fricke, H. (2011) Enhanced Jet Performance Model for High Precision 4D Flight Path Prediction. Proceedings of the International Conference on Application and Theory of Automation in Command and Control Systems (ATACCS), Barcelona, 26-27 May 2011, 38-45.
18. Condor Flight Operations Engineering, HO/E (Bölling, M., et al.), Take off and Landing Flight Performance Calculations, November 2011.
19. DFS: FANOMOS Flight Track Data of Approaches/Departures at Frankfurt/Main Airport, May-October 2011, Langen, November 2011.

NOTES

*An application to a Large Close-in Obstacle at Frankfurt Airport's New Runway System.
1. Obstacle clearance altitude is referenced to mean sea level and obstacle clearance height is referenced to the threshold elevation or in the case of non-precision approaches to the aerodrome elevation or the threshold elevation, ICAO Doc 8168, Vol. II, Section I-1-1-6 [2] . Beyond the OCA/H, the approach surface is extended by a Visual Approach Surface (VSS) assuming the reduced aircraft ANP as guidance is formally limited to visual cues.

2. "New objects or extensions of existing objects should not be permitted above the conical surface and the inner horizontal surface… except when, in the opinion of the appropriate authority, an object would be shielded by an existing immovable object, or after aeronautical study it is determined that the object would not adversely affect the safety or significantly affect the regularity of operations of aeroplanes" [1] .

3. This is the turning departure route BIBTIE 3E.

4. Restricted by other obstacles resp. environmental constraints.

5. [5]Pursuant to ICAO PANS-OPS (Pt. 1-Section 2, Chapter 3, Table I-2-3-1) the maximum bank angle until 305 m (1000 ft.) altitude for departures is given by $\Phi = 15°$.

CHAPTER 5

Development of an Aircraft Routing System for an Air Taxi Operator

F.M. van der Zwan[1], K. Wils[1] and S.S.A. Ghijs[2]

[1] Delft University of Technology, The Netherlands

[2] Fly Aeolus, Belgium

1. INTRODUCTION

Due to increasing congestion of the road network and the major airports, the established modes of transport like car and scheduled air travel are having difficulty to fulfil the need for efficient business travel. Therefore, recent years have seen a growing demand for business aviation services like business aircraft charter and (fractional) aircraft ownership. These services enable improved time efficiencies, access to a wider range of airports, and the ability to control flight scheduling (Budd & Graham, 2009). However, business aviation is characterized by high prices and due to the economic downturn of 2008 corporate travel departments now seek less costly and more efficient business travel solutions (American Express, 2009; Gall & Hindhaugh, 2009). This has paved the way for the development of a new phenomenon in the air transport industry, namely the on-demand air taxi service. Air taxi operators offer travellers a low fare, on-demand travel service (Budd & Graham, 2009). As such, they provide the flexibility and time efficiency of existing on-demand business aviation, giving them the same time-efficiency advantage over car and scheduled air transport, but at a lower price (Bonnefoy, 2005).

One of the key enablers that allows air taxi operators to offer lower fares and still remain profitable is, apart from their smaller and thus cheaper aircraft, the optimization of resources, particularly during the employment of the aircraft during the aircraft routing phase. This chapter therefore sets out to develop an aircraft routing system which allows a full on-demand per-aircraft air taxi company to generate aircraft routing plans which adhere to its planning objectives, during the first operational years. Firstly, the air taxi business model is discussed. Then the aircraft routing problem and modelling approaches are described, followed by the design, testing and validation of the aircraft routing model, after which conclusions on the approach taken are drawn.

2. THE AIR TAXI BUSINESS MODEL

This section discusses the air taxi business model, the planning process for on-demand air transport (ODAT) and how this differs from scheduled air transport services.

2.1. AIR TAXI CHARACTERISTICS

Existing air taxi services can be differentiated with respect to their offer. Both per-seat and per-aircraft air taxi services exist. In the former the customer buys a single seat on a flight, while in the latter the customer charters a whole aircraft. Also 'semi on-demand' and 'full on-demand' air taxi services exist. In the former the air taxi service only operates from a predetermined set of airports, while in the latter the customer can freely choose the departure and arrival airport. Another difference lies in the type of aircraft used; operators either fly very light jets (VLJs) or small piston aircraft. A VLJ typically seats between four and seven passengers, can be flown by a single pilot, has a maximum take-off weight of less than 10,000 lb (4536 kg), and an average range of around 2,000 km (Budd & Graham, 2009, p.289). Piston aircrafts have one or more piston-powered engines connected to the propeller(s). Piston aircraft used for business typically seats one to six passengers and fly relatively short missions of 300-400 miles, using very small general aviation airports that are often without air traffic control towers (NBAA, 2010). The characteristics of the air taxi business model are summarised in Table 1 below.

TABLE 1. Characteristics of the air taxi business model (after Dyson, 2006)

Focused outside hub airports
Faster than scheduled air carriers for most point-to-point travel
One to six passengers per aircraft (though some take more)
High utilization of aircraft (via optimization of resources)
Cheaper than a charter or a fractionally owned aircraft
On-demand or semi on-demand (no fixed or published schedules)
Sold by the plane or per-seat
Clear, all-in pricing, by distance, time or zone
Using small piston or VLJ aircraft

As mentioned at the beginning of this chapter, air taxi services offer the flexibility and time efficiency of existing on-demand business aviation but at a lower price. Prices for per-aircraft air taxi (PAAT) operators in Europe range from € 2,500 per flight hour (London Executive Aviation, 2010) to € 800 per flight hour (Fly Aeolus, 2010). Bonnefoy (2005) indicates that air taxi operators are able to offer these lower fares due to their use of a small aircraft single type fleet and optimization of resources. The small aircraft used by air taxi operators have lower acquisition and operating costs than jets used by existing business aviation (Bonnefoy, 2005). Also, Budd and Graham (2009) and Espinoza et al. (2008a) note that the landing and take-off characteristics of these smaller aircraft enable access to an even greater range of airports, which possibly lie closer to where business travellers want to go. Apart from the smaller aircraft, optimization of resources is also key to the cost feasibility of the air taxi model (Dyson, 2006; Mane & Crossley, 2009). The resource optimization challenges faced by air taxi operators present themselves during the operational planning and management of the day-to-day operations.

2.2. PLANNING PROCESS FOR AIR TAXI MODEL

The airline planning process for scheduled air transport typically consists of five phases: flight scheduling, fleet assignment, aircraft routing, crew scheduling and crew rostering (Bazargan, 2004). The planning problem for on-demand air travel has a different nature than that of scheduled airlines (Hicks et al., 2005; Yao, 2007). The primary cause for this is the demand mechanism (Yao, 2007): scheduled airlines decide on their flight schedule months in advance, while in contrast, on-demand air transport operators like air taxis and fractional management companies sometimes know their flight requests only several hours in advance since customers can book up to six or eight hours before departure (Hicks et al., 2005; Yang et al., 2010). In addition, the flight legs flown by on-demand operators are less predictable and differ from day to day and week to week (Ronen, 2000; Yao et al., 2008). However, one can still distinguish the different phases of the scheduled airline's planning process in the planning process of on-demand air travel operators.

The **flight scheduling** phase for on-demand air transportation is customer driven. The customer contacts the air taxi operator a couple of days or hours in advance with his flight request. This process leads to an unpredictable and non-repeatable flight schedule. The customer's flight request specifies a departure location and departure time and an arrival location. Air taxi companies have the option to reject a request. The accept/reject decision depends on a number of factors, for example whether it is possible to fly the request and also whether is worthwhile flying the flight, both financially and strategically. This assessment should be performed with respect to the flight request itself, with respect to expected demand (Fagerholt et al., 2009), and also future customer demand as a customer denied service may choose not to request future trips (Mane & Crossley, 2007a). During the flight scheduling phase it must also be decided when and where aircraft will undergo scheduled maintenance (Keysan et al., 2010). If the request is accepted, the flight leg(s) of the customer's flight request are added to the flight schedule.

The second phase of the scheduling process considers **fleet assignment**, where fleet types (not specific aircraft) are assigned to each flight leg in the schedule (Bazargan, 2004). For an air taxi company, this phase is virtually non-existent as they mostly operate a single type fleet (Bonnefoy, 2005; Dyson, 2006).

The next phase is **aircraft routing**. During this phase each individual aircraft is assigned a routing, which is a sequence of flight legs, so that each leg is covered exactly once while ensuring that the aircraft visits maintenance stations at regular intervals thereby fulfilling maintenance requirements (Barnhart et al., 2003). The aircraft routing phase for on-demand air transport operators consists of assigning specific aircraft to the customer requested flight legs (Yao, 2007). During aircraft routing the operator thus aims to create aircraft routes that cover all the flight requests while minimizing the operational cost and adhering to the operational constraints, related to the aircraft's maintenance limit, crew regulations and the availability of fuelling facilities at airports (Martin et al., 2003; Yao et al., 2008). As departure and arrival locations can be chosen freely by the customer it is unlikely that an aircraft is always directly available at the departure location. Therefore an ODAT operator will have to conduct repositioning flights, also called 'deadhead' or 'non revenue' flights, to reposition their aircraft to the departure locations. Yao et al. (2008) mentions that deadhead flights may represent over 35% of the total flying conducted by a fractional management company (FMC). Air taxis and FMCs mostly need to bear the cost of repositioning themselves since the customer only pays for his actual flight request (Yang et al., 2008). Therefore, as a major driver during the aircraft routing phase is minimizing operational cost, it is important to minimize repositioning flights, both in number and in length (Bonnefoy, 2005; Mane and Crossley, 2007b; Yao, 2007). It must be noted that per-seat on-demand operators face an additional challenge during the aircraft routing phase. Apart from planning so that all customer requests are served, customer requests should be interleaved where possible such that that different customers can be put on the same aircraft, thereby obtaining a load factor sufficiently high to remain profitable (Espinoza et al., 2008a). This greatly increases the complexity of the underlying aircraft routing problem.

The two last phases in the scheduled airline's planning process involve the planning of crew resources. The fourth phase is **crew scheduling** or crew pairing. This treats the process of identifying sequences of flight legs that start and end at the same crew base, i.e. trip pairings (Bazargan, 2004). While creating trip pairings one strives for a minimum-cost set of pairings so that every flight leg is covered and that every pairing satisfies the applicable crew work rules (Bazargan, 2004). Finally, in the **crew rostering**phase, crew pairings are combined into monthly schedules which are then assigned to individual crew members (Bazargan, 2004). The crew pairing and crew rostering phase for ODAT operators are different from those used by scheduled airlines. As the definitive flight schedule is not known beforehand, no accurate monthly crew schedules can be created. Therefore crew of ODAT operators work with on-duty periods, during which they must be ready to operate flight legs which may not yet be known at the start of the on-duty period, and off-duty periods (Hicks et al., 2005; Yao et al., 2008). Yao et al. (2008) mention that, as aircraft are not always stationed at crew bases, crews coming on duty may need to travel to the aircraft's location, which counts as on-duty time and (possibly) needs to be paid for by the ODAT operator. Needless to say, ODAT crew schedules need to meet crew regulations as stipulated by the air transport authorities.

With respect to the demand uncertainty of ODAT, Hicks et al. (2005) report, based on the operations of an FMC, that approximately 80% of the trips are requested 48 hours or more prior to departure and 20% with as little as 4 hours notice. In addition, 30% of trips are changed at least once within 48 hours of the requested departure time (Hicks et al., 2005). Historical data analysis on FMC operations shows that on average 5% of demand is unknown for the first day for which the schedule is created, and 10-20% and 25-40% for the second and the third day respectively (Yao et al., 2007 , 2008). Despite the uncertainty in demand, ODAT operators need to create advance flight schedules, aircraft routings and crew schedules. There are a number of reasons for this: some airports require arrival information 24 hours in advance, the crews need to be given enough time to relocate to where they are required to come on duty, and maintenance locations need to be booked for scheduled maintenance (Yang et al., 2010). However, these advance schedules need to be adapted as more demand information gets known. For example, Yang et al. (2008) reports that around 25% of the requested trips of an FMC arrived after the original schedule was created, which causes the need to re-optimize the schedule.

Because of this dynamic process of creating and adapting the schedules it is desirable to both have a flexible and persistent schedule (Karaesmen et al., 2007; Yang et al., 2010): a flexible schedule to be able to cost-effectively cater extra demand, and a persistent schedule to avoid that the planning completely changes every time a change in demand occurs. Next to that, the schedule needs to be robust. Robustness denotes the ability of the schedule to cope with and cost-effectively recover from unforeseen changes in supply (e.g. aircraft break-down) or other disruptions (e.g. adverse weather) (Ball et al., 2007). As Bian et al. (2003) state, in a robust schedule enough slack is built in, such that a single disruption does not cause later flights to be delayed. However, a robust schedule comes at a price as schedule slack basically means that aircraft will be standing idle. Finally, the ODAT schedule needs to achieve a sufficient customer service level. This means on the one hand carefully looking at accepting or rejecting of flight requests, but also considering whether to schedule their customer's flights as requested or not (Fagerholt et al., 2009). Delaying the departure time by 30 minutes may enable a more cost-effective aircraft routing but this may have a negative impact on the service level as perceived by the client. In summary, the ODAT operator has to balance minimising costs whilst keeping a high service level and developing an appropriate schedule (flexible, persistent, and robust).

3. THE AIRCRAFT ROUTING PROBLEM

In this section the aircraft routing problem as it applies to the air taxi operator is laid out. As explained before, the aircraft routing problem arises during the aircraft routing phase, in which the ODAT operator aims to create aircraft routes that cover all flight legs and satisfy operational restrictions while minimizing operational cost. Below, the different aspects of the aircraft routing problem are treated:

- **The scope:** planning horizon
- **The input:** flight legs (trips) and aircraft
- **The main objective:** to create an aircraft routing plan that covers all flight legs while minimizing the operating cost
- **The secondary objectives:** robustness, flexibility and persistence
- **The constraints:** operational requirements

Please note that this work has been undertaken in collaboration with Fly Aeolus, a Belgian full on-demand per-aircraft air taxi operator. This routing problem and subsequent model has been kept as generic as possible. However, in some instances the specifics of the Fly Aeolus business model have been taken as an input for the routing problem. In the text it is clearly stated how the routing problem might be (slightly) different for ODAT operators with a different business model.

An important parameter in the planning process is the **planning horizon** that is used. This determines the size, and thus complexity, of the planning problems that need to be solved during the different phases. If a 24 hour horizon is chosen the aircraft routing system will find a local optimum for that day, but the ending locations of the aircraft will not be optimized towards the start of the next day. Theoretically, the more days that are included in the planning horizon, the closer the solution will be to a global optimum. However, due to the computational complexity and the uncertainty of demand, there is a limit on the number of days that can be incorporated in the planning horizon. Ronen (2000) has tested the effect of the planning horizon by solving aircraft routing problems for 24, 36 and 48 hour horizons for a fractional management company (FMC). For his application a 24 to 36 hour planning horizon proved most effective, as planning decisions taking place beyond that horizon were always changed again later. On the other hand, Martin et al. (2003) state that the operations of the FMC they worked with allowed a two to three day planning horizon. Overall, in the ODAT literature it is generally accepted that planning beyond 72 hours is highly speculative. Therefore, the planning horizon is set to a maximum of 72 hours.

At each instance the operations control centre (OCC) wants to create an aircraft routing plan, the aircraft routing system takes the known flight schedule (output of the flight scheduling phase) and extracts the booked flight requests. Each flight request consists out of at least one **flight leg** (trip). Each of these flight legs has a departure airport, a requested time of departure, an arrival airport and a specified number of passengers for that flight. Customers also have the possibility to make a booking consisting of multiple flight legs. For these requests they are given the option to have the aircraft and crew remaining on standby at the airport in between the coupled flight legs, such that all their separate trips are served by the same aircraft. Apart from the flight requests, the aircraft routing system also retrieves the current status of all the **aircraft** available within the available fleet. For each aircraft the home base and its lease cost per hour are known. In addition to this, it is known when and where the aircraft will become available, and the aircraft's fuel status and flight hours remaining till maintenance at that time. An aircraft can be unavailable at the start of the planning horizon because it is still in-flight at that time, or because maintenance is performed on the aircraft.

The objective of the aircraft routing system is now to assign to each aircraft a feasible sequence of flight legs in a manner that **minimizes the cost of operating all the flights** while meeting the operational requirements. Of the total operating costs, the variable direct operating costs (DOCs) are the only costs that are controlled by the aircraft routing system. This is because these are the costs that are directly related to the amount of flying that is undertaken (Doganis, 2002), which in turn depends on the aircraft routing plan is created.

The following DOC items incurred by the air taxi operator are directly influenced by the aircraft routing plan: variable aircraft costs, passenger and non-passenger related airport charges, air traffic service charges, fuel costs, and variable flight crew costs. With respect to the fuel costs, these costs depend on the amount of flying that is planned. In addition, they also rely on where and how much the aircraft is planned to be fuelled. Because fuel prices vary significantly between airports (Flyer Forums, 2010), so-called fuel ferrying can in certain cases be a valid strategy to reduce fuel cost. When adopting this approach, excess fuel is carried on a flight leg to decrease the amount of fuel that needs to be bought at the next airport where fuel prices are high. The crew costs consist of the pilot wages which are directly related to how much flying the pilot carries out (as measured by the Hobbs meter of the aircraft, which registers the time that the engine is running and thus also includes taxiing and idling (e.g. when waiting for take-off clearance)). Therefore, this part of crew costs is also controlled during the aircraft routing phase.

As explained in the previous section, achieving **robustness, persistence and flexibility** while creating operational plans are additional objectives to consider during the planning process. These three secondary planning objectives are also to be considered during the design of the aircraft routing system. However, these planning objectives are interrelated and cannot be fully achieved together. For example, requiring a very robust schedule (thus a large amount of slack time), decreases the ability of the air taxi operator to accept future flight requests. During the development of the aircraft routing system it is therefore ensured that the system parameters that control robustness, persistence and flexibility can be altered by the OCC such that they can prioritize these objectives as they wish.

The aircraft routing plans that are created also have to fulfil a number of **operational requirements**. Firstly, aircraft should not exceed their maintenance flight hour limit when flying the assigned flight sequence. Secondly, flight sequences should be fuel feasible. Both because there are airports where it is not possible to fuel and because the number of passengers that is carried influences the amount of fuel that can be taken. For example, an SR22 carrying a pilot and two passengers can only carry 38% of its maximum usable fuel (Cirrus, 2007). In addition, aircraft routing plans must be crew feasible, i.e. they should not violate the crew work rules. For this purpose, it is customary in the ODAT industry to couple an aircraft with its crew for scheduling purposes (Martin et al., 2003; Yao et al., 2007 ; Yang et al., 2008). Hence, an aircraft is viewed as an operational unit which has to meet both its own restrictions (e.g. maintenance) and crew duty time limits. Because of this, the aircraft routing system will generate routing plans that do not necessarily require crew swaps during the day as a single crew can (legally) stay with a single aircraft throughout the whole day.

In summary, during the routing phase the air taxi operator's OCC needs to create aircraft routing plans covering all flight legs taking place during the scheduling horizon with the aim to minimize operating cost while adhering to the aircraft's maintenance flight hour limits, fuel feasibility, aircraft availability and crew duty rules, as shown in Figure 1 above.

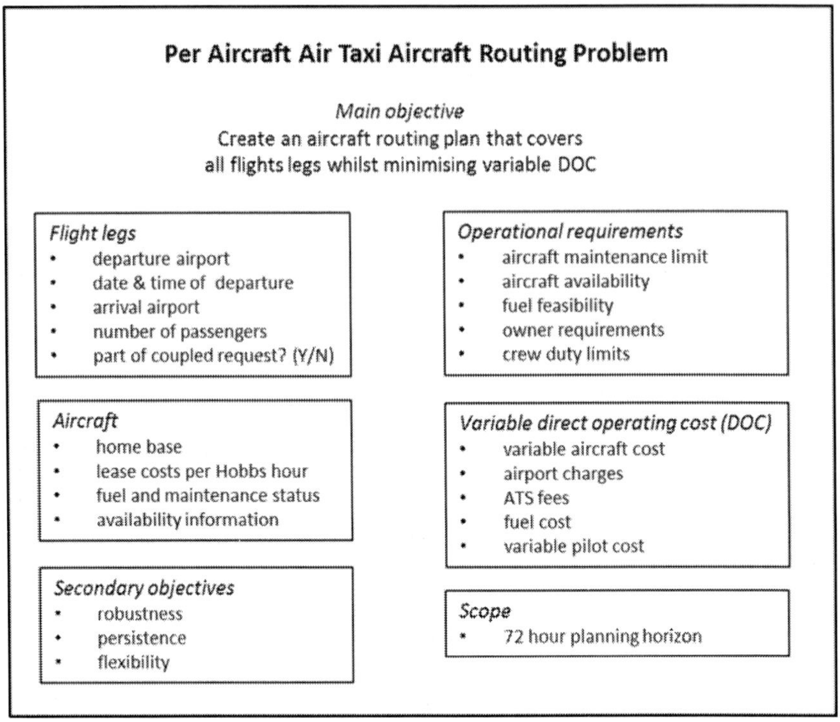

Figure 1. Full on-demand per-aircraft air taxi routing problem

4. CHOICE OF MODELLING APPROACH

In this section the modelling approach for the aircraft routing problem is chosen, after which the chosen methodology is further explained.

4.1. Choosing the Appropriate Modelling Approach

To the authors' best knowledge there exists no literature on the aircraft routing problem faced by a per-aircraft air taxi (PAAT) operator. However, their aircraft routing problem is very similar to that of a FMC or charter. Both aim to minimize the operating cost while serving all the customer requests on time, and satisfying the applicable constraints. However, there are some differences between the aircraft routing problem of an FMC and that of a typical air taxi operator. First of all, since most air taxi operators operate a single type of aircraft, all its aircrafts are in principle compatible with every request. For an FMC, which operates different fleet types, there is a specific aircraft type that fits each request and this requirement must be taken into account during the aircraft routing. In addition, an FMC does not incur the flight hour related aircraft costs as a part of its variable operating costs because they either own or lease their aircraft for a continuous period of time. So their aircraft cost is irrespective of how much they use the aircraft. However, despite these differences the core of the aircraft routing problem for an FMC and a PAAT operator remains the same. Therefore, this section focuses on studies related to FMCs to develop a system that solves the PAAT aircraft routing problem. These are listed in Table 2below.

Table 2. Modelling and solution approaches for the aircraft routing problem

Modelling approach	Author(s)	Maintenance constraints	Crew constraints	Planning horizon [h]	Mathematical formulation*	Solution method#
M1	Keskinocak & Tayur (1998)	N	N	72	0 – 1 IP	CPLEX
	Keskinocak & Tayur (1998)	N	N	72	0 – 1 IP	Heuristic
M2	Ronen (2000)	Y	Y	48	SP	Heuristic
M3	Martin et al. (2003)	Y	Y	48 – 72	0 – 1 IP	CPLEX
M4	Hicks et al. (2005)	Y	Y	24 – 72	Mixed IP	CG, GENCOL
M5	Karaesmen et al. (2005)	Y	N	24	0 – 1 IP	CPLEX
M6	Karaesmen et al. (2005)	N	N	24	Mixed IP	CPLEX
	Yang et al. (2008)	N	N	24	Mixed IP	CPLEX
M7	Karaesmen et al. (2005)	Y	Y	24	0 – 1 IP	Heuristic, CPLEX
	Yang et al. (2008)	Y	Y	24	0 – 1 IP	Heuristic CPLEX
M8	Karaesmen et al. (2005)	N	Y	24	Mixed IP	CPLEX
	Yang et al. (2008)	N	Y	24	Mixed IP	CPLEX
M9	Karaesmen et al. (2005)	Y	Y	24	SP	BP, CPLEX
	Yang et al. (2008)	Y	Y	24	SP	BP, CPLEX
	Yang et al. (2008)	Y	Y	72	SP	BP, CPLEX
	Yang et al. (2008)	Y	Y	96	SP	BP, CPLEX
M10	Yao et al. (2005)	Y	Y	72	SP	CG
	Yao et al. (2008)	Y	Y	72	SP	CG
M11	Yao & Zhao (2006)	Y	Y	72	SP	na
	Yao et al. (2007)	Y	Y	72	SP	CG

* IP = integer programming, SP = set-partitioning   # CG = column generation, BP = branch-and-price

To choose a modelling approach for a PAAT aircraft routing problem one must first specify the decision criteria that are used during the trade-off. These criteria are derived from the requirements that are posed upon the aircraft routing system and from the nature of the PAAT aircraft routing problem. The modelling approach should allow for both crew feasible and maintenance feasible routings. It should be able to accommodate a multiple day planning horizon and multiple fleet types. It should also have the flexibility to incorporate additional operational rules and finally, the flight leg cost should be dependent on the aircraft route. In Table 2 above the specifics of the different aircraft modelling approaches deployed in FMC literature are given. Since the criteria that schedules must be crew and maintenance feasible need to be fulfilled a number of modelling approaches can already be discarded. These are model 1 proposed by Keskinocak and Tayur (1998) and models 5, 6 and 8 described byKaraesmen et al. (2005) and Yang et al. (2008). Furthermore, model 7 can also be discarded since it is noted by Karaesmen et al. (2005) and Yang et al. (2008) that this modelling approach is not flexible enough to handle planning horizons longer than 24 hours. This leaves us with six models (2, 3, 4, 9, 10 and

11). All these models produce schedules that are crew and maintenance feasible and can be used in multiple day planning horizons.

However, model 10 and 11 do not incorporate the cost of flying customer trips in the operational cost of a schedule. The authors, Yao et al. (2005, 2008); Yao and Zhao (2006), state that the customer of an FMC pays for the fuel and the crew costs of their trip, thus they do not consider these trip costs as a part of operational cost. For the PAAT aircraft routing problem this is not a valid assumption. The cost for one specific flight leg namely differs depending on which aircraft routing the leg is in. Air taxi customers always pay a fixed price per flight hour, which is independent of what total cost the operator has to bear to operate this flight.

Models 2 and 9 utilize a set-partitioning formulation. In an overview of vehicle routing models, Bunte and Kliewer (2010) state that an advantage of the set-partitioning formulation, in contrast to other formulations, is that additional constraints or operational rules can be easily incorporated. Also, Ronen (2000) and Karaesmen et al. (2005) note that the set-partitioning approach is very appropriate to problems where costs are nonlinear and discrete, and complicated rules are imposed, as is the case in the PAAT aircraft routing problem. A direct comparison with a 0-1 and mixed IP formulation is made by Karaesmen et al. (2005) and Yang et al. (2008). They conclude that the set-partitioning formulation is the only one that has adequate flexibility to cover long planning horizons and to incorporate complex operational rules (Karaesmen et al., 2005). Therefore, in order to model the PAAT aircraft routing problem the set-partitioning formulation is chosen.

4.2. The Set-Partitioning Model

For the set-partitioning formulation of the PAAT aircraft routing problem, which is adopted from Yang et al. (2008), the following notation is used:

Ω_i = the set of feasible routes for aircraft i,		$i \in N$
p = a feasible route,		$p \in \Omega_i$
c_{ip} = the total cost of taking route p by aircraft i,		$i \in N$ and $p \in \Omega_i$
a^i_{jp} = 1 if trip j is on route p; 0 otherwise,		$i \in N, p \in \Omega_i$ and $j \in M$
θ_{ip} = 1 if aircraft i takes route p; 0 otherwise,		$p \in \Omega_i$ and $i \in N$

Where $N = \{1, 2;\dots n + m\}$ is the set of all aircraft such that $\{1,\dots, n\}$ represents the own fleet, and $\{n+1,\dots, m\}$ represents any charter aircraft for subcontracting the trips (in case the own fleet proves insufficient to operate all the flights). For each i $(n + 1, \dots, m)$, i consists of a single route that only takes trip j with a cost b_j (the cost that is incurred when subcontracting this trip). Furthermore, M denotes the set of all trips. Using this notation, the aircraft routing problem of a PAAT operator can be formulated as the following set-partitioning problem (SP):

$$(SP) \quad \min \sum_{i \in N} \sum_{p \in \Omega_i} c_{ip}\theta_{ip}$$

$$\text{s.t.} \quad \sum_{i \in N} \sum_{p \in \Omega_i} a^i_{jp}\theta_{ip} = 1, \qquad\qquad j \in M$$

$$\sum_{p \in \Omega_i} \theta_{ip} \le 1, \qquad\qquad i \in N$$

$$\theta_{ip} \in \{0, 1\}, \qquad\qquad i \in N \text{ and } p \in \Omega_i$$

This way aircraft are assigned to feasible routes while minimizing the total cost of serving all trips (1). Constraints (2) ensure that each trip is covered exactly once in the solution. Constraints (3) make sure every aircraft is assigned at most one route. Also, either a route is part of the solution or it is not, therefore parameters $_{ip}$ must be either 0 or 1 (4).

To solve the set-partitioning model of the aircraft routing problem it is possible to use either an exact solution method or a heuristic (Hillier and Lieberman, 2008). As Hillier and Lieberman (2008) andSilver (2004) note heuristics are often used when the time required to find an optimal solution for an accurate model of the problem would be very large. For the aircraft routing system under consideration computational time is indeed an important requirement. However, the FMC studies mentioned in Table 2 that use exact solution methods to solve a set-partitioning model already achieve solution times in the order of seconds. The second performance measure for a solution method, apart from computational speed, is the degree of optimality of the solution (Silver, 2004). In practice, and even for exact solution methods, there is a maximum optimality gap given which is the maximum percentage that the obtained solution differs from the optimal solution. For the exact methods computational time or memory limits can cause the model not to be solved to optimality. The maximum optimality gap for the exact solution methods for the set-partitioning models has been found to be smaller than half a percent. Also, in the considered literature on the aircraft routing problem for on-demand air transport no heuristics have been used to solve the set-partitioning model. In addition, when designing a heuristic or adapting an existing one, both computational performance and solution quality can only be assessed after implementation. So when opting for a heuristic solution method for this problem no performance guarantees can be given beforehand. Because exact solution methods have proven to yield an adequate computational performance and solution quality for solving the set-partitioning model of the aircraft routing problem for on-demand air transport, the authors have chosen to use an exact solution method to solve the set-partitioning model that is created by the model creator.

To implement an exact solution method for the solver of the aircraft routing system, there are two options. First of all, one can program and implement an exact solution method, like branch-and-price or column generation, themselves. However, as Feillet (2010) has recently noted, carrying out an accurate implementation of an exact solution algorithm is a long and difficult task due to the inherent complexity of these methods and the lack of simple and comprehensive descriptions of these methods. The second method is to use a solver software package that embeds one or more of the exact solution methods needed to solve the set-partitioning problem. There already exist a number of solver packages that are applicable for the problem at hand, as can be seen in Table 2. Therefore, in this project a solver software package is used to solve the set-partitioning formulation of the PAAT aircraft routing problem. The solver package that is used is the IBM ILOG CPLEX Optimizer, since it has already proven its worth in ODAT applications. To solve the set-partitioning problem with CPLEX use is made of the built-in "mipopt" optimizer module. This optimizer uses a branch-and-cut algorithm, which is a hybrid method of branch-and-bound and cutting plane methods (Smith and Taşkin, 2008).

5. DESIGNING THE AIRCRAFT ROUTING SYSTEM

To construct the set-partitioning model of the aircraft routing problem for a PAAT operator, the model creator receives the following input data via the I/O interface of the aircraft routing system: flight leg (trip) data of customer and dummy flight requests, parameters of the real and dummy aircraft, airport parameters, a list of persistent aircraft-trip assignments, and operational parameters. This is shown inFigure 2 below.

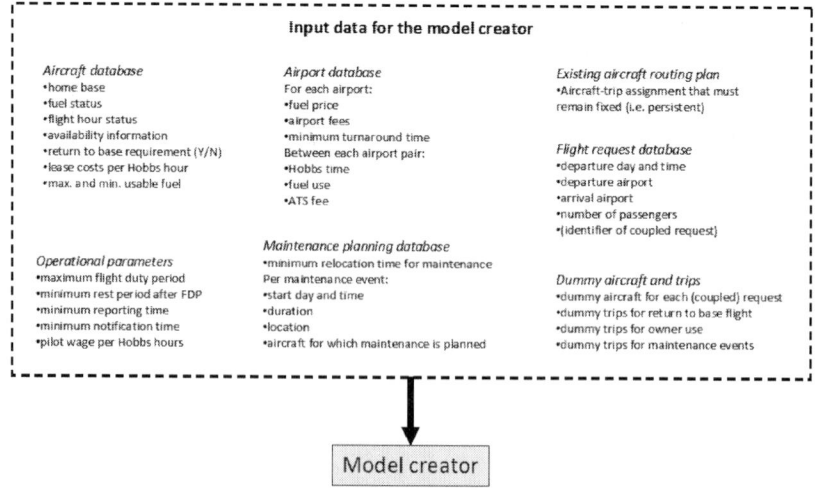

Figure 2. Input data of the aircraft routing system.

Dummy requests are created for each return to base trip, for owner use and for maintenance events. Dummy aircraft on the other hand are added to represent the aircraft of another company when a trip is subcontracted. The model creator uses the above input data to generate the set of feasible routes for each aircraft (Ω_i), the total cost of operating each of these routes (c_{ip}) and information as to which trips are contained in each route (a^i_{jp}). With these parameters the set-partitioning formulation of PAAT aircraft routing problem is constructed and written to an LP file. This file is read by CPLEX which then solves the set-partitioning model. The obtained solution specifies which routes are part of the aircraft routing plan (i.e. which θ_{ip}'s are 1). This information is processed by the solver module which uses it to generate the aircraft routing plan that forms the solution of the aircraft routing problem. The model creator and the solver module of the aircraft routing system described in this chapter are programmed in MATLAB. The model creator uses the input data provided by the I/O interface to generate an LP file of the set-partitioning model that represents the aircraft routing problem. This LP file serves as input for the solver module. Figure 3 provides a detailed overview of this process.

First, the preprocessor module of the model creator uses the input data to create matrices in which it is specified whether aircraft i is initially compatible with trip j and whether aircraft i can fly trip j_2 after trip j_1. Together with the input data these two matrices are passed to the next module of the model creator, namely the route generator. The route generator module uses the information stored in the matrices to create the set of possible routes for each aircraft.

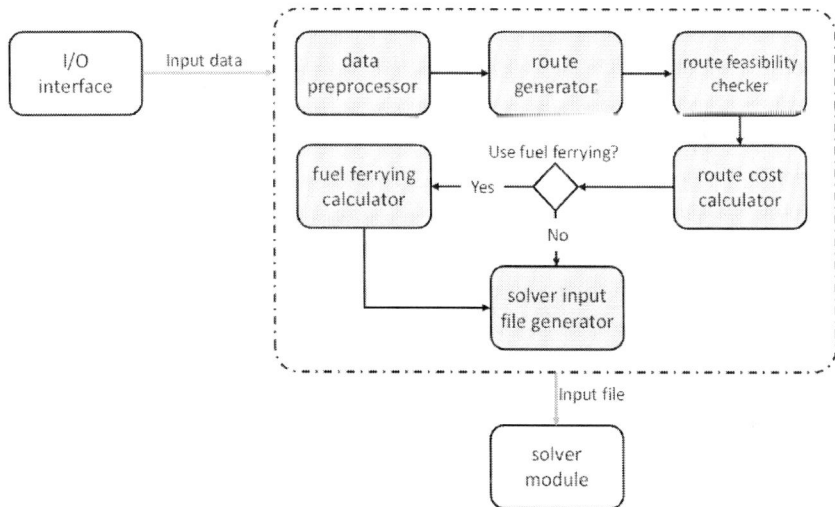

Figure 3. The model creator module of the aircraft routing system.

Each of these routes consists of a sequence of flight legs. This set of routes for each aircraft is passed to the route feasibility checker. The route feasibility checker processes the route list for each aircraft generated by the route generator module and removes the routes that are infeasible. The list of feasible routes is then forwarded to the route cost calculator. The route cost calculator processes each of the feasible routes and calculates the total route cost for each of the feasible routes. The total route cost consists of the aforementioned variable aircraft cost, airport charges, ATS charges, fuel costs, and variable crew cost. If the fuel ferrying strategy is used, the fuel cost is calculated using the fuel ferrying calculator. This module uses the assumed percentage burn-off fuel ferrying calculation method. The final module of the model creator is the solver input file generator. This module writes an LP file of the set-partitioning model that is created using the previous modules. This LP file can be directly read and solved by CPLEX. The solver module of the aircraft routing system consists out of a MATLAB script that solves the created set-partitioning formulation, as written to the LP file, using CPLEX and retrieves the solution, i.e the aircraft routing plan.

6. TESTING AND VALIDATING THE MODEL

In this section the aircraft routing model is both tested and validated.

6.1. Testing The Model

To test and validate the developed aircraft routing system operational data is needed as input. However, at the time of writing Fly Aeolus is not yet fully operational and as such the system can not be tested in the environment where it will eventually be used. Therefore a virtual test environment is created. This environment contains a virtual airport network in which a per-aircraft air taxi operates. The virtual environment is created by an input data generator which creates all the input data needed by the model creator. The input data generator utilizes operational parameters that reflect the (expected) PAAT operations of Fly Aeolus, see Table 3.

Table 3. Parameters used to generate input test data

Parameter	Value	Unit
Parameters used to generate airport test data		
Farthest point-to-point distance in the network	800	km
Number of airports	225	-
Average cruise speed	310	km/h
Average fuel burn	1.12	litre per minute
ATS fee	0.00	€ per km
Passenger arrival fee	0 – 10	€ per pax
Passenger departure fee	0 – 30	€ per pax
Aircraft arrival fee	10 – 150	€
Aircraft departure fee	0 – 60	€
Turnaround time	45	minutes
Avgas price	0.90 – 2.70	€ per liter
Number of maintenance bases	6	-

Table 3. (Continued)

Parameters used to generate trip test data		
Planning horizon	72	hours
Minimum Hobbs time of flight request	50	minutes
Average Hobbs time of flight request	90	minutes
Departure time distribution	-	xx
Average number of passengers	1.3	-
Percentage of coupled flight requests	10	% of flight requests
Number of trips in a coupled request	2	-
Waiting time between each trip of a coupled request	6 – 8	hours
Aircraft that have maintenance events	10	% of aircraft
Planned maintenance duration	6	hours
Minimum relocation for maintenance	120	minutes
Aircraft that have owner use	30	% of aircraft
Start time of owner use, between	5 – 20	hour
Aircraft that have return to base requirement	30	% of aircraft
Aircraft has to be back at base at	22	hour
Parameters used to generate aircraft test data		
Time when aircraft becomes available, between	0 – 17	hours of the 1st day
Initial fuel status, between	52 – 87	litres
Maximum useable fuel for 0, 1, and 2 pax	348; 272; 132	litres
Minimum reserve fuel	52	litres
Flight hours remaining till maintenance, between	6 – 50	flight hours
Lease costs per Hobbs hour, between	n/a	€
Charter cost per Hobbs hour, between	n/a	€
Parameters used to in the aircraft routing system		
Maximum flight duty period	13	hours
Minimum rest period after an FDP	12	hours
Time pilots need to report for duty	60	minutes
Pilot wage per Hobs hour	50	€ per hour
Surplus fuel burn (sfb_{hour})	4	% per hour
Allowance for unknown costs (a_u)	0.05	€ per litre
Number of closest routes to append (k)	10	-

In addition, to test the use of persistent aircraft-trip assignments a demand uncertainty simulator is programmed. The input generator that is created to generate the test environment and corresponding data consists of two MATLAB scripts: an 'airport data generator' and a 'trip and aircraft data generator'. The airport data generator creates a virtual world containing a number of airports, i.e. the network in which a virtual per-aircraft air taxi service operates. The trip and aircraft data generator in turn creates flight requests that need to be served by that virtual operator and the data of the virtual operator's aircraft. Both the virtual world and the virtual PAAT operator are modelled on our Belgian air taxi company and its operations to ensure that the order of magnitude of the generated input data is the same as that of the environment where the system will eventually be employed.

The final parameter, k, determines the number of closest trips that are appended to a route during the route generation process. To choose the value for k, tests are conducted with the aircraft routing system with k ranging from 3 to 15. The results of these experiments are given in Tables 4 and 5. Note that each line of results that is reported in the results tables represents the average result of ten tests with independent random data sets. Furthermore the number of flight requests per day in the planning horizon is indicated in the results by j and the number of available aircraft by i. The data sets are generated using the virtual environment and virtual air taxi operator as described in this section.

Table 4. The effect of k on the aircraft routing solution

1. k	2. (j, i)	3. Routes	4. ct_{MC}	5. ct_{CPLEX}	6. Solution cost
3	(7, 19)	1,729	5.3	0.01	18,649
5	(7, 19)	3,427	8.0	0.03	18,491
10	(7, 19)	8,090	15.6	0.08	18,401
15	(7, 19)	10,799	20.3	0.12	18,395
3	(15, 25)	6,086	17.7	0.13	38,469
5	(15, 25)	16,631	37.2	0.22	38,149
10	(15, 25)	72,260	177.9	1.21	37,849
15	(15, 25)	180,389	608.9	5.05	37,832

The first column in Table 4 indicates the value of parameter k that is used for the experiments. Column 2 gives the number of flight requests per day that are generated and the number of aircraft in the fleet (excluding subcontractor aircraft). In the next column the total number of feasible routes present in the set-partitioning model is stated (thus for all the aircraft). Column 4 and 5 respectively denote the computational time in seconds of the model creator and CPLEX. Finally, in the last column the total cost of the aircraft routing plan is given. Table 5 below gives the results of the same experiments as shown in Table 4, but this time relative to the results for k = 15.

Table 5. The effect of k on the aircraft routing solution (relative to k = 15)

1. k	2. (j, i)	3. Routes	4. ct_{MC}	5. ct_{CPLEX}	6. Solution cost
3	(7, 19)	16%	27%	13%	101.4%
5	(7, 19)	33%	40%	33%	100.5%
10	(7, 19)	77%	78%	71%	100.0%
15	(7, 19)	100%	100%	100%	100.0%
3	(15, 25)	4%	3%	4%	101.8%
5	(15, 25)	10%	7%	5%	100.8%
10	(15, 25)	40%	30%	24%	100.0%
15	(15, 25)	100%	100%	100%	100.0%

It can be seen in column 3 of Table 4 that decreasing k causes the number of feasible routes that is generated by the model creator to decrease. This is because fewer of the closest trips are appended during the route generator loop. Due to the lower number of routes that are generated, both the model creator and CPLEX need less time to create and solve the set-partitioning model. This effect can be observed in column 4 and 5 of Table 5. While the number of created routes and the computational time for k = 3 can be as little as 4% of those for k = 15, the average increase in total cost of the aircraft routing solution is maximum 1.8%. When changing k from 15 to 10, the increase in solution cost is less than 0.1% while the computational time is decreased by at least 22% and in the (15, 25) case even by 70%. The latter can be explained by the fact that the number of routes that is created grows exponentially with the number of flight requests per day. A change in parameter k therefore has an increasingly larger effect on the change in the number of generated routes, and thus in computational time, when increasing the flight requests per day. In summary, varying parameter k from 15 to 3 has a large effect on the number of feasible routes that are created and, consequently, on the computational time needed by the aircraft routing system. However, the conducted experiments show that the effect on the solution cost is limited to an increase by maximum 1.8%. These findings are similar to the ones reported in the research of Ronen (2000) who varied k from 11 to 5 using 47 aircraft and 50 trips and reported a maximum cost increase of less than 0.1%. It can therefore be concluded that while having a high value of k causes the system to generate a larger number of feasible routes and thus possible solutions, the chosen aircraft routing solution mostly contains those routes that minimize the repositioning cost. In other words, when creating an aircraft routing it is a good rule of thumb to only consider those trips that are close to the arrival airport of the previous trip. As shown by the results in Table 5 only considering the five trips for which the departure airport is closest to the arrival airport of the previous trip, causes the routing solution to be at most 0.8% more costly compared to when considering the fifteen closest ones. Because changing k from 15 to 10 only increases the solution cost by less than 0.1% while reducing the solution time sometimes by as much as 70%, a k value of 10 is adopted as the standard value.

When creating the virtual airport network, demand and PAAT operator it is attempted to create input data for the aircraft routing system that has the same order of magnitude of the expected real input data. However, there are a number of differences between the virtual test environment and the air taxi operator expected operations. First of all, the virtual test environment uses a square airport grid in which airports are uniformly distributed instead of the actual locations of the airports in Western Europe. In addition, the airport charges the virtual environment uses are therefore not the ones encountered in reality. Also related to the airports is the fact that the turnaround time is taken to be 45 minutes for all airports. In reality the turnaround time depends on the airport.

The trip data generator randomly chooses a departure and arrival airport for all the trips. It therefore does not simulate demand hot spots, i.e. airports where a lot of demand originates or arrives. In reality, certain airports do attract significantly more business flight traffic than others. For example, Geneva Cointrin airport, which is the second busiest business airport in Europe, has almost 50% more business flight departures than the fifth busiest airport, Nice (Eurocontrol, 2010). In addition, the trip data generator only uses the hourly business flight departure pattern of France. In reality, however, the hourly departure pattern is dependent on the country the flight departs from and on the month and the day of the week (Eurocontrol, 2010). Also the trip data generator creates few day returns, i.e. customers that fly from A to B in the morning and return from B to A in the evening. Furthermore, the data generator only earmarks aircraft unavailable for a maximum of a single day. However, it is possible that the aircraft is unavailable for a longer period of time. Yao et al. (2008) report that a FMC with 35 aircraft encountered 49 mid-day unscheduled maintenance events during a one-month period. It is therefore probable that unplanned maintenance will also occur during air taxi operator's operations. However, this virtual environment does not account for this.

It is apparent from the above that there exist differences between the virtual test environment that is created and the expected operational environment of the air taxi operator. However, as explained before, the virtual test environment is only used in this project to test and validate the workings of the aircraft routing system, and not for example to find out what the exact effect of the fleet size is on the air taxi operations. Therefore, it is of main importance that all the necessary input elements (like airport turnaround times, owner use trips and coupled requests) are specified in the virtual environment and that these have the same order of magnitude as in reality. As such it can be assessed whether the aircraft routing system is capable of handling all these inputs correctly. It is in this stage not important that the virtual test environment does exactly represent reality. The main aim is the development of the aircraft routing system and therefore the virtual test environment described in this section is adequate for its purpose, namely validating the aircraft routing system.

6.2. Validating The Model

Validation is an important step in model development because through validation the level of confidence in and the credibility of the aircraft routing system is established. To carry out the validation in this project five types of validation are conducted, namely conceptual model validation, computerized model verification, data validation, experimental validation and operational validation (Landry et al., 1993; Sargent, 2007).

In the **conceptual validation** phase, in which it is determined whether the underlying theories and assumptions of the system are correct, face validation and desk checking are the used validation techniques. It is argued that the theory of the aircraft routing system is indeed correct. However, there are some assumptions of which the exact effect on the accuracy of the conceptual model could not be determined as no real operational data is available, namely the assumptions that a fixed percentage of ferried fuel is burned per hour, that the Hobbs time and fuel use between airport pairs is specified for the one passenger case, and that it is acceptable to impose crew constraints on the aircraft. It is therefore advised that these assumptions and their effect are checked as soon as real operational data is available. Next is the **computerized model verification**. This process ensures that the implementation of the conceptual model in programming code is correct. To make sure this holds true for the MATLAB implementation of the aircraft routing system the authors used bottom-up development and testing, debugging and desk checking. In addition, experiments show that the results produced by the routing system are consistent. Next, **data validation** is performed to ensure that the data used to build and evaluate the model is adequate and correct. For building the model the mental and written databases provided by the air taxi operator and found in ODAT literature are considered appropriate. In addition, the virtual test environment and corresponding numerical input data created to support the evaluation of the aircraft routing system is deemed adequate for its purpose. **Experimental validation** aims to check whether or not the model contradicts qualitative, expert knowledge. To this end both the convergent validation and face validation techniques are utilized. For the first technique the results produced by the aircraft routing system are checked against six expert statements. In all cases the results of the routing system adhered to the expert knowledge. In addition face validation is carried out with Fly Aeolus' director. He is given an example aircraft routing problem to assess whether the solution produced by the routing system is correct, which he indeed found to be true. Finally, **operational validation** assesses the usefulness and timeliness of the solutions and cost of implementing the system. To carry out the routing plans produced by the aircraft routing system the air taxi operator must first decide on and then apply the waiting strategy for the routing plans. Furthermore, as mentioned in the conceptual validation, it must be assessed what the exact effect on the accuracy of the routing plans is of the fuel ferrying, and Hobbs time and fuel use assumptions before implementing the routing system. With respect to the timeliness it is found that the aircraft routing system is capable of solving routing problems with sizes that correspond to air taxi operator's first operational years in under five minutes. Noteworthy in this respect is that the routing system is capable of producing a solution to a routing problem in little over one second that takes a human dispatcher 30 minutes to solve.

7. DISCUSSION & CONCLUSIONS

In this section important limitations of the work are discussed and conclusions are drawn.

7.1. Discussion

As this aircraft routing system has been developed for an air taxi operator who – at time of writing – is not yet fully operational, no real operational input data can be used for system validation and testing and instead a virtual test environment that approximates the air taxi operator's operations is utilized. Therefore, this research has some limitations.

The air taxi operator needs to check some assumptions underlying the routing system before putting it to use. The exact effect of these assumptions on the routing plans could not be assessed in this project because no real operational data is available. First, it needs to be assessed whether the increase in routing costs caused by imposing crew constraints upon the aircraft is acceptable. In addition, the air taxi operator must check whether specifying Hobbs time and fuel use between airport pairs for the one passenger load case is accurate enough. Finally, the relation to reality of the assumption that a fixed percentage of ferried fuel is burned per hour must be checked. Once operational data becomes available, the air taxi operator is advised to check these assumptions. This is deemed a necessary step in the implementation of the aircraft routing system. However do note that with respect to the Hobbs time and fuel use assumption, the air taxi operator can also opt to remove this assumption from the system altogether by specifying these input values for each passenger load case.

The aircraft routing plans created with the aircraft routing system do not contain a waiting strategy for repositioning flights, only the earliest and latest start time for deadheads are indicated. If the air taxi operator wants to adopt a 'fly-first' or 'wait-first' waiting strategy, this can be directly applied to the created routing plans and as such this limitation is easily removed from the system. However, the authors suggest that the aircraft routing system is augmented with a module which calculates the optimal waiting strategy per aircraft route instead of adopting the 'fly-first' or 'wait-first' approach as the latter two are suboptimal approaches. Once the waiting strategy is determined, also aircraft parking fees, crew waiting wages, and airport opening hours can be incorporated into the aircraft routing system, as these all depend on the strategy that is adopted.

With respect to the secondary planning objectives, the degree of robustness, flexibility and persistence of the aircraft routing plans are not measured. The developed system does allow the OCC to control these parameters, but only to a limited extent. Robustness is solely controlled by slack time, and no other means of enforcing robustness are currently provided. The same holds for persistence and flexibility, whose balance is only altered by using the persistent aircraft- trip assignments. But the aircraft routing system does not provide information as to exactly how robust, flexible or persistent the created routing plans are. Though the dispatcher knows he is increasing/decreasing robustness, flexibility or persistence when he is adapting the corresponding control parameters. The authors would like to further explore and implement other control parameters to enforce these three secondary planning objectives of an aircraft routing plan in an on-demand air transport (ODAT) context.

7.2. Conclusions

This paper has aimed to be the first study that provides a detailed description of the aircraft routing problem as faced by a per-aircraft air taxi (PAAT) operator and also the first study that treats the development of an aircraft routing system for a per-aircraft air taxi operator. The developed aircraft routing system, which consists out of the model creator, CPLEX and the solution module, is capable of solving the aircraft routing problem faced by the air taxi operator in under five minutes for the first operational years. This will allow the air taxi operator to achieve an estimated cost reduction of 12% on their routing plans with respect to using a human dispatcher. This amounts to estimated savings of 10 % on variable direct operating costs during the air taxi's first operational year. By altering the slack time parameter the OCC can control the robustness, i.e. the capability to cope with and recover from external disruptions, of the aircraft routing plans it creates. Furthermore, via the persistent aircraft-trip assignments input, the system allows the OCC to control the balance between persistence, the degree to which the aircraft routing plan deviates from the previous one when it is re-optimized, and flexibility, the ability of the schedule to cost-effectively deal with future demand, of the generated aircraft routing plans. In addition, by setting aircraft-trip assignments as input the OCC can specify 'hard' input decisions. The aircraft routing system also allows the air taxi operator to incorporate fuel ferrying in the aircraft routing phase as a strategy to reduce the operational cost of the aircraft routing plan.

The next stage for our project is for our Belgian air taxi operator to become fully operational, further develop the routing system and eventually implement it at their OCC. Then the various databases in the aircraft routing system can be filled with real life operational data and help the air taxi operator to provide affordable and reliable full on-demand personal air transport.

REFERENCES

1. A. Abdelghany, K. Abdelghany, 2009 Modeling Applications in the Airline Industry, Ashgate, 978-0-75467-874-8Farnham, United Kingdom
2. American Express 2009Global business travel forecast. In: American Express, 02.11.2011, Available at: http://corp.americanexpress.com/gcs /travel/news/pressreleases.aspx
3. M. Ball, C. Barnhart, G. Nemhauser, A. Odoni, 2007Chapter 1 Air Transportation: Irregular operations and control, In: Transportation. 14of Handbooks in Operations Research and Management Science, Barnhart, C. & Laporte, G. (Eds.), 1 67Elsevier, 978-0-44451-346-5Amsterdam, the Netherlands
4. C. Barnhart, P. Belobaba, A. Odoni, 2003Applications of operations research in the air transport industry, Transportation Science, 37 4 368 391 0041-1655
5. M. Bazargan, 2004 Airline Operations And Scheduling, Ashgate Publishing, 978-0-75469-772-5Aldershot, United Kingdom
6. F. Bian, E. Burke, S. Jain, G. Kendall, G. Koole, Silva. J. Landa, J. Mulder, M. Paelinck, C. Reeves, I. Rusdi, M. Suleman, 2003Making airline schedules more robust In: Proceedings of the 1st Multidisciplinary International Conference on Scheduling: Theory and Applications (MISTA 2003), 678 693
7. P. Bonnefoy, 2005Simulating air taxi networks. Proceedings of 37th Winter Simulation Conference, 1586 1595 0-78039-519-0
8. L. Budd, B. Graham, 2009Unintended trajectories: liberalization and the geographies of private business flight. Journal of Transport Geography, 17 4July 2009), 285 292 0966-6923
9. S. Bunte, N. Kliewer, 2010An overview on vehicle scheduling models. Public Transport, 1 4March 2010), 299 317
10. Cirrus 2007Cirrus SR22 weight and balance loading form, In: Airplane information manual for the Cirrus Design SR Cirrus, 04.05.2011, Available on: http://servicecenters.cirrusdesign.com/TechPubs/pdf/POH/sr22 /pdf/20880 001 InfoManual.pdf
11. R. Doganis, 2002 Flying Off Course, 3rd Edition, Routledge, 0-41521-323-1York
12. E. Dyson, 2006Visible demand: The new air-taxi market, In: Fly Your Jet website, 17.11.2010, Available at: http://www.flyyourjet.com/Portals/0 /DysonRelease.pdf
13. D. Espinoza, R. Garcia, M. Goycoolea, G. Nemhauser, M. Savelsbergh, 2008aPer-seat, on-demand air transportation part i: problem description and

an integer multicommodity ow model. Transportation Science, 42 3August 2008), 263 278 0041-1655

14. D. Espinoza, R. Garcia, M. Goycoolea, G. Nemhauser, M. Savelsbergh, 2008bPer-seat, on-demand air transportation part ii: Parallel local search. Transportation Science, 42 3August 2008), 279 291 0041-1655

15. Eurocontrol 2010Business aviation in Europe 2009, In: Eurocontrol website, 04.05.2010, Available at: http://www.eurocontrol.int/statfor/public /subsitehomepage/homepage.html

16. K. Fagerholt, B. A. Foss, O. J. Horgen, 2009A decision support model for establishing an air taxi service: A case study. Journal of the Operational Research Society, 60 9September 2009), 1173 1182 0160-5682

17. D. Feillet, 2010A tutorial on column generation and branch-and-price for vehicle routing problems. 4OR, A Quarterly Journal of Operations Research, 8 4June 2010), 407 424 1619-4500

18. Fly Aeolus 2010Private Jet Price Private Plane Charter Prices, In: Fly Aeolus website, 03.05.2011, Available at http://www.flyaeolus. com/affordable-and-low-price/

19. Flyer Forums 2010Avgas price table, In: Flyer Forums website, 25.10.2010, Available at: http://forums.yer.co.uk/viewtopic.php?t=51535

20. C. Gall, J. Hindhaugh, (2009). Communicating change to sr. management economic & industry outlook 2010, In: National Business Travel Association website, 12.07.2010, Available at: http://www2.nbta.org/Lists/ Resource%20Library/F103%20Communicating%20Change%20   - %20Industry%20Outlook.pdf

21. R. Hicks, R. Madrid, C. Milligan, R. Pruneau, M. Kanaley, Y. Dumas, B. Lacroix, J. Desrosiers, F. Soumis, 2005Bombardier flexjet significantly improves its fractional aircraft ownership operations. Interfaces, 35 1January- February 2005), 49 60 0092-2102

22. F. S. Hillier, G. Lieberman, 2008 Introduction to Operations Research. 8th Edition, McGraw Hill, 007123828New York

23. I. Karaesmen, P. Keskinocak, S. Tayur, W. Yang, 2005Scheduling multiple types of time-shared aircraft: Models and methods for practice. In: Proceedings of the 2nd Multidisciplinary International Conference on Scheduling: Theory and Applications. Kendall, G., Lei, L. & Pinedo, M. (Eds.), New York (July 2005), 1 19 38

24. I. Karaesmen, W. Yang, P. Keskinocak, 2007Optimization vs. persistence in scheduling: heuristics to manage uncertainty in on-demand air travel. In: Proceedings of the 3rd Multidisciplinary International Conference on Scheduling: Theory and Applications, Paris (August 2007), 1 242 250

25. P. Keskinocak, S. Tayur, 1998Scheduling of time-shared jet aircraft, Transportation Science, 32 3 277 294 1526-5447

26. G. Keysan, G. Nemhauser, M. Savelsbergh, 2010Tactical and operational planning of scheduled maintenance for per-seat, on-demand air transportation. Transportation Science, 44 3August 2010), 291 306 0041-1655

27. M. Landry, O. Muhittin, 1993In search of a valid view of model validation for operations research. European Journal of Operational Research, 66 2April 1993), 161 167 0377-2217

28. London Executive Aviation 2010In: London Executive Aviation website, 23.11.2010, Available at: http://www.flylea.com/

29. M. Mane, W. A. Crossley, 2007aAn approach to predict impact of demand acceptance on air taxi operations. In: 7th AIAA Aviation Technology, Integration, and Operations Conference. 1Belfast, Northern Ireland, 954 961

30. M. Mane, W. A. Crossley, 2007bProbabilistic approach for selection of maintenance facilities for on-demand operations. In: 7th AIAA Aviation Technology, Integration, and Operations Conference. 1Belfast, Northern Ireland, 944 953

31. M. Mane, W. A. Crossley, 2009Importance of aircraft type and operational factors for air taxi cost feasibility. Journal of Aircraft, 46 4 1222 1230 0021-8669

32. C. Martin, D. Jones, P. Keskinocak, 2003Optimizing on-demand aircraft schedules for fractional aircraft operators. Interfaces, 33 5September-October 2003), 22 35 0092-2102

33. NBAA 2010Piston Engine Aircraft, In: National Business Aviation Association website, 03.05.2011, Available at: http://www.nbaa.org/businessaviation/aircraft/pistons/

34. D. Ronen, 2000Scheduling charter aircraft. The Journal of the Operational Research Society, 51 3March 2000), 258 262 0160-5682

35. R. Sargent, 2007Verification and validation of simulation models. WSC '07 Proceedings of the 39th conference on Winter simulation: 40 years! The best is yet to come, IEEE Press Piscataway, NJ, 124 137 1-42441-306-0

36. E. A. Silver, 2004An overview of heuristic solution methods. The Journal of the Operational Research Society, 55 9May 2004), 936 956 0160-5682

37. J. C. Smith, Z. C. Taşkin, 2008A tutorial guide to mixed-integer programming models and solution techniques. In: Optimization in Medicine and Biology. Lim, G. J., Lee, E. K. (Eds.), 521 548Taylor and Francis, Auerbach Publications, 978-0-84930-563-4New York

38. W. Yang, I. Karaesmen, P. Keskinocak, S. Tayur, 2008Aircraft and crew scheduling for fractional ownership programs. Annals of Operations Research, 159December 2007), 415 431 1572-9338

39. W. Yang, I. Z. Karaesmen, P. Keskinocak, 2010Managing uncertainty in on-demand air travel. Transportation Research Part E: Logistics and Transportation Review, 46 6November 2010) 1169 1179 1366-5545

40. Y. Yao, W. Zhao, O. Ergun, E. Johnson, 2005Crew Pairing and Aircraft Routing for On-Demand Aviation with Time Window. In: Social Sciences Research Network eLibrary., 04.05.2011, Available from: http://ssrn.com/abstract=822265

41. Y. Yao, W. Zhao, 2006Optimization problem in fractional ownership operations. Collection of Technical Papers- 6th AIAA Aviation Technology, Integration, and Operations Conference, 1 319 323

42. Y. Yao, 2007 Topics in fractional airlines. Georgia Institute of Technology, School of Industrial and Systems Engineering, PhD thesis, May 2007

43. Y. Yao, O. Ergun, E. Johnson, 2007Integrated model for the dynamic on-demand air transportation operations. In: Dynamic Fleet Management. Sharda, R., Vo, S., Zeimpekis, V., Tarantilis, C. D., Giaglis, G. M. &Minis,

I. (Eds.), 38of Operations Research/Computer Science Interfaces Series, 95 111Springer US, New York, 978-0-38771-721-0

44. Y. Yao, O. Ergun, E. Johnson, W. Schultz, J. M. Singleton, 2008Strategic planning in fractional aircraft ownership programs. European Journal of Operational Research, 189 2September 2008), 526 539 0377-2217

CHAPTER 6

On Cup Anemometer Rotor Aerodynamics

Santiago Pindado *, Javier Pérez and Sergio Avila-Sanchez

IDR/UPM, ETSI Aeronáuticos, Universidad Politécnica de Madrid, Pza. del Cardenal Cisneros 3, Madrid 28040, Spain

ABSTRACT

The influence of anemometer rotor shape parameters, such as the cups' front area or their center rotation radius on the anemometer's performance was analyzed. This analysis was based on calibrations performed on two different anemometers (one based on magnet system output signal, and the other one based on an opto-electronic system output signal), tested with 21 different rotors. The results were compared to the ones resulting from classical analytical models. The results clearly showed a linear dependency of both calibration constants, the slope and the offset, on the cups' center rotation radius, the influence of the front area of the cups also being observed. The analytical model of Kondo *et al.* was proved to be accurate if it is based on precise data related to the aerodynamic behavior of a rotor's cup.

KEYWORDS:

anemometer calibration; cup anemometer; cup aerodynamics; rotor aerodynamics

1. INTRODUCTION

Rotation anemometers, such as cup and propeller anemometers, are the most commonly used instruments for wind speed measurements. Thanks to their linearity and accuracy they are optimal for a large number of applications in the wind energy sector, from routine observations to field measurements. Cup anemometers have been widely studied since the first half of the twentieth century, with the early works devoted to studying the optimal number of cups and arm length [1,2]. Additionally, the cups' rotor aerodynamics and its behavior, especially in turbulent wind, was the subject of great interest throughout the twentieth century. Wyngaard [3] used a 2-cup model to illustrate the "overspeeding" effect of the anemometer, that is, the cup anemometer responds quicker to wind accelerations than to wind decelerations. Other interesting models were developed by Ramachandran [4] and Kondo et $al.$ [5], to analyze the cups' rotor dynamics as a function of the aerodynamic drag of the cups. Both authors referred to the research by Brevoort and Joyner [6,7] for this cup aerodynamic drag.

In a previous study at the IDR/UPM Institute [8] large series of calibrations were analyzed, the calibration coefficients of the anemometers' transfer function being studied as a function of the anemometers' shape. The transfer function of an anemometer is represented by the following expression:

$$V = A \cdot f + B$$

where V is the wind speed, f is the anemometer's rotation frequency output, and A (slope) and B (offset) are the calibration coefficients. This linear equation, which correlates the wind speed and the anemometer's output frequency [9], has to be defined by means of a calibration process. The aforementioned transfer function can be rewritten in terms of the anemometers' rotation frequency, f_r, instead of the output frequency f, as:

$$V = A_r \cdot f_r + B$$

where A_r is the result of multiplying the calibration constant A by the number of pulses per revolution given by the anemometer, N_p. The number of pulses is different depending on the anemometer's inner system for translating the rotation into electric pulses. Magnets-based systems give from 1 to 3 pulses per revolution, whereas opto-electronics-based systems normally give higher pulse rates per revolution, from 6 to 44 [8].

This research, performed on more than 20 models of commercial cup anemometers, showed a linear correlation between the coefficients A_r and the cups' center rotation radius (defined in Figure 1), R_{rc}, $A_r = 0.012R_{rc} + 0.546$ (R_{rc} expressed in mm). However, this linear fitting changed to $A_r = 0.019R_{rc} + 0.196$, with a better regression coefficient ($R^2 = 0.753$ instead of $R^2 = 0.485$, in the previous fitting) leaving aside some anemometers, those that have very different shape from the others, when calculating the linear fitting. Finally, a brief calculation using the 2-cup analytical method showed a very close result to the mentioned fitting, $A_r = 20.7R_{rc}$ (R_{rc} expressed in meters). On the other hand, based on the analysis, B calibration coefficients of the studied anemometer models did not seem to correlate to the cups' center rotation radius, R_{rc}, nor to the cups' front area, S_c.

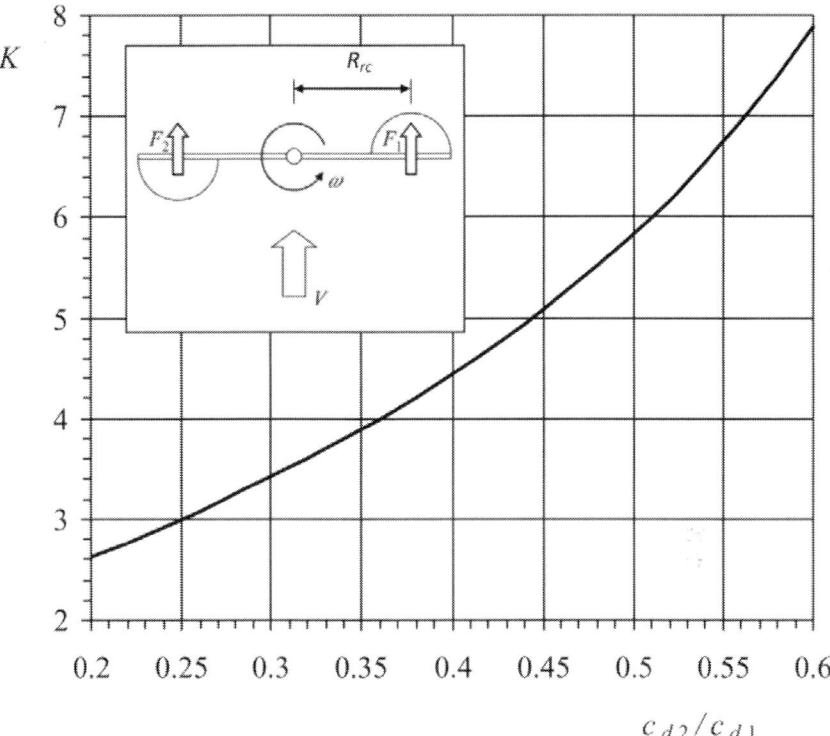

Figure 1. 2-cup anemometer model: anemometer factor (see Section 1.1), K, as a function of the ratio between the aerodynamic drag coefficient of the cups at $0°$ wind angle, c_{d1}, and at $180°$ wind angle, c_{d2}.

1.1. The 2-Cup Model

This simple model was used in the past to study the aerodynamics of the cup anemometers' rotor (see a sketch of the model in the Figure 1). This model is based on the perfect rotor equilibrium assumption, that is, the aerodynamic torque is neglected (see [1,3]). The response of a cup anemometer can be derived from the following expression [10]:

$$I\frac{d\omega}{dt} = Q_A + Q_f$$

where I is the moment of inertia, Q_A is the aerodynamic torque, and Q_f is the frictional torque. The frictional torque, Q_f, can be neglected as it is usually (for wind speeds larger than 1 m·s^{-1}) very small in comparison to the aerodynamic torque, Q_A[11,12]. So finally, the response can be obtained if the aerodynamic torque is assumed to be equal to zero, that is, $F_1 = F_2$, where F_1 and F_2 are the aerodynamic forces on the cups (see Figure 1). This condition can be expressed as:

$$\frac{1}{2}\rho\left(V-\omega R_{rc}\right)^2 c_{d1}S_c = \frac{1}{2}\rho\left(V+\omega R_{rc}\right)^2 c_{d2}S_c$$

where V is the wind speed, ω is the rotor's angular velocity, R_{rc} is the cups' center rotation radius, S_c is the front area of the cups, and c_{d1} and c_{d2} are the drag coefficients of the cups, respectively at 0° and 180° regarding the wind direction (the wind direction angle with respect to the cup is indicated in Figure 2). This equation can be simplified to obtain the transfer function of the anemometer:

$$V = 2\pi\left(\frac{1+\sqrt{c_{d2}/c_{d1}}}{1-\sqrt{c_{d2}/c_{d1}}}\right)R_{rc}f_r = A_r f_r$$

The ratio between the wind speed, V, and the rotation speed of the cups' center, ωR_{rc}, is called the anemometer factor, K:

$$K = \frac{V}{\omega R_{rc}} = \frac{A_r}{2\pi R_{rc}} = \left(\frac{1+\sqrt{c_{d2}/c_{d1}}}{1-\sqrt{c_{d2}/c_{d1}}}\right)$$

From an early study by Patterson, this factor was found to be between 2.5 and 3.5 [9]. Based on the data from [8], most common commercial anemometers have factors between 2.97 and 3.54 (calculated leaving aside the offset of the transfer function). In Figure 1, the K factor estimated with this 2-cup model is shown as a function of the ratio c_{d2}/c_{d1}.

1.2. The Ramachandran Model

Ramachandran [4] derived the rotor's behavior from the normal aerodynamic force coefficient of a single cup, c_N, (see inFigure 2, a sketch regarding the aerodynamic normal force on a cup in relation to the wind speed direction). Taking into account the three cups of a rotor, the Equation (3) can be rewritten as:

$$I\frac{d\omega}{dt} = \frac{1}{2}\rho S_c R_{rc}V_r^2(\theta)c_N(\alpha(\theta)) + \frac{1}{2}\rho S_c R_{rc}V_r^2(\theta+120°)c_N(\alpha(\theta+120°))$$

$$+\frac{1}{2}\rho S_c R_{rc}V_r^2(\theta+240°)c_N(\alpha(\theta+240°))$$

where V_r is the relative wind speed to the cups, α is the wind direction with respect to the cups, and θ is the angle of the rotor with respect to a reference line (see inFigure 4, a more detailed sketch with regard to these variables). This approximation leaves aside the aerodynamic moment of the cups, as its contribution to the rotor's torque is, in principle, less important (this effect was checked in the present research). The relative wind speed, V_r, to the cup at θ rotor angle with respect to the reference line can be expressed as:

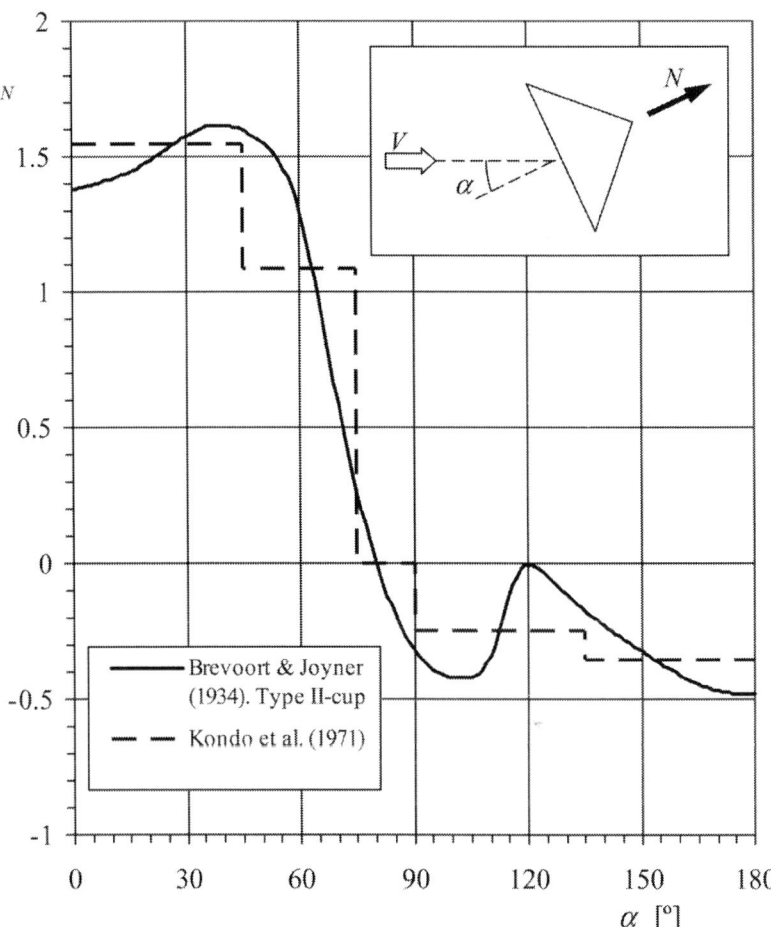

Figure 2. Normal aerodynamic force coefficient, c_N, of the Brevoort and Joyner Type-II cup plotted as a function of the wind direction with respect to the cup, α.

$$V_r(\theta) = \sqrt{V^2 + (\omega R_{rc})^2 - 2V\omega R_{rc} \cos(\theta)}$$

whereas the wind direction with respect to the cup, α, can be derived from these two equations:

$$V_r(\theta)\sin(\alpha) = V \sin(\theta)$$

$$V_r(\theta)\sin(\alpha-\theta) = \omega R_{rc}\sin(\theta)$$

to the following expression:

$$\tan(\alpha) = \frac{K\sin(\theta)}{K\cos(\theta)-1}$$

Ramachandran made two important assumptions in his calculations:

$$V_r(\theta) \approx V - \omega R_{rc}\cos(\theta)$$

and:

$$\alpha \approx \theta$$

Taking into account the aforementioned ratios between the wind speed and the rotation speed (that is, the anemometer factor, K), it can be observed that Equation (12) does not deviate excessively from the exact Equation (8) (up to 8.3%, 5.7% and 4.2%, respectively for factors $K = 2.5$, 3 and 3.5). However, the second assumption, indicated with the Equation (13), involves more important deviations (up to 23.6°, 19.5° and 16.6° for the mentioned values of K, see Figure 3).

The anemometer's behavior in steady state can be obtained from Equation (7) by making its average value equal to zero in one turn, that is:

$$V = 4\pi R_{rc}\frac{b_m}{a_m}f_r = A_r f_r$$

where the ratio b_m/a_m can be simplified as:

$$\frac{b_m}{a_m} = \frac{\int_0^{2\pi} c_N(\alpha)\cos(\alpha)d\alpha}{\int_0^{2\pi} c_N(\alpha)d\alpha}$$

As mentioned, in order to calculate the last expression Ramachandran suggested the use of Brevoort and Joyner results [6,7]. These authors measured the normal aerodynamic force coefficient of anemometer cups, c_N, as a function of the wind angle, α, for five cup shapes (named in the referred reports: Type I-hemispherical, without bead, 4.03 in diameter-, Type II-conical, without bead, 4.56 in diameter-, Type III-hemispherical, with bead, 2.03 in diameter-, Type IV-conical, with bead, 4.70 in diameter-, and Type V-hemispherical, without bead, 6.00 in diameter-). In Figure 2, the normal force coefficient, c_N, related to the Type II cup, is shown as a function of the wind direction with respect to the cup, α. Solving Equations (14) and (15) with this normal force coefficient distribution, the anemometer's factor becomes $K = 2.64$, and the calibration coefficient $A_r = 16.6R_{rc}$ (R_{rc} expressed in meters). This value of the calibration coefficient is lower than the one resulting from the previous research ($A_r = 19 \cdot R_{rc}$) [8].

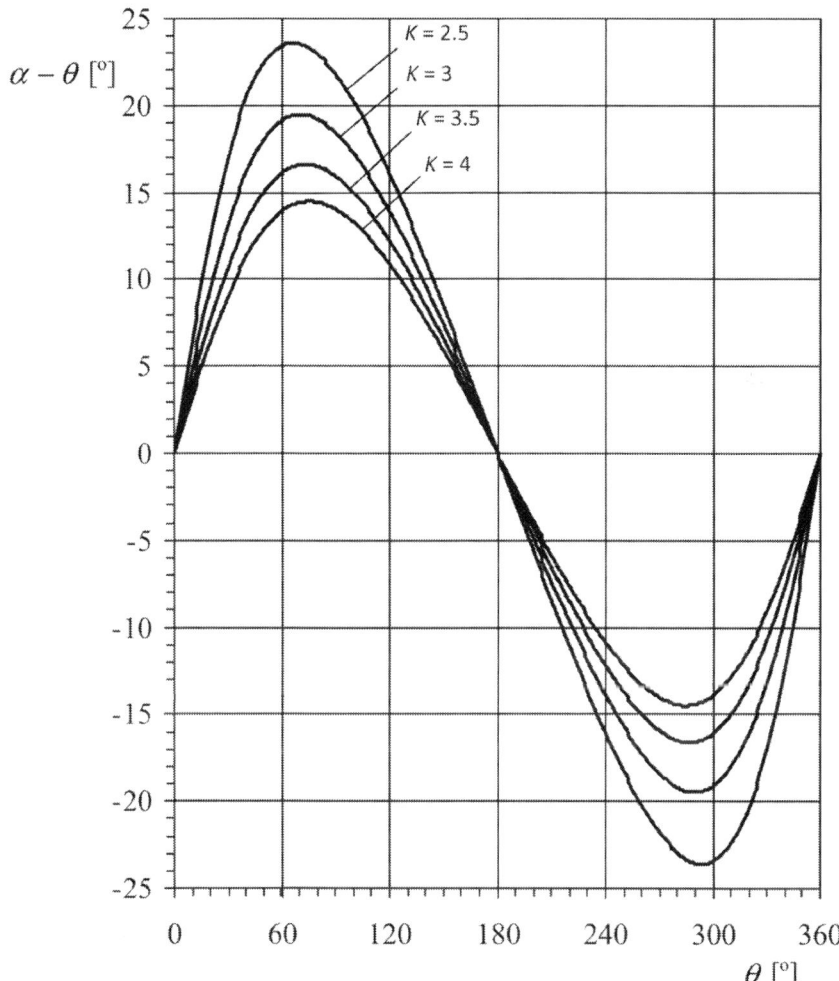

Figure 3. Difference between the local wind angle with respect to the cup (commonly known as angle of attack), α, and the anemometer rotor's rotation angle, θ, as a function of the latter, for different anemometer factors, K. See also the sketch included in Figure 4.

Kondo *et al.* [5] proposed a solution to the problem considering the anemometer's rotor position, θ, different from the local wind angle with respect to the cup, α. See in Figure 4, the normal aerodynamic force coefficient of the Brevoort and Joyner Type-II cup plotted as a function of both angles, α and θ, together with a sketch regarding the geometric relations between V, ωR_{rc}, α, and θ. Again, averaging Equation (7) and making the result equal to zero, it is possible to obtain the following equation:

$$0 = \left(V^2 + (\omega R_{rc})^2\right) \int_0^{2\pi} c_N(\alpha(\theta)) d\theta - 2V \omega R_{rc} \int_0^{2\pi} c_N(\alpha(\theta)) \cos(\theta) d\theta$$

that leads to the solution of the anemometer's steady state:

$$K = \frac{V}{\omega R_{rc}} = \frac{b}{a} + \sqrt{\left(\frac{b}{a}\right)^2 - 1}$$

where:

$$\frac{b}{a} = \frac{\int_0^{2\pi} c_N(\alpha(\theta)) \cos(\theta) d\theta}{\int_0^{2\pi} c_N(\alpha(\theta)) d\theta}$$

Kondo et al. simplified the calculations by using a stepwise line for the coefficient $c_N(\alpha)$ (see Figure 2). However, it is possible to express this coefficient as a function of θ using the Equation (11) together with a reasonable value of the factor K, and then solve Equations (17) and (18) in order to obtain a more accurate value of this factor. Using this method it is possible to get the solution in a few iterations. In the present case $K = 3.5$, and the calibration coefficient $A_r = 22.02 R_{rc}$ (R_{rc} expressed in meters).

The aim of the present work is to analyze the correlation between the cups' rotor dynamics in steady state (that is, the anemometer's transfer function) and the rotor's shape (more specifically, the cups center rotation radius, R_{rc}, and the cups' front area, S_c), by means of a specific testing research. Other effects such as the cups' aerodynamic drag coefficient measured both on one isolated cup and on one cup surrounded by the other two inside the rotor, are also taken into account. The testing campaign was divided into two parts, the first one included calibrations performed on two different anemometers with 21 different rotors (varying both the cups' size and the rotation radius), whereas the second one included the aerodynamic forces measurements on a cup, isolated and in a rotor (that is, surrounded by the other two). In order to have a better understanding of the anemometers' rotor dynamics, the results were compared to the aforementioned classic analytical models.

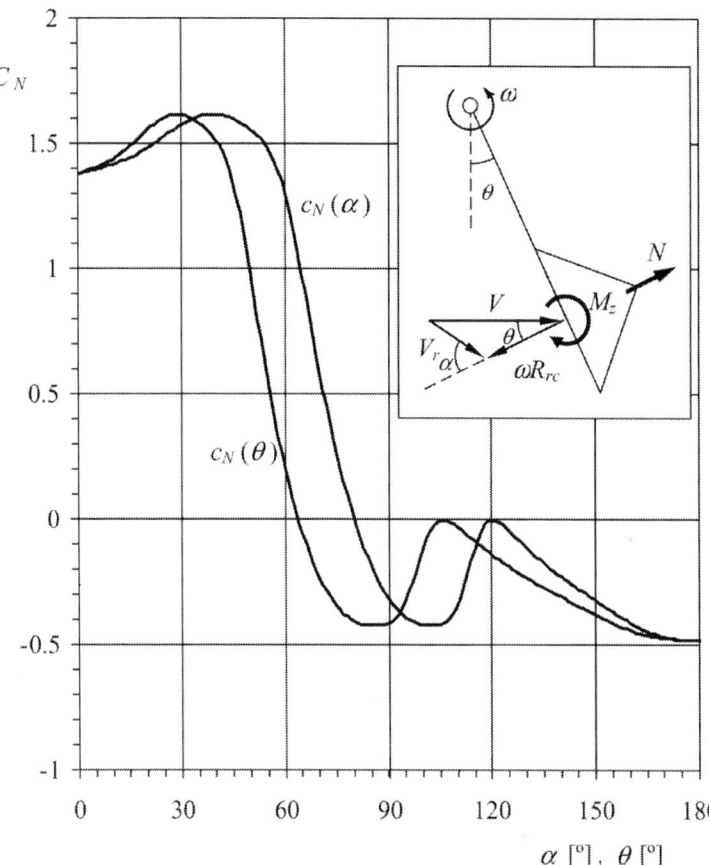

Figure 4. Normal aerodynamic force coefficient, c_N, of the Brevoort and Joyner Type-II cup plotted as a function of the wind direction with respect to the cup, α, and the rotor's rotation angle, θ. Calculated for anemometer factor $K = 3.5$.

2. TESTING CONFIGURATION AND CASES STUDIED

Two anemometers, Climatronics 100075 (also known as F460 model, by Climatronics Corp.: Bohemia, New York, USA), and Ornytion 107A (Ornytion: Bergondo, A Coruña, Spain) were used in the testing campaign (see Figure 5). These anemometers will be referred hereinafter in the text as Cl-100075 and Ory-107 respectively. As said, 21 different rotors were tested on both anemometers (see Figure 6), varying the cups' front section, S_c, and the cups' center rotation radius, R_{rc}, from one another (see Table1). All cups tested were conical (90° cone-angle). The cups were made in a 3D printer of ABS plastic, and the arm of each cup was made of 5 mm diameter aluminum tube. Each set of three cups were attached to the Cl-100075 anemometer rotor's head. In order to use this head with the Ory-107 anemometer, a special piece had to be designed and manufactured as an interface to the Ory-107 shaft. These anemometers have different electronic systems, the Cl-100075 is equipped with an opto-electronic system that gives 30 squared pulses per rotation, while the Ory-107 has a magnets-based system that gives two harmonic pulses per rotation.

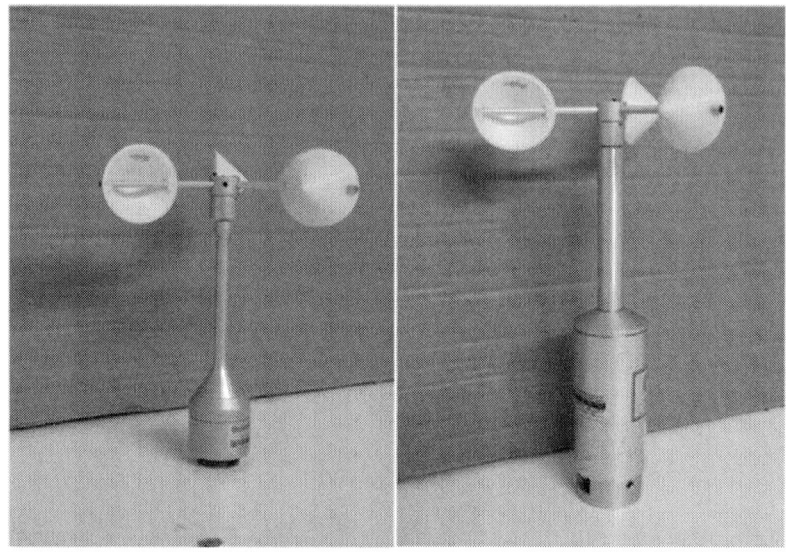

Figure 5. Ornytion 107A (**left**) and Climatronics 100075 (**right**) anemometers, both with 50/60 rotor (see in Table 1 the characteristics of each rotor tested).

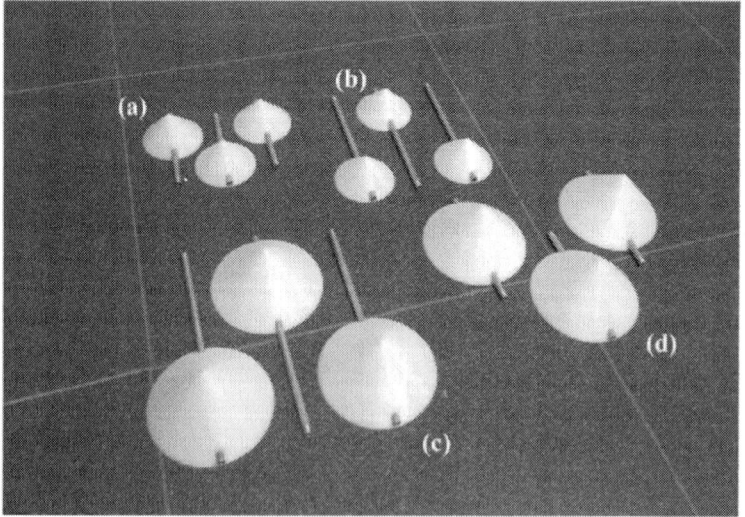

Figure 6. Cups sets correspondent to the 50/60 (**a**), 50/100 (**b**), 80/120 (**c**) and 80/60 (**d**) rotors (see in Table 1 the characteristics of each rotor tested).

Table 1. Results of the calibration performed on the Climatronics 100075 and Ornytion 107A anemometers. The calibration constants, A, B and A_r, are indicated together with the cups' diameter, D_c, and the cups' center rotation radius, R_{rc}, correspondent to each rotor tested.

Climatronics 100075					
Rotor	D_c [mm]	R_{rc} [mm]	A	B	A_r
40/40	40	40	0.0310	0.2593	0.9295
40/50	40	50	0.0420	0.3010	1.2592
40/60	40	60	0.0518	0.3562	1.5526
50/40	50	40	0.0293	0.2867	0.8777
50/60	50	60	0.0495	0.2567	1.4850
50/80	50	80	0.0697	0.3387	2.0909
50/100	50	100	0.0890	0.5447	2.6692
60/40	60	40	0.0279	0.1900	0.8361
60/60	60	60	0.0481	0.1559	1.4425
60/80	60	80	0.0682	0.2167	2.0464
60/100	60	100	0.0866	0.3731	2.5991
60/120	60	120	0.1064	0.4731	3.1922
70/60	70	60	0.0461	0.1642	1.3833
70/80	70	80	0.0674	0.1754	2.0210
70/100	70	100	0.0873	0.2003	2.6200
70/120	70	120	0.1067	0.2997	3.1997
80/60	80	60	0.0454	0.1376	1.3633
80/80	80	80	0.0653	0.1864	1.9601
80/100	80	100	0.0859	0.2221	2.5781
80/120	80	120	0.1052	0.2539	3.1557
80/140	80	140	0.1248	0.3040	3.7455
Ornytion 107A					
Rotor	D_c [mm]	R_{rc} [mm]	A	B	A_r
40/40	40	40	0.4809	0.2739	0.9617
40/50	40	50	0.6396	0.3707	1.2792
40/60	40	60	0.7827	0.5387	1.5654
50/40	50	40	0.4477	0.1479	0.8954
50/60	50	60	0.7584	0.3513	1.5168
50/80	50	80	1.0571	0.5058	2.1141
50/100	50	100	1.3489	0.6509	2.6978
60/40	60	40	0.4313	0.0702	0.8625
60/60	60	60	0.7363	0.1824	1.4727
60/80	60	80	1.0397	0.2957	2.0795
60/100	60	100	1.3143	0.4646	2.6285
60/120	60	120	1.6139	0.5543	3.2278
70/60	70	60	0.7074	0.1848	1.4148
70/80	70	80	1.0275	0.2796	2.0549
70/100	70	100	1.3173	0.2182	2.6347
70/120	70	120	1.6022	0.3444	3.2044
80/60	80	60	0.6794	0.1500	1.3588
80/80	80	80	0.9936	0.1902	1.9873
80/100	80	100	1.3001	0.2174	2.6003
80/120	80	120	1.5906	0.2359	3.1812
80/140	80	140	1.8830	0.3191	3.7659

The calibrations were carried out at the IDR/UPM Institute, in the S4 wind tunnel. This facility is an open-circuit wind tunnel with a closed test section measuring 0.9 by 0.9 m. It is served by four 7.5 kW fans with a flow uniformity under 0.2% in the testing area. More details concerning the facility and the calibration process are included in references [8,13]. The calibrations analyzed in the present paper were performed following the MEASNET [14,15] recommendations (over 13 points and from 4 to 16 m·s^{-1} wind speed).

The aerodynamic forces on the cups were measured in the wind tunnel of the Department of Mechanical Engineering of the Vrije Universiteit Brussel (Belgium). This facility is also an open-circuit wind tunnel with a 2 by 1 m closed test section. The facility is served by a 55 kW centrifugal fan. The testing section is equipped with a 6-component balance made by TEM Engineering Limited (Sussex, UK). Three larger-scale cups (0.2 m diameter) were manufactured to measure the aerodynamic forces. These cups were also made in a 3D printer of ABS plastic, and they are an exact scale-replica of the rotors' ones. As stated in the introduction, two different tests were carried out. In the first one, the forces on an isolated cup were measured by varying the wind direction, whereas in the second one the cup was surrounded by the other two in order to better simulate the anemometer's rotor (see Figure 7). These tests were carried out in smooth flow (with low turbulence, 1–1.5%), with around 15 m·s^{-1} wind speed. 12,000 samples were taken at 50 Hz in each measurement. The forces were made non dimensional with the dynamic pressure directly measured by a BnC-Lambrecht 630a (Goettingen, Germany) pitot tube located at the ceiling of the testing chamber, upstream to the point where the models are allocated, and connected to a SETRA Model 239 (Boxborough, MA, USA) differential pressure sensor.

3. RESULTS AND DISCUSSION

The calibration results, together with the characteristics of each tested rotor, are included in Table 1. The calibration constants A_r, that is, the anemometers' transfer function slopes, are plotted in Figure 8 as a function of the cups' center rotation radius, R_{rc}, for the two anemometers tested. A linear fit has been added to both graphs included in the figure. The results seem to be identical for both anemometers, although there is some dispersion due to the different front areas of the cups. In Figure 9, the calibration constants A_r measured in both anemometers equipped with 50 mm and 80 mm diameter cup rotors, are shown as a function of the cups' center rotation radius. The same linear tendency is observed, with better correlation coefficients to the data, R^2. In Table 2, the linear fittings of A_r:

Figure 7. Both cup configurations measured in the wind tunnel of the Vrije Universiteit Brussel: isolated cup (**top**), and cup inside a 3-cup rotor (**bottom**).

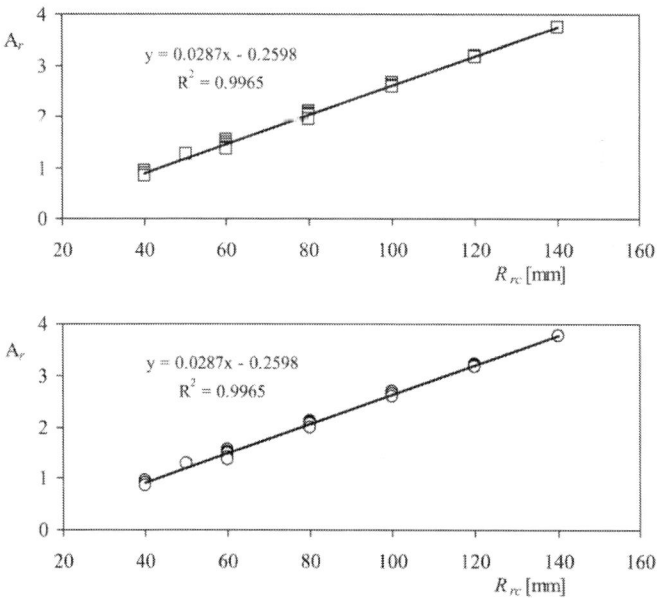

Figure 8. Calibration coefficients, A_r, as a function of the cups' center rotation radius, R_{rc}, for the Cl-100075 (**top**) and the Ory-107 (**bottom**) anemometers tested each one with all the rotors prepared for this research (see also Table 1).

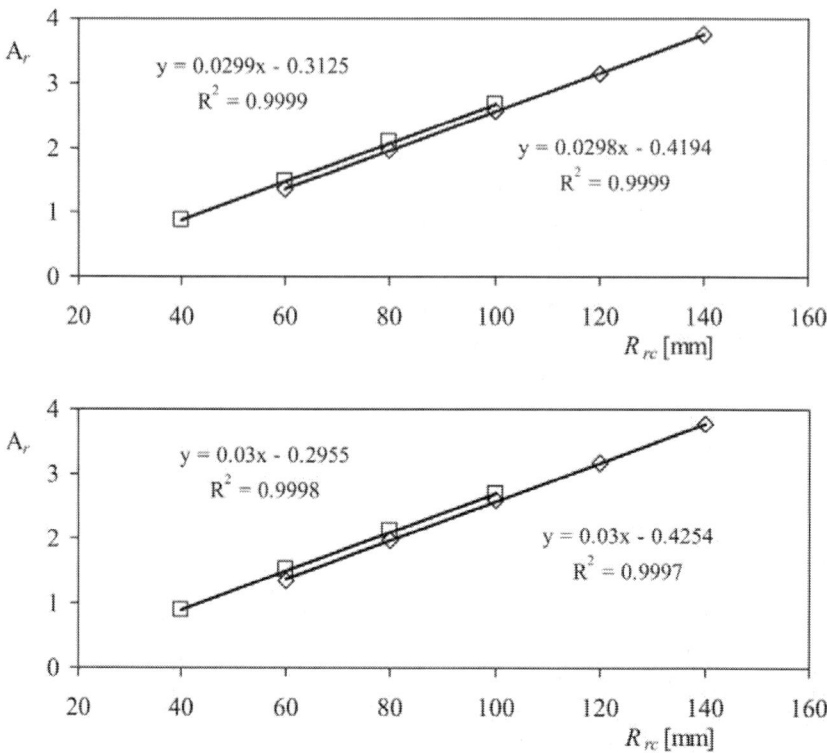

Figure 9. Calibration coefficients, A_r, as a function of the cups' center rotation radius,R_{rc}, for the Climatronics 100075 (**top**) and the Ornytion 107A (**bottom**) anemometers. Anemometers equipped with 50 mm diameter cups (squares), and 80 mm diameter cups (rhombi).

Table 2. Linear fittings (slope, dA_r/dR_{rc}, offset, A_{r0}, and correlation coefficient, R^2) of calibration constants A_r as a function of the cups center rotation radius, R_{rc}, with regard to calibrations performed on Climatronics 100075 and Ornytion 107A anemometers, equipped with the same rotors. The diameter, D_c, and front area of the rotors' cups, S_c, are also included.

Climatronics 100075					
Rotors tested	D_c [mm]	S_c [mm^2]	dA_r/dR_{rc}	A_{r0}	R^2
40/40, 40/50, 40/60	40	1,256.6	3.116E−02	−3.107E−01	0.99888
50/40, 50/60, 50/80, 50/100	50	1,963.5	2.990E−02	−3.125E−01	0.99986
60/40, 60/60, 60/80, 60/100, 60/120	60	2,827.4	2.934E−02	−3.243E−01	0.99974
70/60, 70/80, 70/100, 70/120	70	3,848.5	3.024E−02	−4.157E−01	0.99953
80/60, 80/80, 80/100, 80/120, 80/140	80	5,026.5	2.980E−02	−4.194E−01	0.99989
Ornytion 107A					
Rotors tested	D_c [mm]	R_{rc} [mm]	dA_r/dR_{rc}	A_{r0}	R^2
40/40, 40/50, 40/60	40	1,256.6	3.019E−02	−2.406E−01	0.99911
50/40, 50/60, 50/80, 50/100	50	1,963.5	3.002E−02	−2.955E−01	0.99980
60/40, 60/60, 60/80, 60/100, 60/120	60	2,827.4	2.943E−02	−3.003E−01	0.99968
70/60, 70/80, 70/100, 70/120	70	3,848.5	2.974E−02	−3.497E−01	0.99923
80/60, 80/80, 80/100, 80/120, 80/140	80	5,026.5	3.004E−02	−4.254E−01	0.99970

$$A_r = \frac{dA_r}{dR_{rc}} R_{rc} + A_{r0}$$

are included for all cups' diameters. It can be observed that the slopes of all fittings are quite the same, $dA_r/dR_{rc} = 0.03$ (with R_{rc} expressed in mm, hereinafter all dimensions concerning cups and rotor geometry will be expressed in mm), whereas the offsets, A_{r0}, are different from one case to another. However, once divided by the cups' front area, S_c, there seems to be a quite good power fitting to these coefficients, when they are plotted against this parameter (see Figure 10). After all these considerations, it has been found that the calibration constant A_r can be expressed as:

$$A_r = \frac{dA_r}{dR_{rc}} R_{rc} - S_c \left(\delta + \eta S_c^{-\xi} \right)$$

where δ, η, and ξ, are coefficients that depend on the anemometer (see in Figure 10 the values correspondent to each coefficient for both anemometers tested). The last expression indicates that there is a small contribution to the slope of the anemometers' transfer function that depends on the cups' front area. And this contribution makes the anemometer more efficient in terms of transforming the wind speed into shaft rotation.

The same analysis can be performed on the anemometers' transfer function offset, B. Let's assume that this coefficient has a linear behavior with R_{rc} as the transfer function slopes:

$$B = \frac{dB}{dR_{rc}} R_{rc} + B_0$$

In Figure 11 the behavior of this coefficient is shown as a function of the cups' center rotation radius, R_{rc}, for the two anemometers equipped with the rotors of 50 mm and 80 mm diameter cups. It can be observed in the figure that the linear fitting is confirmed as quite accurate for describing the coefficient's behavior, despite the fact that some fittings are better than the others (see in Table 3 the linear fittings correspondent to this coefficient). Unlike what happened with the constant, A_r, both the slope, dB/dR_{rc}, and the offset, B_0, are quite different from one fitting to the other one. Nonetheless, it is possible to find some fitting to describe them as a function of the other important parameter of the rotors' shape, the cups' front area, S_c. In Figure 12, the power fittings to the data concerning the slope and offset of Equation (21) are shown. It can be observed that the power fittings represent the behavior of the aforementioned slope and offset quite well. Based on this fact, Equation (21) can be rewritten as:

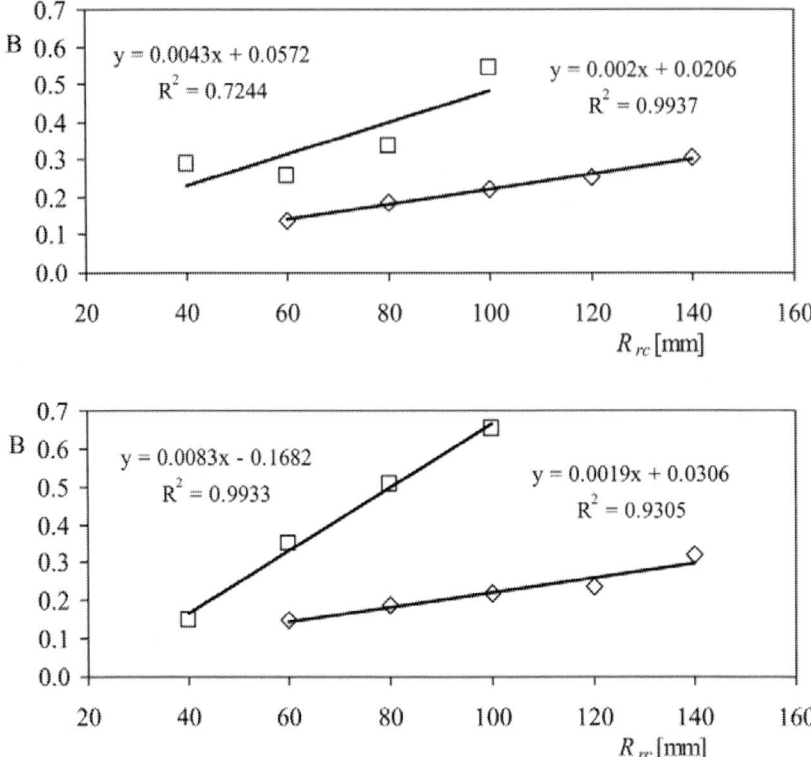

Figure 11. Calibration coefficients, B, as a function of the cups' center rotation radius, R_{rc}, for the Climatronics 100075 (**top**) and the Ornytion 107A (**bottom**) anemometers. Anemometers equipped with 50 mm diameter cups (squares), and 80 mm diameter cups (rhombi).

$$B = \left(\varepsilon + \phi S_c^{-\gamma} \right) R_{rc} - \mu S_c^{-\psi}$$

where ε, ϕ, γ μ, and ψ, are coefficients that depend on the anemometer (the values correspondent to each coefficient can be derived from the power fittings in Figure 12, for both anemometers tested).

The previous analysis illustrates the relation between both calibration constants, A_r and B, and the two most important shape parameters of the anemometers, the cups' center rotation radius, R_{rc}, and the cups' front area, S_c. The comparison between the calibration results and the constants estimated with Equations (20) and (22), together with the measured wind speed percentage deviation at $V = 7$ m·s^{-1}, is shown in Table 4. The variations with regard to the slope of the transfer function, A_r, are up to 4.9% (Cl-100075) and 2% (Ory-107), whereas with regard to the offset, B, they are larger, up to 43% (Cl-100075) and 211% (Ory-107). These large differences, also between both anemometers, are explained as the offset is mainly produced by the friction [16,17]. It should also be said that the friction term in the expression that describes the anemometer's behavior (3) is affected by changes in the temperature [10]. Therefore, changes in the air temperature during the calibration, together with the different bearings systems of both anemometers can explain the aforementioned differences.

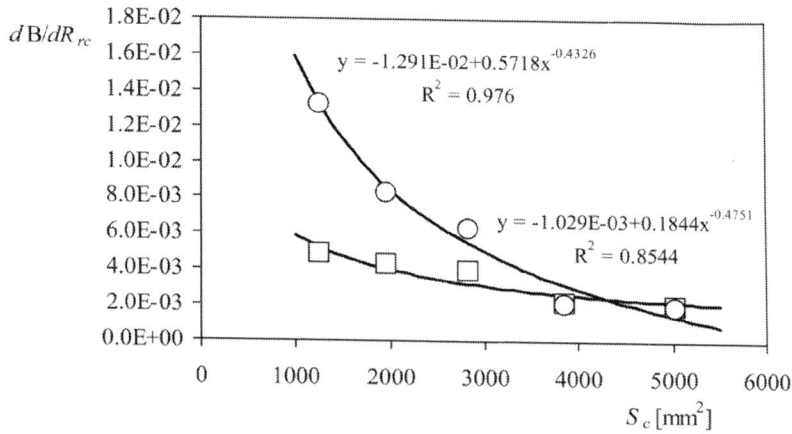

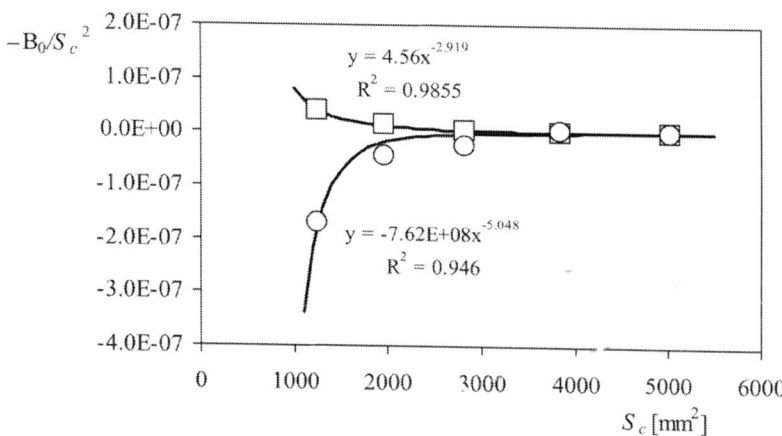

Figure 12. Slope and offset (divided by the squared front area of the cups) of the linear fittings from Table 3, as a function of the cups' front area, S_c. Data correspond to the Climatronics 100075 (squares) and the Ornytion 107A (circles) anemometers. Power fittings have also been added to the graph.

To compare these results with the ones from the classical models it must be taken into account that the anemometer's factor depends on both calibration constants:

$$K = \frac{V}{\omega R_{rc}} = \frac{A_r}{2\pi R_{rc}} + \frac{B}{2\pi R_{rc} f_r}$$

However, it can be accurate enough to leave aside the second term of this expression, assuming an average deviation up to 8.6% (based on the calibration results in Table 1, the average deviations are 7.4%, 4.1% and 2.8% -Cl-100075-, and 8.6%, 4.7% and 3.2% -Ory-107-, respectively for V = 4, 7, and 10 $m \cdot s^{-1}$ wind speed). The results are then K = 4.77 and dA_r/dR_{rc} = 0.03, which are a bit far from the results of the 2-cup model (K = 3.88 and dA_r/dR_{rc} = 0.0244), Ramachandran model (K = 2.64 and dA_r/dR_{rc} = 0.0166), Kondo et al. model (K = 3.50 and dA_r/dR_{rc} = 0.022), and the previous research [8] (K = 3.02 and dA_r/dR_{rc} = 0.019).

Table 3. Linear fittings (slope, dB/dR_{rc}, offset, B_0, and correlation coefficient, R^2) of calibration constants B as a function of the cups center rotation radius, R_{rc}, with regard to calibrations performed on Climatronics 100075 and Ornytion 107A anemometers, equipped with the same rotors. The diameter, D_c, and front area of the rotors' cups, S_c, are also included.

Climatronics 100075					
Rotors tested	D_c [mm]	S_c [mm²]	dB/dR_{rc}	B_0	R^2
40/40, 40/50, 40/60	40	1,256.6	4.8445E−03	6.3273E−02	0.99370
50/40, 50/60, 50/80, 50/100	50	1,963.5	4.2794E−03	5.7171E−02	0.72444
60/40, 60/60, 60/80, 60/100, 60/120	60	2,827.4	3.9178E−03	2.0341E−02	0.83575
70/60, 70/80, 70/100, 70/120	70	3,848.5	2.1567E−03	1.5786E−02	0.81394
80/60, 80/80, 80/100, 80/120, 80/140	80	5,026.5	2.0019E−03	2.0607E−02	0.99374
Ornytion 107A					
Rotors tested	D_c [mm]	R_{rc} [mm]	dB/dR_{rc}	B_0	R^2
40/40, 40/50, 40/60	40	1,256.6	1.3239E−02	−2.6754E−01	0.97653
50/40, 50/60, 50/80, 50/100	50	1,963.5	8.3167E−03	−1.6821E−01	0.99334
60/40, 60/60, 60/80, 60/100, 60/120	60	2,827.4	6.2523E−03	−1.8674E−01	0.99262
70/60, 70/80, 70/100, 70/120	70	3,848.5	2.0861E−03	6.9016E−02	0.58547
80/60, 80/80, 80/100, 80/120, 80/140	80	5,026.5	1.9197E−03	3.0577E−02	0.93054

The results of the normal aerodynamic force coefficient measured on a cup are included in Figure 13. In this figure the two cases are included, one isolated cup and the measurement cup surrounded by other ones simulating a 3-cup rotor. In this last case, the scaled cup's center rotation radius corresponds to a rotor with R_{rc} = 60 mm. In the figure, the yawing moment coefficient, c_{mz}, multiplied by the ratio D_c/R_{rc} has also been included (see the sketch in Figure 4). As can be observed, the effect of the aerodynamic moment in terms of torque, is much less important than the normal aerodynamic force on the cup. Except for angles between α = 120° and α = 240°, the experimental results follow the Brevoort & Joyner early results quite well. Nevertheless, a quite large discrepancy is found in that bracket. This can be explained due to the cups' geometries, as the Type II-cup tested by Brevoort and Joyner has rounded edges instead of the sharp edges from the cups tested, and a lower cone-angle (80° instead of 90°).

The use of these experimental results in combination with the described analytical models gives more accurate values of K and dA_r/dR_{rc}, when compared to the ones calculated with the Brevoort & Joyner measurements. The results are summarized in Table 5. Notwithstanding, it is possible to go one step further.

In Figure 14, the standard deviation of the normal force on the cup divided by the mean value of this force, σ_N/N, is shown as a function of the wind angle with respect to the cup, for the two cases tested (the isolated cup, and the cup in a 3-cup rotor). In this graph the effect of the wake produced by the upstream cups can be observed. Between $\alpha = 70°$ and $\alpha = 130°$ the cup in which the forces are measured is clearly in the wake of another, the effect being a dispersion up to 22 times the average force measured. Furthermore, the same effect can be observed between $\alpha = 255°$ and $\alpha = 280°$, although in this last case the effect of the non steady aerodynamic effects is less important. It is well known from basic aerodynamics that when a body is in the wake of another, the aerodynamic forces are highly affected. Supposing that in the mentioned intervals the cup is affected by a sudden loss of drag (so the normal force is neglected), the calculations of the anemometer's factor give a quite good result, $K = 4.75$ ($dA_r/dR_{rc} = 0.0298$), when compared to the result from the calibrations.

Table 4. Percentage variations of the anemometer transfer function slope and offset, A_r and B, between the measured values and the estimated ones (based on the geometry parameters, Equations (20) and (22). The measured wind speed variations at $V = 7$ m·s^{-1} wind speed are also included in the table.

Rotor	Climatronics 100075			Ornytion 107A		
	ΔA_r [%]	ΔB [%]	ΔV [%]	ΔA_r [%]	ΔB [%]	ΔV [%]
40/40	−3.47%	4.93%	−3.16%	−0.83%	92.54%	2.82%
40/50	−4.92%	7.59%	−4.38%	−1.99%	77.80%	2.24%
40/60	−3.57%	5.49%	−3.11%	−0.75%	46.84%	2.91%
50/40	−0.72%	−29.27%	−1.89%	1.46%	132.63%	4.23%
50/60	−0.92%	10.13%	−0.51%	−0.55%	46.92%	1.83%
50/80	−0.93%	7.08%	−0.55%	−0.27%	36.06%	2.36%
50/100	0.08%	−18.74%	−1.38%	0.40%	32.16%	3.35%
60/40	1.21%	−16.50%	0.73%	0.31%	211.19%	2.43%
60/60	0.25%	42.76%	1.20%	−0.50%	79.73%	1.59%
60/80	−0.01%	32.19%	0.98%	−0.68%	47.77%	1.36%
60/100	1.81%	−6.06%	1.39%	1.40%	17.58%	2.47%
60/120	1.69%	−12.41%	0.74%	1.16%	18.25%	2.51%
70/60	2.72%	9.90%	2.89%	0.61%	2.86%	0.67%
70/80	0.00%	32.78%	0.82%	−1.53%	−9.35%	−1.84%
70/100	0.03%	42.43%	1.25%	−0.43%	45.23%	1.00%
70/120	0.66%	12.71%	1.18%	0.59%	10.40%	1.08%
80/60	2.35%	8.49%	2.47%	1.76%	−43.43%	0.80%
80/80	1.80%	3.59%	1.85%	−0.23%	−40.53%	−1.32%
80/100	0.67%	6.60%	0.86%	−0.67%	−34.94%	−1.74%
80/120	1.26%	10.49%	1.59%	0.05%	−28.06%	−0.90%
80/140	1.33%	6.65%	1.56%	0.45%	−37.96%	−1.30%

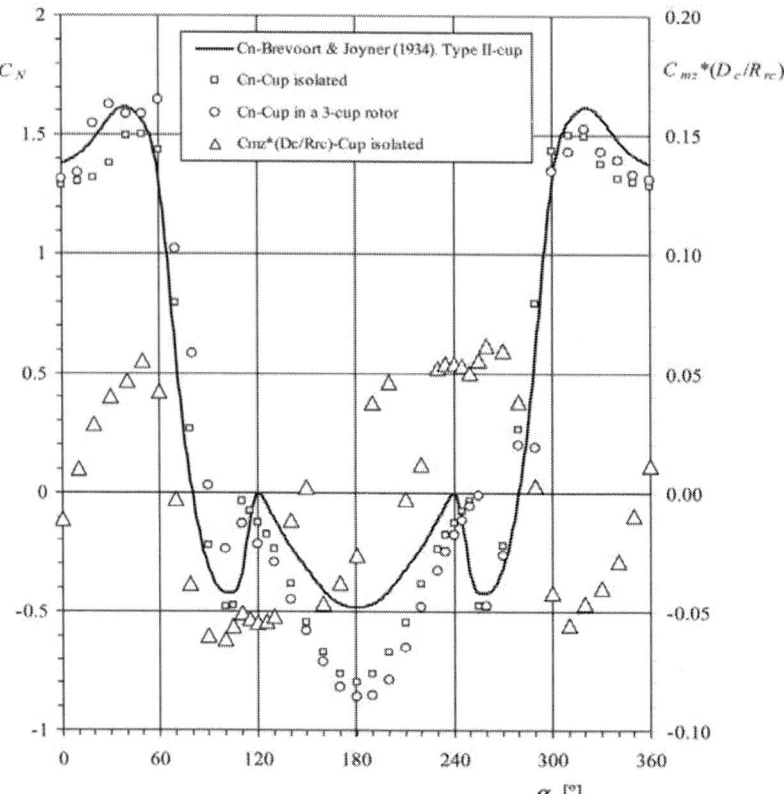

Figure 13. Normal aerodynamic force coefficient, c_N, measured on cup, isolated and surrounded by the other two simulating an anemometer's rotor as a function of the wind direction with respect to the cup, α. The aerodynamic yawing moment coefficient, c_{mz}, multiplied by the ratio D_c/R_{rc} (D_c = 50 mm and R_{rc} = 60 mm were selected to calculate this ratio as normal values from commercial anemometers), has been also included in the graph.

4. CONCLUSIONS

In the present study, the influence of the rotor's shape (cups' front area, S_c, and cups' center rotation radius, R_{rc}) on the anemometers' performance, has been analyzed by wind tunnel calibrations performed on two different anemometers, tested with 21 different rotors. The results have been compared to those from analytical models improved with data from specific cup aerodynamics wind tunnel testing.

Table 5. Anemometer factor, K, and slope of the A_r calibration constant as a function of the rotation radius, dA_r/dR_{rc}, calculated with the different methods explained in the text. The experimentally measured results are also included in the table.

Analytical models based on Brevoort & Joyner results [6,7]		
Model	**K**	**dA_r/dR_{rc}**
2-cup model [3]	3.88	0.0244
Ramachandran [4]	2.64	0.0166
Kondo *et al.* [5]	3.5	0.022
Analytical models based on drag force measured on one isolated cup		
Model	**K**	**dA_r/dR_{rc}**
2-cup model	8.39	0.0527
Ramachandran	3.37	0.0212
Kondo *et al.*	4.8	0.0302
Analytical models based on drag force measured on one cup in a 3-cup rotor		
Model	**K**	**dA_r/dR_{rc}**
2-cup model	9.51	0.0598
Ramachandran	3.46	0.0218
Kondo *et al.*	4.98	0.0313
Kondo *et al.* with shielding effects	4.75	0.0298
Experimental results		
Testing campaign	**K**	**dA_r/dR_{rc}**
Based on calibrations on commercial anemometers [8]	3.02	0.019
Present research	4.77	0.03

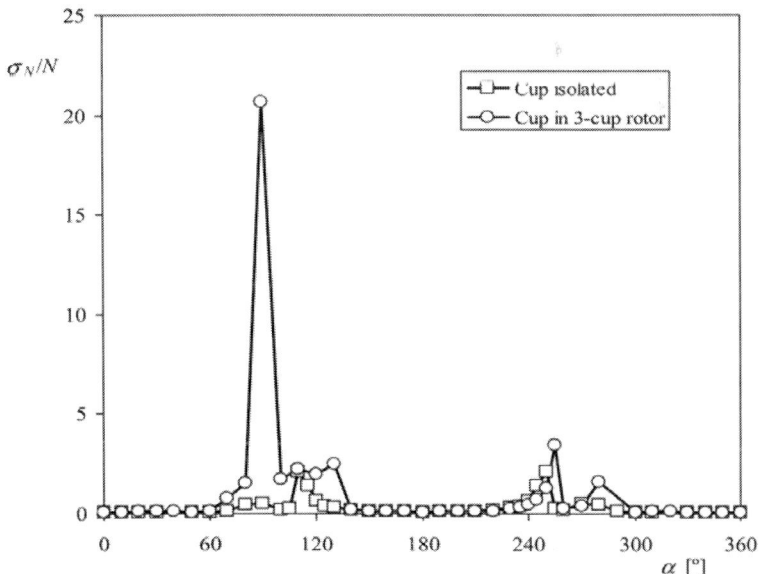

Figure 14. Standard deviation of the normal force on the cup divided by the mean value of this force, σ_N/N, as a function of the wind angle with respect to the cup, α, in the two cases tested, cup isolated and cup in a 3-cup rotor.

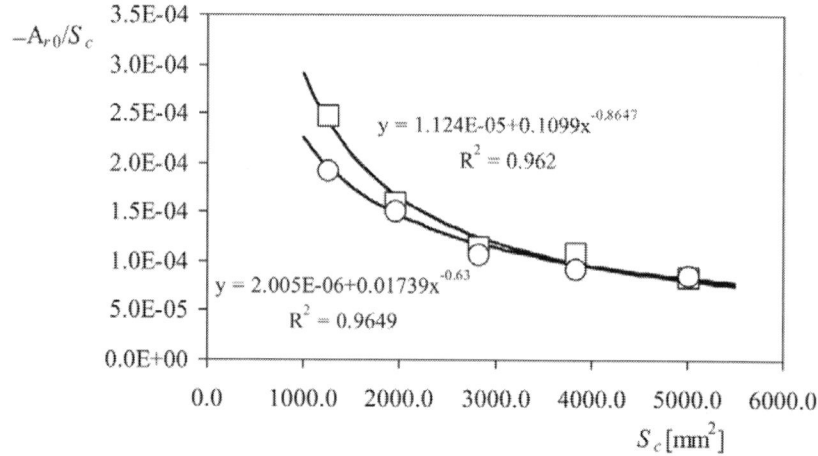

Figure 10. Offset of the linear fittings from Table 2 divided by the cups' front area, A_{r0}/S_c, as a function of the cups' front area, S_c. Data correspond to the Climatronics 100075 (squares) and the Ornytion 107A (circles) anemometers. Power fittings have been also added to the graph.

The major conclusions resulting from this work are:

1. Both calibration constants, A_r and B, of a cup anemometer depend on the aforementioned shape parameters. Both constants show a linear relationship with the cups' center rotation radius, R_{rc}. In the case of constant A_r, the slope of this linear behavior, dA_r/dR_{rc}, only depends on the cups' aerodynamics (and does not depend on any rotor shape parameter), whereas the offset, A_{r0}, seems to depend only on the cups' front area, S_c. In case of constant B, both the slope, dB_r/dR_{rc}, and the offset, B_{r0}, seem to depend only on the cups' front area.

2. The slope of an anemometer's transfer function, that is, the calibration constant A_r, can be accurately estimated with Kondo et al. analytical model if it is based on precise aerodynamic data related to the aerodynamics of the rotor cups.

ACKNOWLEDGMENTS

The authors are indebted to Enrique Vega, Alejandro Martínez, Pedro López, Luis García and Eduardo Cortés, from the IDR/UPM Institute for their friendly help and support in the present research. This work was done in collaboration with the Department of Mechanical Engineering of the Vrije Universiteit Brussel. In this sense, the authors are truly indebted to Chris Lacor and Alain Wery, for the opportunity to carry out part of the testing campaign in Brussels, and also for their friendly support. The authors would also like to thank Angel Sanz for sharing his ideas and giving his encouraging support. The authors are indebted to Alfonso Rosende from Ornytion for his constant and kind support for the authors' research. Finally, the authors are grateful to Brian Elder and Tania Tate for their kind help in improving the style of the text.

REFERENCES

1. Sanz, A.A. *Trabajo de Investigación*; Research Report; Competitive Exam for Head of Department Position at the Polytechnic University of Madrid: Madrid, Spain, 2001.
2. Kristensen, L.; Jensen, G.; Hansen, A.; Kirkegaard, P. *Field Calibration of Cup Anemometers*; Risø–R–1218(EN); Risø National Laboratory: Roskilde, Denmark; January; 2001.
3. Wyngaard, J.C. Cup, propeller, vane, and sonic anemometers in turbulence research. *Ann. Rev. Fluid Mech.* **1981**, *13*, 399–423.
4. Ramachandran, S.A. Theoretical study of cup and vane anemometers. *Quart. J. Roy. Meteorol. Soc.* **1969**, *95*, 163–180.
5. Kondo, J.; Naito, G.; Fujinawa, Y. Response of cup anemometer in turbulence. *J. Meteorol. Soc. Jpn.* **1971**, *49*, 63–74.
6. Brevoort, M.J.; Joyner, U.T. *Aerodynamic Characteristics of Anemometer Cups*; NACA Technical Note 489; Washington, DC, USA, 1934.
7. Brevoort, M.J.; Joyner, U.T. *Experimental Investigation of the Robinson-Type Cup Anemometer*; NACA Technical Note 513; Washington, DC, USA, 1935.
8. Pindado, S.; Vega, E.; Martínez, A.; Meseguer, E.; Franchini, S.; Sarasola, I. Analysis of calibration results from cup and propeller anemometers. Influence on wind turbine Annual Energy Production (AEP) calculations. *Wind Energy* **2011**, *14*, 119–132.
9. Kristensen, L. Cup anemometer behavior in turbulent environments. *J. Atmos. Ocean. Technol.* **1998**, *15*, 5–17.
10. Wind Turbines. *Part 12-1: Power Performance Measurements of Electricity Producing Wind Turbines*, 1st ed.; International Standard IEC-61400-12-1; International Electrotechnical Comission: Geneva, Switzerland, 2005.
11. Solov'ev, Y.P.; Korovushkin, A.I.; Toloknov, Y.N. Characteristics of a cup anemometer and a procedure of measuring the wind velocity. *Phys. Oceanogr.* **2004**, *14*, 173–186.
12. Kaganov, E.I.; Yaglom, A.M. Errors in wind speed measurements by rotation anemometers. *Bound.-Lay. Meteorol.* **1976**, *10*, 1–11.
13. Pindado, S.; Sanz, A.; Wery, A. Deviation of cup and propeller anemometer calibration results with air density. *Energies* **2012**, *5*, 683–701.
14. *Cup Anemometer Calibration Procedure, Version 1*; Measuring Network of Wind Energy Institutes (MEASNET): Madrid, Spain; September; 1997.
15. *Anemometer Calibration Procedure, Version 2*; Measuring Network of Wind Energy Institutes (MEASNET): Madrid, Spain; October; 2009.
16. Wind speed measurement and use of cup anemometry. In *International Energy Agency Programme for Research and Development on Wind Energy Conversion Systems. Recommended Practices for Wind Turbine Testing and Evaluation*, 1st ed.; Renewable Energy Systems Ltd.: Glasgow, UK, 2003.
17. Lockhart, T.J. Some cup anemometer testing methods. *J. Atmos. Ocean. Technol.* **1985**, *2*, 680–683.

CHAPTER 7

An Adaptive Altitude Information Fusion Method for Autonomous Landing Processes of Small Unmanned Aerial Rotorcraft

Xusheng Lei * and Jingjing Li

Science and Technology on Inertial Laboratory, Beijing 100191, China

ABSTRACT

This paper presents an adaptive information fusion method to improve the accuracy and reliability of the altitude measurement information for small unmanned aerial rotorcraft during the landing process. Focusing on the low measurement performance of sensors mounted on small unmanned aerial rotorcraft, a wavelet filter is applied as a pre-filter to attenuate the high frequency noises in the sensor output. Furthermore, to improve altitude information, an adaptive extended Kalman filter based on a maximum *a posteriori* criterion is proposed to estimate measurement noise covariance matrix in real time. Finally, the effectiveness of the proposed method is proved by static tests, hovering flight and autonomous landing flight tests.

KEYWORDS:

small unmanned aerial rotorcraft; wavelet filter; altitude information fusion; adaptive extended Kalman filter

1. INTRODUCTION

With the ability to land vertically, small unmanned aerial rotorcraft (SUAR) have an irreplaceable role in civil applications [1]. Thus, they have been widely used in many areas, including road traffic monitoring, city building surveillance and power line inspection, *etc.* [2,3].

SUAR is a complex multi-input and multi-output (MIMO) system. Compared with the hovering and straight flight processes, there exists land disturbance in the landing process [4]. High performance altitude information is the basis factor for SUARs to realize stable landing control [5]. Due to the constraints of weight and size, sensors with low size and low performance are often used by SUARs, including micro-electronic mechanic system (MEMS) accelerometers and gyroscopes, barometers, the global positioning system (GPS) and ultrasonic sensors.

Integrated by the Euler equations, or quaternions, SUAR can get the corresponding aircraft attitude angles and position information, however, inertial sensors, especially gyroscopes, have fixed bias, drift bias, asymmetric scale factor errors and temperature-varying biases, causing the integration results to drift from true attitude [6]. GPS can provide absolute position and velocity information [7], but GPS information is easily affected by sources of interference [8]. Furthermore, GPS has low data frequency to get position and velocity information for a SUAR system [9]. Based on the relationship between the air pressure and the altitude, barometers can provide altitude information [10], but they are easily affected by wind disturbances, air fluctuations, and temperature [11]. Ultrasonic sensors are also often used in SUAR systems. Although they can provide high performance altitude measurements, they have measurement region limitations. When the altitude of SUAR surpasses the upper bound of ultrasound, the measurement results will have errors, therefore all the current sensors have limitation for SUAR to realize stable landing control.

Using filter methods, system can get high performance information based on different sensors. The most used filtering method is the extended Kalman filter (EKF) [12]. With the predict and update theory, Beard has used EKF to realize attitude acquisition for a unmanned aerial vehicle [13]. Nevertheless, poor performance or even divergence arising from the linearization implicit in EKF has led to the development of other filters [14]. The unscented Kalman filter (UKF) is also used in UAV systems [15]. Based on second or higher-order approximations of nonlinear functions, UKF can estimate the mean and covariance of state vectors [16]. With UKF, Seung realized target relative position and velocity determinations for follower UAV systems [17]. However, UKF is sensitive to the statistical distribution of the stochastic processes [18]. Based on the concept of sequential sampling and Bayesian theory, particle filtering (PF) is also used in dealing with nonlinear and non-Gaussian noise in SUAR systems. Kamrani used PF for efficient path planning of a UAV [19], but its computational demands are too complex. Wavelet analysis has also been widely used for its time and frequency domain convenience and it can effectively eliminate high frequency noise. Tsiotras used a wavelet transform to construct an approximation of the environment at different levels for small UAVs [20].

Inspired by the discussion above, an adaptive extended Kalman filter (AEKF) method based on the wavelet filter is proposed to get high performance altitude information for a SUAR during the autonomous landing process. The wavelet decomposition and reconstruction method is used to restrain the high frequency noise in the barometer, ultrasonic and GPS sensor information. Since the measurement noise is greatly changed after wavelet filtering, an AEKF based on a maximum *a posteriori* criterion is proposed to estimate the measurement noise matrix in real time to get high performance altitude information.

The paper is organized as follows: the dynamic model of the SUAR system is described in Section 2. The wavelet decomposition and reconstruction method is presented in Section 3. In Section 4, an AEKF based on a maximum *a posteriori*criterion is proposed to improve altitude information. The simulation and test results confirm the effectiveness of the proposed method in Section 5. Finally, conclusions are drawn in Section 6.

2. THE DYNAMIC MODEL OF SUAR

2.1. The State Model

For SUAR, altitude information is mainly controlled by the main rotor speed and longitudinal cyclic input. Therefore, the simple altitude dynamic model for SUAR can be defined as:

$$\begin{cases} \dot{x}_1 = f_1 = x_2 \\ \dot{x}_2 = f_2 = a_0 + a_1 x_2 + a_2 x_2^2 + \left(a_3 + a_4 x_4 - \sqrt{a_5 + a_6 x_4}\right)x_3^2 \\ \dot{x}_3 = f_3 + u_1 = a_7 + a_8 x_3 + (a_9 \sin x_4 + a_{10})x_3^2 + u_1 \\ \dot{x}_4 = f_4 = x_5 \\ \dot{x}_5 = f_5 + u_2 = a_{11} + a_{12}x_4 + a_{13}x_3^2 \ \sin x_4 + a_{14}x_5 + u_2 \end{cases}$$

where $x_i = (h, \dot{h}, w, \theta, \dot{\theta})^T$, $i = 1,\cdots,5$. The term h is the estimated altitude measured by barometer, GPS, and ultrasonic sensors,w is rotor speed of main rotor, θ is the longitudinal angle of rotor blade speed, $u = (u_1, u_2)^T$ is the corresponding input, $u_1$and u_2 are throttle and collective input respectively, playing an important role in longitudinal cyclic input, lateral cyclic input and blade speed. $a_j(j = 1,2,\cdots,15)$ are unknown identification parameters, obtained by the adaptive genetic method [21]. Therefore, the altitude state model of SUAR can be defined as follows:

$$\dot{h} = f\left(h, w, \theta, \dot{\theta}\right)$$

2.2. The Measurement Model

The output accuracy of a barometer is mainly affected by the high frequency noise and constant error which is related to air pressure and temperature. The high frequency noise can be restrained largely by a wavelet filter. Thus, the barometer output h_b mainly includes a constant error ε_b and measurement noise v_1. The function of the h_b can be defined as follows:

$$h_b = h + \varepsilon_b + v_1$$

where h is the altitude of the SUAR system.

With the location method of the ranging interchange theory, DGPS can provide position information for SUAR systems with sub-meter performance. The output of DGPS can be defined as follows:

$$h_g = h + v_2$$

where h_g is the output of DGPS, and v_2 is measurement noise

Ultrasonic sensors can provide high performance altitude information from 0.15 m to 6.05 m, and the error is less than 1 millimeter. When the altitude surpasses the upper limitation, the output of ultrasonic sensor fluctuates greatly. Therefore, the output of the ultrasonic sensor can be defined as follows:

$$h_u = \begin{cases} \omega_{lu} & h \geq 6.05m \\ \omega_u & 0 \leq h < 6.05 \end{cases}$$

where h_u is the output of ultrasonic sensor, w_u is the random error within the bounds and w_{lu} is random error without bounds.

When a SUAR finishes a certain task at low altitude, there exists land disturbance causing an increase of barometer error. Ultrasonic sensors can provide high precision altitude information for SUARs at low altitude, therefore the measurement matrix can be constructed with inputs from different sensors.

Case 1

If the integrated navigation altitude is larger than 6 m, the SUAR is beyond the range of the ultrasonic sensor. The output of barometer and DGPS are fused. Thus, the constant error of barometer sensor can be revised by the DGPS. The measurement equation can be defined as:

$$\begin{bmatrix} h_b \\ h_g \end{bmatrix} = \begin{bmatrix} 1 & 1 \\ 1 & 0 \end{bmatrix} \begin{bmatrix} h \\ \varepsilon_b \end{bmatrix} + \begin{bmatrix} v_1 \\ v_2 \end{bmatrix}$$

Case 2

If the integrated navigation altitude is less than 6 m, the barometer is easily affected by land disturbance. The output of DGPS and ultrasonic sensor are used to construct the measurement vector. The measurement equation can be defined as follows:

$$\begin{bmatrix} h_g \\ h_u \end{bmatrix} = h + \begin{bmatrix} v_2 \\ \omega_u \end{bmatrix}$$

Therefore, the measurement equation can be expressed as:

$$Z_k = H_k X_k + V_k$$

where H_k is the state vector, Z_k is the measurement vector, H_k is the measurement matrix and V_k is the measurement noise vector. With the wavelet filter, the high frequency noises can be largely eliminated. Then, using the AEKF based on maximum *a posteriori* criterion, the altitude h and the constant error ε_b can be estimated unbiasedly. The whole procedure is shown in Figure 1.

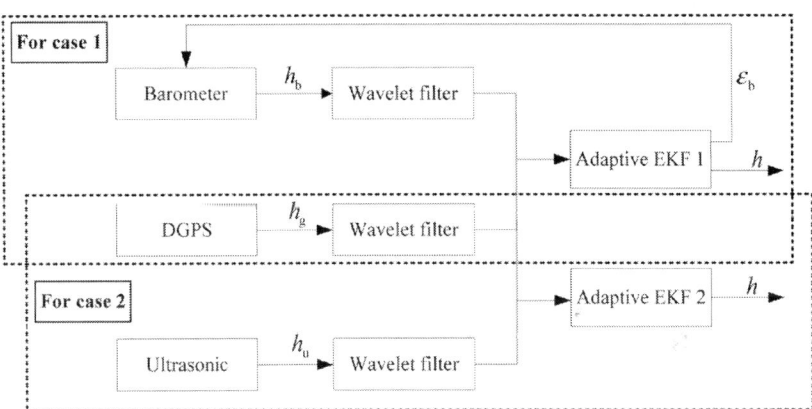

Figure 1. The scheme of altitude fusion.

3. THE WAVELET DECOMPOSITION AND RECONSTRUCTION METHOD

To get high precision altitude information, it is necessary to use a data filter to deal with high frequency noises in the output of barometer sensor, DGPS, and ultrasonic sensor. Wavelet analysis is a time and frequency domain method, having good representation for partial signal characteristics, therefore, a wavelet filter is used here as a tool to reduce high frequency noises in the sensor information. Lifting-based wavelet transform implementation has shown high potential in reducing the number of computations, so it is used to reduce computation burden in real tasks. It includes three steps:

1. Split: splitting the original signal $s^j = \left\{ s^j, k \mid 0 \leq k < 2^j \right\}$ into even and odd ones. That is:

2.
$$\begin{cases} Split(s^j) = (E_{j-1}, O_{j-1}) \\ E_{j-1} = \{E_{j-1,k} = s_{j,2k}\}, O_{j-1} = \{O_{j-1,k} = s_{j,2k+1}\} \end{cases}$$

3. Predict: defining the detailed representation characteristics by choosing a predictor:

$$O_{j-1} \mathrel{-}= P(E_{j-1})$$

4. Update: averaging the signal of rough representation against original signal:

$$E_{j-1} \mathrel{+}= U(O_{j-1})$$

The basic principle of lifting scheme is to factorize the polyphase matrix of a wavelet filter into a sequence of alternating upper, lower triangular matrices and a diagonal matrix with constants. The factorization is obtained by using an extension of the Euclidean algorithm. The resulting formulation can be implemented by means of banded matrix multiplications.

Suppose that the z-transform of wavelet filter $h = \{h_k, k \in Z\}$ can be defined as $h(z)$. Let $\hat{h}(z)$ and $\hat{g}(z)$ be the low and high pass analysis filters, and $h(z)$, $g(z)$ be the low and high pass synthesis filters. $\hat{h}(z)$, $\hat{g}(z)$, $h(z)$ and $g(z)$ are biorthogonal filters. The filters can be divided into even and odd parts as:

$$\begin{cases} \hat{h}(z) = \hat{g}_e(z^2) + z^{-1}\hat{g}_o(z^2) \\ \hat{g}(z) = \hat{g}_e(z^2) + z^{-1}\hat{g}_o(z^2) \\ h(z) = h_e(z^2) + z^{-1}h_o(z^2) \\ g(z) = g_e(z^2) + z^{-1}g_o(z^2) \end{cases}$$

The polyphase matrices are then defined as:

$$\begin{cases} \hat{P}(z) = \begin{bmatrix} \hat{h}_e(z) & \hat{h}_o(z) \\ \hat{g}_e(z) & \hat{g}_o(z) \end{bmatrix} \\ P(z) = \begin{bmatrix} h_e(z) & h_o(z) \\ g_e(z) & g_o(z) \end{bmatrix} \end{cases}$$

If the $(\hat{h}, \hat{g})$ is a complementary filter pair, then $\hat{P}(z)$ can be factored as follows:

$$P(z) = \prod_{i=m}^{m} \begin{bmatrix} 1 & s_i(z) \\ 0 & 1 \end{bmatrix} \begin{bmatrix} 1 & 0 \\ t_i(z) & 1 \end{bmatrix} \begin{bmatrix} K & 0 \\ 0 & K^{-1} \end{bmatrix}$$

where K is a constant value.

Therefore, the low pass samples are multiplied by the time domain equivalent of $s_i(z)$, and are added to the high pass samples. Then, the updated high-pass samples are multiplied by the time domain equivalent of $t_i(z)$ and are added to the low-pass samples. If a diagonal matrix is present in the factorization, the low pass coefficients are multiplied by K and the high-pass coefficients are multiplied by K^{-1}. The polyphase-based wavelet transform in lifting scheme is shown in Figure 2.

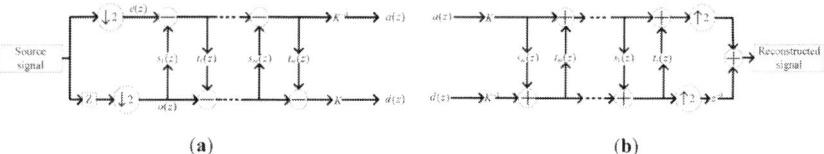

(a)   (b)

Figure 2. The polyphase-based wavelet transform in lifting scheme. (**a**) The wavelet analysis process. (**b**) The wavelet reconstruction process.

In this paper, the wavelet "db4" is utilized to construct the wavelet method. The coefficients of the filter are shown in Table 1.

Table 1. The coefficients of the "db4" filter.

h_0	**0.48296291314453**	g_0	**0.12940952255126**
h_1	0.83651630373780	g_1	0.22414386804201
h_2	0.22414386804201	g_2	−0.83651630373780
h_3	−0.12940952255126	g_3	0.48296291314453

The comparisons of original data and the wavelet filtered data of barometer and DGPS are shown in Figure 3. Obviously, the wavelet method can filter out the high frequency noise effectively.

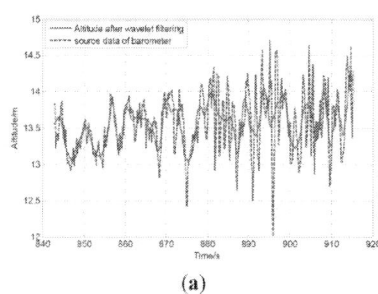
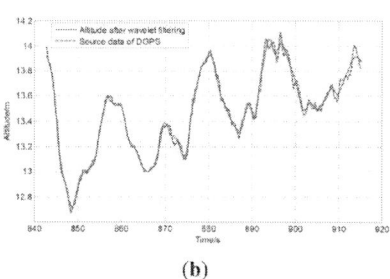

(a)   (b)

Figure 3. (**a**) The comparison between the original barometer data and the wavelet filtered data. (**b**) The comparison between the original DGPS data and the wavelet filtered data.

4. THE ADAPTIVE EXTENDED KALMAN FILTER

Since the measurement noise structure has changed greatly after wavelet filtering, experiential value or the statistics of partial noise cannot be used to provide a good description of measurement noise covariance, therefore, an AEKF is proposed to estimate the measurement noise covariance in real time to improve altitude information.

Since the nonlinear dynamic equation of SUAR is continuous and the measurements are a discrete series, a continuous-discrete EKF is proposed to fuse altitude sensor information. In EKF, the state equation and measurement equation can be expressed as:

$$\begin{cases} \dot{X}(t) = f(X(t), t) + B(t)u(t) + W(t) \\ Z_k = H_k X_k + V_k \end{cases}$$

where, k is the number of time step. $X(t)$ and Z_k are the state vector and measurement vector respectively. $f(X(t), t)$ is nonlinear ordinary differential equations, and H_k is the measurement matrix. $B(t)$ is the input matrix, $u(t)$ is the input vector, $W(t)$, V_k are the system noise and measurement noise vector respectively. Besides, the system noise and the measurement noise are uncorrelated, and the system noise can be treated as Gaussian white noise.

In EKF, measurement noise covariance matrix R plays an important role in obtaining a converged filter result. If the value of R is small, unreliable results will be obtained, and a big value of the diagonal elements of R can lead to filter divergence. In traditional EKF, the measurement noise is treated as Gaussian white noise, however, the measurement noise structure has changed greatly after wavelet filtering. Using traditional experiment values or partial statistics of sampling data as measurement noise matrix, the filtering speed will become slow, and filtering performance will become bad, therefore, the sub-optimal and unbiased maximum *a posteriori* method is proposed to estimate R in real time. As shown in Equation (21), the current R is updated by the innovation ε_k and R at previous time. The filter consists of the following stages:

1. The prediction stage:

$$\hat{X}_{k/k-1} = \hat{X}_{k-1} + \left\{ f\left[\hat{X}_{k-1}, t_{k-1}\right] + B(t_{k-1})u(t_{k-1}) \right\}T$$

$$P_{k/k-1} = \emptyset_{k,k-1} P_{k-1} \emptyset_{k,k-1}^T + Q_{k-1}$$

$$\hat{Z}_{k/k-1} = H_k \hat{X}_{k-1}$$

2. The update stage:

$$\varepsilon_k = Z_k - \hat{Z}_{k/k-1}$$

$$\hat{X}_k = \hat{X}_{k/k-1} + K_{k-1}\varepsilon_{k-1}$$

$$R_k = \left(1 - \frac{1}{k}\right)R_{k-1} + \frac{1}{k}(\varepsilon_k\varepsilon_k^T - H_kP_kH_k^T)$$

$$K_k = P_{k/k-1}H_k^T[H_kP_{k/k-1}H_k^T + R_k]$$

$$P_k = (I - K_kH_k)P_{k/k-1}(I - K_kH_k)^T + K_kR_kK_k^T$$

where $\hat{X}_{k/k-1}$ is the predicted measurement vector for the next epoch, $\hat{Z}_{k/k-1}$ and $P_{k/k-1}$ are the predicted measurement vector and the predicted state covariance matrix respectively. $\varnothing_{k,k-1}$ is the transition matrix after discretization. The innovation ε_k is the difference between the real observations and its estimated values. T is the sampling time. Q_k is the system noise covariance matrix. K_k is the gain matrix, P_k is the estimated state covariance matrix, and R_k is the covariance matrix of measurement noise based on the maximum a posteriori adaptive method. Using ε_k, HkPk/k−1HTkand initial experiment value R_0, the measurement noise covariance matrix can be estimated in real time to improve filtering performance.

5. EXPERIMENT

5.1. Hardware System

Experiments were conducted on a radio-controlled Raptor 90 helicopter, shown in Figure 4. The SUAR is 1.3 m length and 1.46 m span. The total weight is 5 kg, including two liter fuels, a light weight DGPS receiver, and radio telemeter system. The SUAR is powered by a piston engine running on a mixture of methanol and oil. Five servos are used to control the tail, the longitudinal cyclic input, the lateral cyclic input, the collective and the throttle. The longitudinal vertical direction can be stabilized by using the collective and pitch cyclic. Meanwhile, the lateral direction can be controlled by using the roll-cyclic and collective. The heading can be controlled by the tail.

Figure 4. The Raptor-90 helicopter.

For SUAR, there exist weight and size constraints for onboard control components. Thus, a micro guidance navigation control (MGNC) system with little weight was self-developed to realize stable control. The MGNC is only 207 g in weight, with a size of 120 mm × 61 mm × 48 mm. It consists of a horizontal main board, housing three angular rate sensors, two 2-axis accelerometers and a barometer. The barometer, DGPS, and ultrasonic sensor are used to provide altitude information for the SUAR system. The MPXA6115 barometer, produced by Freescale Semiconductor Company, has a range of 15 kPa~115 kPa. The DGPS module employs the Novatel RTK, whose position accuracy is about 0.02 m, and the range of the Mini-S electrostatic ultrasonic transducer is from 0.15 m to 6.05 m.

5.2. Static Distance Test

To test the effectiveness of the proposed information fusion method, a static distance test has been done on the stairs. A six-floor building is chosen as the basis for its high precision measurement. Three marking points are selected on the stairs. The distances between points and ground have been tested by flexible rulers and the distances are 5.12 m (first point), 9.32 m (second point) and 13.52 m (third point). The SUAR is stretched to measure the distance between the current point and the ground. Besides, the sampling time is 60 s per point.

The comparison result of the proposed method and the real altitude, the output of the barometer and the real altitude, the output of the DGPS and the real altitude, the output of the ultrasonic sensor and the real altitude are shown in Figure 5. The measurement results of each sensor are shown in Table 2. It is easy to see that the proposed AEKF method has the best performance. Since there exist barriers for GPS in the building, the maximum error reaches to 1.49

m. Without airflow disturbance, the barometer can provide good measurement results, and the standard deviation is 26 percent of the DGPS after initial alignment. The ultrasonic sensor can provide high performance altitude information under 6 m. The mean error at first point is below 0.11 m. When altitude surpasses the 6 m, the performance of ultrasonic is decreased greatly.

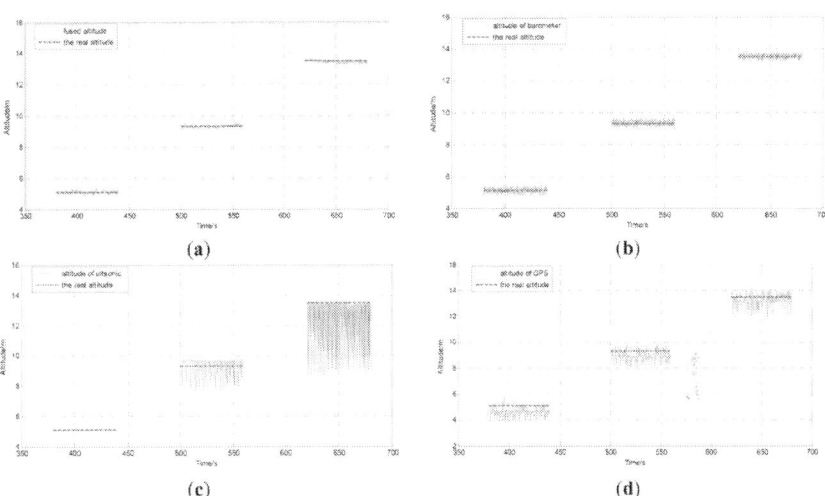

(a) (b) (c) (d)

Figure 5. (a) The comparison between the output of AEKF and the real altitude. (b) The comparison between the output of barometer and the real altitude. (c) The comparison between the output of ultrasonic and the real altitude. (d) The comparison between the output of DGPS and the real altitude.

Table 2. Altitude accuracies for AEKF and each sensor (in meters).

	AEKF	Barometer	Ultrasonic	DGPS
Max. absolute error	0.31	0.32	4.94	1.49
Mean error	0.08	0.14	0.44	0.49
Standard deviation	0.098	0.16	0.94	0.41

5.3. Hovering Flight Test

To test the dynamic performance of the proposed method, a hovering flight test has been done on the SUAR system. Under a 3.4 m/s wind disturbance, the SUAR hovers in the air at 10 m altitude. The LQR control method has been used to adjust altitude and position in real time [22]. Since the planned altitude surpasses the upper limit of ultrasonic sensor, DGPS and barometer are used to provide altitude information for the SUAR system. The altitude generated by the AEKF method, barometer and DGPS are shown in Figure 6. The mean error of the adaptive EKF is only 0.214 m, and the standard deviation is 0.169 m. With the proposed AEKF, the SAUR can realize stable hovering control. Furthermore, it is easy to see that the altitude measured by the barometer fluctuates greatly. Since there exists wind disturbance, the fluctuation of barometer surpasses 1 m. Without shelter, the output of DGPS can provide high performance measurement in short periods. With the fluctuation of the star number, the DGPS output is not so reliable.

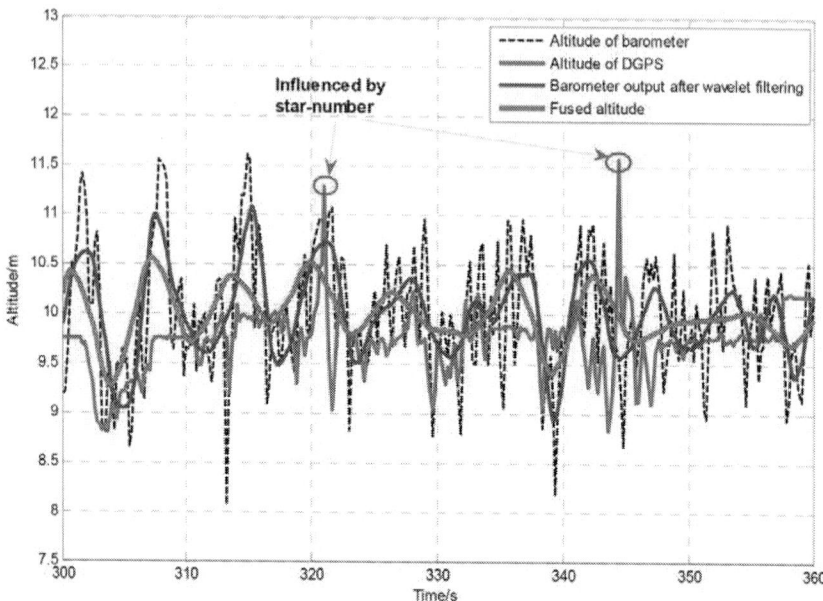

Figure 6. The altitude generated by the AEKF method, barometer and DGPS in a hovering process.

5.4. Autonomous Landing

To test the effectiveness of the proposed method, a series of autonomous landing tests have been done on the SUAR system with the adaptive radial basis function neural network and pilot model. When the SUAR received an autonomous landing command, it changed work station, and flew to the planned hovering point (0,0,10). To satisfy the criteria for position error, speed error and heading error, the SUAR hovered at the planned hovering point. With the constant adjustment for the planned hovering altitude, SAUR descends with hovering stations. Finally, the SUAR landed on the ground. Ten landing tests were conducted from different altitudes, while the wind velocity was less than 3 m/s. The landing results are shown in Figure 7. With the proposed adaptive altitude information fusion method, the SUAR can realize stable autonomous landings, and the average Euclidean distance from the landing target is about 0.67 m. Compared with the navigation system with camera [23,24], the SUAR can get achieve similar landing performance.

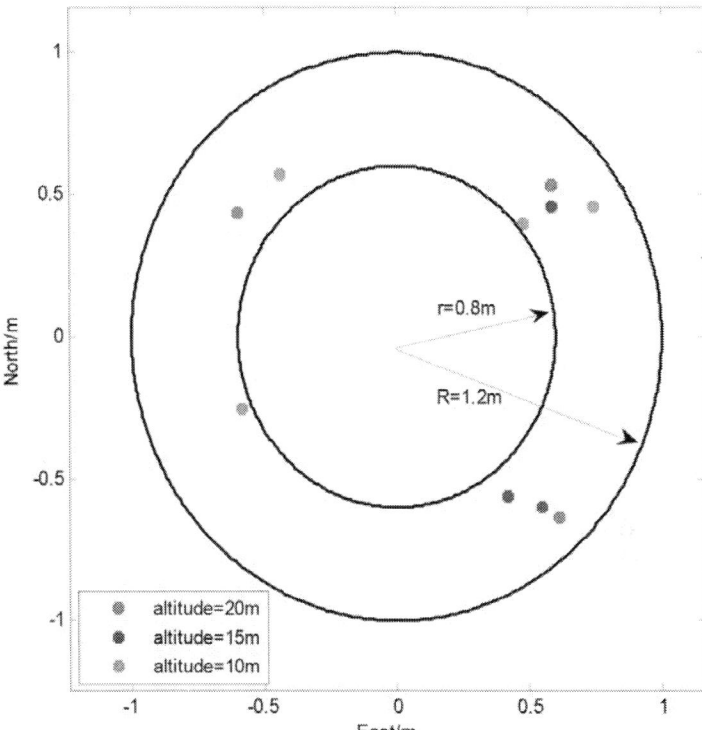

Figure 7. The result of autonomous landing tests from different altitudes.

The comparison of landing performance with AEKF and KF [13] which fuses SINS and DGPS is shown in Figure 8. Using the AEKF, SUAR realized a stable autonomous landing with 0.75 m and 0.45 m error in the East and North directions from the planned landing point. Compared with the KF, the AEFK has much better performance in the autonomous landing process. The altitude, attitude and velocity of the autonomous landing using AEKF are shown in Figure 8(b–d) respectively.

6. CONCLUSIONS

In this paper, an adaptive information fusion method based on wavelet decomposition and reconstruction is proposed to improve the accuracy and reliability of altitude measurement information in the landing process for a SUAR. With the proposed method, the high frequency noises in sensors can be eliminated greatly, and then high performance altitude information can be fused to provide support for SUAR in the autonomous landing process. The effectiveness of the proposed method has been demonstrated by static tests, hovering tests and a series of autonomous landing tests.

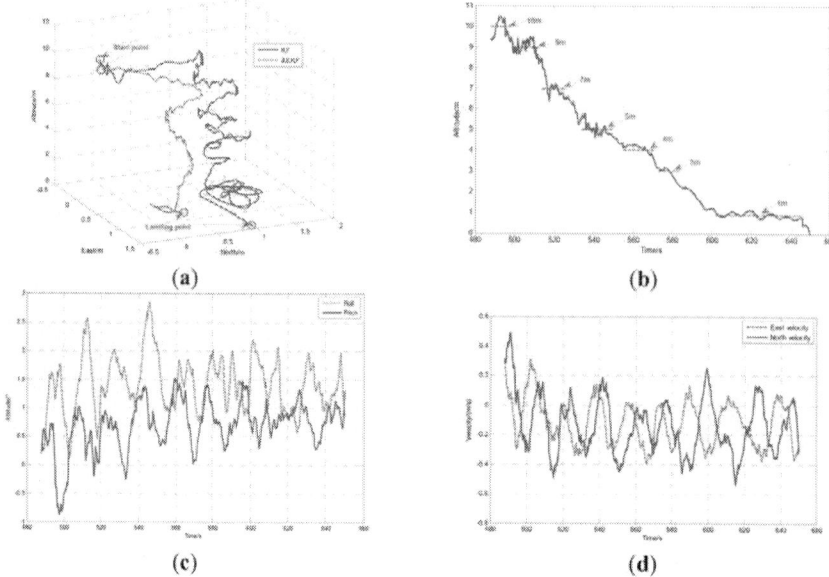

Figure 8. (**a**) The comparison of 3D trajectory of the SUAR with KF and AEKF method in the landing process. (**b**) The altitude trajectory of SUAR in autonomous landing process. (**c**) The pitch and roll angles in autonomous landing process. (**d**) The velocities in two directions.

ACKNOWLEDGMENTS

The research is supported by the National Natural Science Foundation of China (Grant No. 60905056, 60904093, 61121003, 61273033).

REFERENCES

1. Alexander, J.; Sjir, U.; Nick, E. Safety in high-risk helicopter operations: The role of additional crew in accident prevention. *Saf. Sci.* **2009**, *47*, 717–721.
2. George, V.; Liang, T.; Graham, D.; Luis, G. From mission planning to flight control of unmanned aerial vehicles: strategies and implementation tools. *Annu. Rev. Control.* **2005**, *29*, 101–115.
3. Najib, M.; Tarek, H. A UAR for bridge inspection: Visual servoing control law with orientation limits. *Autom. Constr.* **2007**, *17*, 3–10.
4. Cai, G.W.; Chen, B.M.; Lee, T.H.; Dong, M.B. Design and Implementation of a Hardware-in-the-loop Simulation System for Small-scale UAV Helicopters. *Mechatronics* **2009**, *19*, 1057–1066.
5. Fabiani, P.; Fuertes, V.; Piquereau, A.; Mampey, R.; Teichteil, F. Autonomous flight and navigation of VTOL UAVs: From autonomy demonstrations to out-of-sight flights. *Aerosp. Sci. Technol.* **2007**, *11*, 183–193.

6. Tobias, P.; Thomas, R.; Jan, T.G. Modeling of UAV formation flight using 3D potential field. *Simul. Model. Prac. Theory* **2008**, *16*, 1453–1462.
7. Lorenzo, M.; Roboerto, N. Aggressive control of helicopters in presence of parametric and dynamical uncertainties. *Mechatronics* **2008**, *18*, 381–389.
8. Peng, K.M.; Cai, G.W.; Chen, B.M.; Dong, M.B.; Lum, K.; Lee, T.H. Design and implementation of an autonomous flight control law for a UAV helicopter. *Automatica* **2009**, *45*, 2333–2338.
9. Agus, B.; Singgih, S. Optimal tracking controller design for a small scale helicopter. *J. Bionic Eng.* **2007**, *4*, 271–280.
10. Bayraktar, S.; Feron, E. Experiments with small unmanned helicopter nose up landings. *J. Guid. Contr. Dynam.* **2009**,*32*, 332–337.
11. Widyawardana, A.; Adang, S.A.; Jaka, S. Automated flight test and system identification for rotary wing aerial platform using frequency responses analysis. *J. Bionic Eng.* **2007**, *4*, 237–244.
12. Cui, P.L.; Zhang, H.J. QMRPF-UKF master-slave filtering for the attitude determination of micro-nano satellites using gyro and magnetometer. *Sensor* **2010**, *10*, 9935–9947.
13. Randal, B.; Derek, K.; Morgan, Q.; Deryl, S. Autonomous vehicle technologies for small fixed wing UAVs. *J. Aeros. Comp. Inform. Commun.* **2005**, *2*, 92–108.
14. Hong, L.; Wicker, D.A. Spatial-domain multi solution particle filter with threshold wavelets. *Signal Process* **2007**, *87*, 1384–1401.
15. Crassidis, J.L.; Markley, F.L. Unscented filtering for spacecraft attitude estimation. *J. Guid. Contr. Dynam.* **2003**, *26*, 536–542.
16. Carmi, A.; Oshman, Y. Fast particle filtering for attitude and angular-rate estimation from vector observations. *J. Guid. Contr. Dynam.* **2009**, *32*, 70–78.
17. Seung, M.O.; Eric, N.J. Relative Motion Estimation for Vision-Based Formation Flight Using Unscented Kalman Filter. Proceedings of AIAA Guidance, Navigation, Control Conference and Exhibit, Hilton Head, SC, USA, 20–23 August 2007.
18. Zhang, J.Q.; Yan, Y. A wavelet-based approach to abrupt fault detection and diagnosis of sensors. *IEEE Trans. Instrum. Meas.* **2001**, *50*, 1389–1396.
19. Kamrani, F.; Lozano, M.G.; Ayani, R. Path planning for UAVs using symbiotic simulation. Proceedings of European Simulation and Modeling Conference, Toulouse, France, 23–25 October 2006.
20. Tsiotras, P.; Jung, D.; Bakolas, E. Multiresolution Hierarchical path-planning for small UAVs using wavelet decompositions. *J. Intel. Robot. Syst.* **2012**, *66*, 505–522.
21. Lei, X.S.; Du, Y.H. A linear domain system identification for small unmanned aerial rotorcraft based on adaptive genetic algorithm. *J. Bionic. Eng.* **2010**, *7*, 142–149.
22. Lei, X.S.; Bai, L.; Du, Y.H.; Miao, C.X.; Chen, Y.; Wang, T.M. A small unmanned polar research aerial vehicle based on the composite control method. *Mechatronics* **2011**, *21*, 821–830.
23. Farid, K.; Kenzo, N.; Isabelle, F. An Adaptive Vision-based autopilot for mini flying machines guidance; navigation; and control. *Auton. Robot.* **2009**, *27*, 165–188.

24. Hermansson, J.; Gising, A.; Skoglund, M.; Schon, T.B. *Autonomous Landing of an Unmanned Aerial Vehicle*; Technical Report from Automatic Control at Linkopings Universitet: Sweden, 2010.

CHAPTER 8

Nonlinear Adaptive Rotational Speed Control Design and Experiment of the Propeller of an Electric Micro Air Vehicle

Shouzhao Sheng * and Chenwu Sun

College of Automation Engineering, Nanjing University of Aeronautics and Astronautics, 29 YuDao St., Nanjing 210016, China

ABSTRACT

Micro Air Vehicles (MAVs) driven by electric propellers are of interest for military and civilian applications. The rotational speed control of such electric propellers is an important factor for improving the flight performance of the vehicles, such as their positioning accuracy and stability. Therefore, this paper presents a nonlinear adaptive control scheme for the electric propulsion system of a certain MAV, which can not only speed up the convergence rates of adjustable parameters, but can also ensure the overall stability of the adjustable parameters. The significant improvement of the dynamic tracking accuracy of the rotational speed can be easily achieved through the combination of the proposed control algorithm and linear control methods. The experimental test results have also demonstrated the positive effect of the nonlinear adaptive control scheme on the flight performance of the MAV.

KEYWORDS:

nonlinear adaptive control; dynamic tracking; electric propeller; micro air vehicle; hardware-in-loop experiment; flight test

1. INTRODUCTION

In recent years, the study of Micro Air Vehicles (MAVs) driven by electric propellers has gained considerable momentum. Some substantial progress has been made towards designing, building and test-flying remotely piloted high-performance MAVs, and there exist many commercially available MAVs as well as laboratory prototypes [1,2,3]. Such MAVs are intended to lower the total system cost, and are easy enough to operate with minimal training. Moreover, the electric propellers can also reduce the noise signature in practical use.

In general, the thrust force of a MAV largely depends on the rotational speed of the electric propeller while the MAV performs low-speed flight missions [4], and the characteristics of the rotational speed system strongly affect the flight dynamics, stability and reliability of the MAV. Therefore, it is essential to develop a reliable control scheme to ensure an accurate rotational speed of the electric propeller, which can contribute to the significant improvement in the flight performance of the MAV, and new applications for both military and civilian markets.

The studies have been conducted for electric MAVs, as well as other propeller-based propulsion systems, mainly including multidisciplinary design optimization [5,6], system modeling [7,8,9], and control design [10,11,12,13,14,15]. Although the optimization of the electric propulsion systems of such MAVs is extremely crucial, most of the previous investigations were limited to either aerodynamic and structural analyses of propellers [16,17,18], or the optimization of electric motors [19,20]. Until recent times the rotational speed of the electric propeller was mainly regulated according to the altitude control signal without a closed loop speed control [21,22] or it was based on the linearized model around an operation point [23], thus resulting in degrading the flight performance of the MAV, such as the positioning accuracy and stability. Although the conventional method can guarantee the stability of the test plants with moderate flight performance, it largely depends on a lot of information and knowledge derived from the plants, and easily fails to achieve a satisfactory flight performance in the presence of parametric uncertainties.

The purpose of this study is to present a nonlinear adaptive control scheme for the electric propulsion system of a certain MAV, based on which the significant improvement in the dynamic tracking accuracy of the rotational speed can be easily achieved by further using linear control techniques. The control performance of the nonlinear adaptive controller is to be verified by a series of experimental tests.

This paper describes the rotational speed control problem of the propeller of the prototype electric MAV in Section 2. The modeling of the rotational speed system of the propeller is presented in Section 3, and Section 4 provides the nonlinear adaptive rotational speed control of the propeller. The experimental results are shown in Section 5. Finally, conclusions are drawn in Section 6.

2. THE ROTATIONAL SPEED CONTROL PROBLEM OF THE PROPELLER OF THE PROTOTYPE ELECTRIC MAV

The prototype electric MAV, 0.8 m in height and 0.55 m in diameter, is shown in Figure 1. A duct is formed through the fuselage, and a propeller is mounted to the middle portion of the fuselage. Four deflecting vanes are mounted along the longitudinal and lateral axes of symmetry below the propeller to provide pitch, roll and yaw movements. Most of the anti-torque generated by the propeller is compensated by the stators configured inside the duct while the remaining anti-torque is balanced by the deflecting vanes. A center-symmetric landing gear consisting of four legs, *i.e.*, left, right, front, back, made from glass fiber–reinforced plastics is installed on the MAV [2]. The propulsion system mainly consists of an advanced lithium battery, a brushless DC electric motor with an electronic speed controller (ESC), a subminiature encoder, and a propeller. The propeller driven by the electric motor can provide sufficient thrust force to lift the MAV.

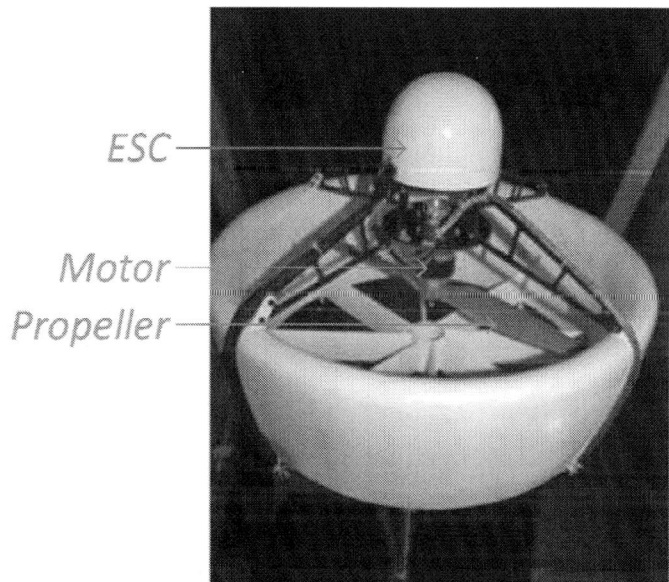

Figure 1. The prototype electric Micro Air Vehicle (MAV).

The rotational speed control of the electric propeller is desired for a high-performance MAV, because the characteristics of the rotational speed are nonlinear, and strongly affect the flight dynamics, stability and reliability of the MAV. It is, however, noted that, until recent times, the rotational speed of the electric propeller is only regulated according to the signal from a command signal generator without a closed-loop rotational speed control structure under fairly stable conditions, thus degrading the flight performance of the MAV.

Therefore, the goal of this study is to design a reliable rotational speed controller for the electric propeller.

3. MODELING OF THE ROTATIONAL SPEED SYSTEM OF THE PROPELLER

The analysis of the basic principle model is crucial for the development of a proper controller to improve the control quality of the rotational speed system. With regard to the aim of this study, the schematic diagram of the motor that conveys detailed information about the electrical components of the motor is shown in Figure 2, where the no-load current is known and is compensated by the ESC.

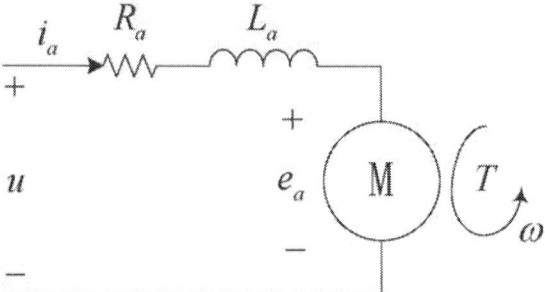

Figure 2. The schematic diagram of the motor.

The equations that describe the motor electrical components are as follows [24]:

$$u = R_a i_a + L_a \frac{di_a}{dt} + e_a$$

$$e_a = \frac{1}{K_V} \omega$$

$$T = K_Q i_a$$

where u (V) is the armature voltage, ia (A) the armature current, Ra (Ω) the armature coil resistance, La (H) the armature coil inductance, ea (V) the back electromotive force, KV (rad/s/V) the motor voltage constant, KQ (Nm/A) the torque constant, T (Nm) the torque generated by the motor, ω (rad/s) the rotational speed of the motor.

Considering that the MAV performs low-speed flight missions, the propeller approximately produces a torque according to [25]

$$T_p = k_p \omega_p^2$$

where kp (Nm s^2/rad^2) is the propeller moment constant, Tp (Nm) the propeller torque, ωp (rad/s) the rotational speed of the propeller.

The following conditions hold

$$\omega_p = \omega, \; T_p = T$$

under the assumption that the propeller is mounted on a propeller shaft which is driven through a rigid connection with the main drive shaft of the motor.

Substituting Equations (2) and (3) into Equation (1) yields

$$a_1 \omega \dot{\omega} + a_0 \omega^2 + b_0 \omega = u$$

where $a_1 = \dfrac{2L_a k_p}{K_O}$, $a_0 = \dfrac{R_a k_p}{K_O}$, and $b_0 = \dfrac{1}{K_V}$. The characteristics of the rotational speed system in Equation (4) are a nonlinear differential equation, and the coefficients of the equation are difficult and expensive to accurately obtain in practice, which results in the fact that most existing autopilots for MAVs with electric propellers, such as the widely used Pixhawk® (Zurich, Switzerland) autopilot [26], only employ a proportional-integral-differential (PID) controller to simply and directly control the altitude without a closed-loop rotational speed control structure under fairly stable conditions. The behavior of the rotational speed directly and greatly affects the flight dynamics, and then indirectly and greatly affects the flight performance, such as the positioning accuracy and stability. The control of the rotational speed is therefore important to the improvement of the flight performance of the MAVs.

In addition, some reasonable assumptions must be made preliminarily to simplify the adaptive control design as follows:

$$a_1 > 0, a_0 > 0, b_0 > 0$$

$$u > 0, \omega > 0$$

It is finally noted that such assumptions are based on the fact that the rotational speed of the propeller remains high enough to lift the MAV.

4. NONLINEAR ADAPTIVE ROTATIONAL SPEED CONTROL OF THE PROPELLER

4.1 Nonlinear Adaptive Control Algorithm

As for the rotational speed system in Equation (4), we first propose the following nonlinear control scheme:

$$u = k_{a_1}\omega\dot{\omega} + k_{a_0}\omega^2 + k_{b_0}\omega + k_b\omega r$$

where r denotes the input signal of the rotational speed system; $k_{a_1}, k_{a_0}, k_{b_0}$ and k_b are adjustable parameters, which need to be tuned adaptively.

Substituting Equation (6) into Equation (4) yields

$$\hat{a}_1\dot{\omega} + \hat{a}_0\omega + \hat{b}_0 = \hat{b}r$$

Where

$$\hat{a}_1 = a_1 - k_{a_1}, \hat{a}_0 = a_0 - k_{a_0}, \hat{b}_0 = b_0 - k_{b_0}, \hat{b} = k_b$$

Thus far, the nonlinear differential equation of the rotational speed system in Equation (4) is found to be inevitably transformed into a linear form by using the nonlinear control scheme, which, however, can contribute to the simplification of the adaptive laws.

Referring to the above transformed model in Equation (7), the ideal model of the rotational speed system can be represented as:

$$a_{m1}\dot{\omega}_m + a_{m0}\omega_m = b_m r$$

where ω_m is the model output; a_{m1}, a_{m0} and b_m are the model parameters determined directly according to the desired performance requirements.

The tracking error between the outputs of the model and the plant

$$e = \omega_m - \omega$$

can then be derived as follows:

$$a_{m1}\dot{e} + a_{m0}e = -\delta_{a_1}\dot{\omega} - \delta_{a_0}\omega - \delta_{b_0} + \delta_b r$$

Where

$$\delta_{a_1} = a_{m1} - \hat{a}_1, \delta_{a_0} = a_{m0} - \hat{a}_0, \delta_{b_0} = -\hat{b}_0, \delta_b = b_m - \hat{b}$$

As for the error system given by Equations (11), we consider the Lyapunov function candidate

$$V = 0.5\left(a_{m1}e^2 + \lambda_{a_1}^{-1}\delta_{a_1}^2 + \lambda_{a_0}^{-1}\delta_{a_0}^2 + \lambda_{b_0}^{-1}\delta_{b_0}^2 + \lambda_b^{-1}\delta_b^2\right)$$

where $\lambda_{a1}, \lambda_{a0}, \lambda_{b0}$ and λ_b are optional positive numbers. From Equations (11) and (13), the derivative of the Lyapunov function candidate can then be derived as

$$\dot{V} = -a_{m0}e^2 + \delta_{a_1}\left(\lambda_{a_1}^{-1}\dot{\delta}_{a_1} - e\dot{\omega}\right) + \delta_{a_0}\left(\lambda_{a_0}^{-1}\dot{\delta}_{a_0} - e\omega\right) + \delta_{b_0}\left(\lambda_{b_0}^{-1}\dot{\delta}_{b_0} - e\right) + \delta_b\left(\lambda_b^{-1}\dot{\delta}_b + er\right)$$

We assume that

$$\begin{cases} \dot{\delta}_{a_1} = \lambda_{a_1}\left(e\dot{\omega} + \mu\dot{\omega}\right), \\ \dot{\delta}_{a_0} = \lambda_{a_0}\left(e\omega + \mu\omega\right), \\ \dot{\delta}_{b_0} = \lambda_{b_0}\left(e + \mu\right), \\ \dot{\delta}_b = -\lambda_b\left(er + \mu r\right), \end{cases}$$

where μ is an optional time-varying parameter and satisfies the condition:

$$\text{sgn}\,(\mu) = \text{sgn}\left(a_{m1}\dot{e} + a_{m0}e\right)$$

The derivative of the Lyapunov function candidate can then be rewritten as follows:

$$\dot{V} = -a_{m0}e^2 - \mu\left(a_{m1}\dot{e} + a_{m0}e\right)$$

Thus, we can conclude that V' is a negative definite.
From Equations (8), (12) and (15), we can obtain the following adaptive laws:

$$\begin{cases} \dot{k}_{a_1} = \lambda_{a_1}\left(e + \mu\right)\dot{\omega}, \\ \dot{k}_{a_0} = \lambda_{a_0}\left(e + \mu\right)\omega, \\ \dot{k}_{b_0} = \lambda_{b_0}\left(e + \mu\right), \\ \dot{k}_b = \lambda_b\left(e + \mu\right)r. \end{cases}$$

It is worth noting that, in contrast with most existing adaptive laws, the term μ introduced here can contribute to the improvement of the convergence rates of adjustable parameters, as discussed below.

However, the nonlinear control algorithm cannot be normally executed at the zero rotational speed because the input signal r will be blocked in this case, as shown in Equation (6). Therefore, we should reset $u = u_{\min}$ if $\omega < \omega_{\min}$ and $u < u_{\min}$, where $u_{\min}$ is a predefined minimum control value, $\omega_{\min}$ a predefined minimum rotational speed.

4.2. Comparative Analysis of the Learning Performance of the Proposed Adaptive Laws

In literatures, most existing adaptive laws for the similar system as that shown in Equation (7) do not contain the optional parameter μ, which is usually disadvantageous to the further improvement of the adaptive learning performance. It is important to note that the optional parameter μ in the proposed adaptive laws can significantly speed up the convergence rates of adjustable parameters. For example, as for the identical Lyapunov function candidate shown in Equation (13), we have

$$\dot{V} = \begin{cases} -a_{m0}e^2, & \mu = 0 \\ -a_{m0}e^2 - \lambda \left(a_{m1}\dot{e} + a_{m0}e\right)^{2k+2}, & \mu = \lambda \left(a_{m1}\dot{e} + a_{m0}e\right)^{2k+1}, \ k = 0, \pm 1, \cdots \\ -a_{m0}e^2 - \lambda\mathrm{sgn}\left(a_{m1}\dot{e} + a_{m0}e\right)\left(a_{m1}\dot{e} + a_{m0}e\right), & \mu = \lambda\mathrm{sgn}\left(a_{m1}\dot{e} + a_{m0}e\right) \end{cases}$$

where $\lambda > 0$. From Equation (19), the introduced terms in $\dot{V}$ for the case with μ, such as $\dot{e}$, can greatly contribute to the convergence rates of adjustable parameters. We can therefore conclude that the learning performance of the proposed algorithm with μ is superior to that without μ, such as most existing adaptive laws.

However, due to the unpredictability of the input signal, large amounts of experimental test results suggest that $\dot{k}_{a1}, \dot{k}_{a0}, \dot{k}_{b0}$ and $\dot{k}_{b}$ should be restricted in a bounded range by means of soft limiting strategies in order to avoid the over-learning problem of adjustable parameters and suppress the oscillation of adjustable parameters, as shown in Section 5. Here, k_{a1}, k_{a0}, k_{b0} and k_{b} are restricted as follows:

$$\begin{cases} -\delta_{a_1} \leqslant \dot{k}_{a_1} \leqslant \delta_{a_1}, \\ -\delta_{a_0} \leqslant \dot{k}_{a_0} \leqslant \delta_{a_0}, \\ -\delta_{b_0} \leqslant \dot{k}_{b_0} \leqslant \delta_{b_0}, \\ -\delta_{b} \leqslant \dot{k}_{b} \leqslant \delta_{b}. \end{cases}$$

where δ_{a1}, δ_{a0}, δ_{b0} and δ_{b} are threshold values.

From an application perspective, the optional parameter μ, together with soft limiting strategies, can not only speed up the convergence rates of adjustable parameters, but can also ensure the overall stability of adjustable parameters, which is especially crucial for the flight safety of the MAV. It is finally important to note that, in fact, the determination of the optional parameter μ and the threshold values of soft limiting strategies according to the constraint condition in Equation (16) are dependent on some *a priori* information of the controlled system, such as the range of the input signal.

4.3. The Nonlinear Adaptive Control Diagram

For the controlled system given by Equation (7), the final design goal is generally to achieve a high-quality tracking performance. The schematic block diagram of the nonlinear adaptive control system, shown in Figure 3, is then constructed, where ωc denotes the command signal of the control system, εm the error between the command signal ωc and the model output ωm.

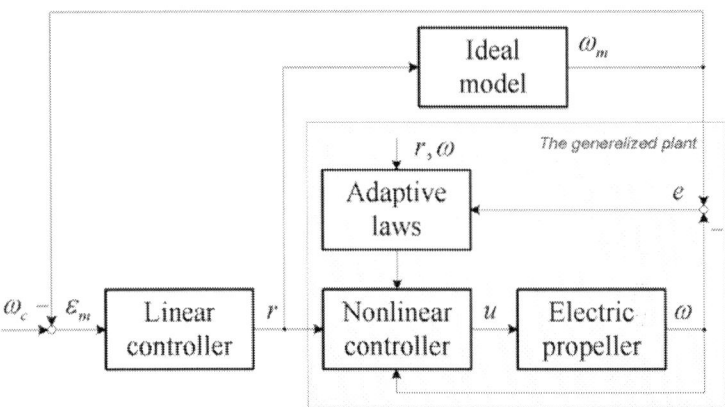

Figure 3. The nonlinear adaptive control diagram.

As mentioned above, the proposed adaptive laws are mainly used to accommodate parametric uncertainties and eliminate the tracking error e as much as possible, which can theoretically ensure that the generalized plant, mainly including the electric propeller and the nonlinear controller, asymptotically has the same characteristics as the ideal model in Equation (9), and based on this, the linear controller with the model feedback is designed to minimize the error εm by means of linear control techniques. As a consequence, the rotational speed ω can precisely track the command signal ωc.

5. EXPERIMENTAL TESTS

In this section, we will first describe the details of the real-time hardware-in-loop (HIL) experimental setup used for validation purposes. Finally, extensive experimental tests will be carried out along with necessary discussions and evaluations.

5.1. The HIL Experimental Setup and Description

The component parts of the rotational speed system, mainly including a propeller, a DC motor with an ESC (*DUALSKY*®XM6355DA and XC9036HV, Shanghai, China), a subminiature encoder with resolution of 20 (JL25, Changchun Institute of Optics, Changchun, Jilin, China), and a digital signal processor (DSP)-based hardware platform with an optional high resolution external timer module, are shown in Figure 4. The software of the nonlinear adaptive control system runs on the hardware platform that can provide reliable support for high precision timer and synchronization operations. The subminiature high resolution incremental encoder is mounted on the end of the main drive shaft to measure the actual rotational speed of the propeller. The command signal generator, including a MAV mathematical model and an autopilot, is normally used to generate the command signal used as input to the rotational speed system according to the flight missions. The schematic diagram of the experimental setup is shown in Figure 4, but to be safe, the MAV is fixed to a post.

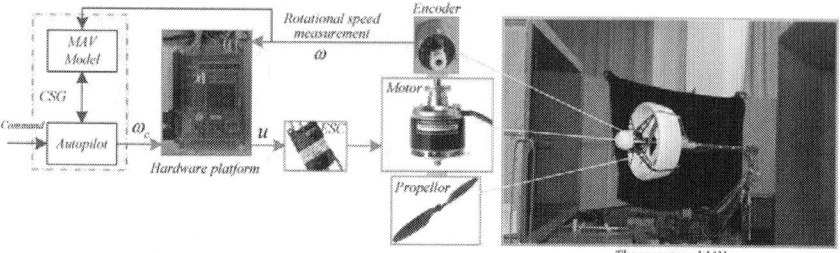

Figure 4. The hardware-in-loop (HIL) experimental setup.

Considering that the propeller only rotates in a fixed direction, namely a clockwise direction or a counter-clockwise direction, the actual rotational speed of the propeller can be determined by scanning the interval time between two neighboring A-phase or B-phase pulses from the incremental encoder, which is implemented based on the high resolution external timer module on the hardware platform.

The predefined parameters are described as follows:
1. Model parameters:

$$a_{m1} = 1.0 \times 10^{-4}, \ a_{m0} = 5.0 \times 10^{-3}, \ b_m = 7.5 \times 10^{-2};$$

2. Initial values of adjustable parameters:

$$k_{a_1}(0) = 4.0 \times 10^{-4}, \ k_{a_0}(0) = 0, \ k_{b_0}(0) = 5.5 \times 10^{-3};$$

3. Adjustable parameter: $k_b(t) \equiv 1.5 \times 10^{-2};$
   Adaptive learning algorithm:

$$\lambda_{a_1} = 1.0 \times 10^{-5}, \ \lambda_{a_0} = 2.0 \times 10^{-7}, \ \lambda_{b_0} = 1.0 \times 10^{-4},$$
$$\mu = 100 \left(a_{m1}\dot{e} + a_{m0}e \right);$$

4. Soft limiting elements: $\dot{k}_{a_1}, \dot{k}_{a_0}, \dot{k}_{b_0}, \dot{k}_b \in (-1.0 \times 10^{-3}, 1.0 \times 10^{-3});$
5. Linear controller: PID controller.

On the one hand, the proper choice of the initial values of adjustable parameters can speed up the convergence rates of the adjustable parameters, while, on the other hand, from Equations (7) and (9), k_b can be artificially fixed to reduce the learning variables and simplify the adaptive learning task without any modification of the adaptive laws, both of which can contribute to the adaptive learning performance. Otherwise, the adaptive learning process of adjustable parameters probably lasts longer. Then, a series of real-time HIL experimental tests are conducted to assess the control performance of the proposed nonlinear adaptive control scheme based on the above assumption.

5.2. Experimental Tests and Discussions

Figure 5 shows the experimental results of the proposed nonlinear control scheme during take-off and hovering flight. It is illustrated from Figure 5a–d that the resulting controlled system can achieve a good tracking performance within about 10 s even without any reliable and accurate *a priori* information about the behavior of the rotational speed system, where ε denotes the tracking error between the command signal ω_c and the actual rotational speed ω. Thereafter, the tracking error typically remains in the range of 0.75 rad/s, which is sufficient to meet the application requirements of a high-performance MAV. Meanwhile, the adjustable parameters of the nonlinear controller also have good convergence properties as shown inFigure 5e, although they finally fluctuate more or less around their mean values. It is, however, to note that the control signal from the nonlinear controller changes rapidly with time, as shown in Figure 5f.

Figure 6 shows the learning process of adjustable parameters with different μ or without soft limiting elements. The increase of μ can obviously contribute to the convergence of adjustable parameters; however, this results in the significant fluctuations of adjustable parameters in the later stage of the learning process mainly because of measurement noises. It is even worse that the adjustable parameters exhibit sharp fluctuations for a longer period of time if removing the soft limiting elements from the nonlinear adaptive control scheme, as shown in Figure 6b, which can easily render the over-learning problem of adjustable parameters.

Figure 7 presents the experimental results of the optimized PID controller, with the nonlinear controller removed from the control scheme, where ω_p is the rotational speed, u_p the controller output, ε_p the tracking error between ω_c and ω_p in this test case. In contrast to what has been shown in Figure 5, the tracking error increases significantly from 0.75 rad/s to 8.0 rad/s, although the control performance is basically acceptable in practice, which also comparatively demonstrates that the proposed nonlinear control scheme can contribute to the significant improvement in the control performance, as shown in Figure 8.

Figure 9 shows the comparisons of some experimental results of two types of control schemes in a wide range of rotational speeds, where the initial values of the adjustable parameters of the nonlinear controller are set to the final values obtained in the first experimental test shown in Figure 5e. The results have once again verified that the dynamic tracking accuracy of the proposed nonlinear adaptive control scheme is much more satisfactory than that of the optimized PID controller, as shown in Figure 9a–c. It is of further note that the achievements largely depend on the fast control capability of the proposed nonlinear adaptive controller, as shown in Figure 9d.

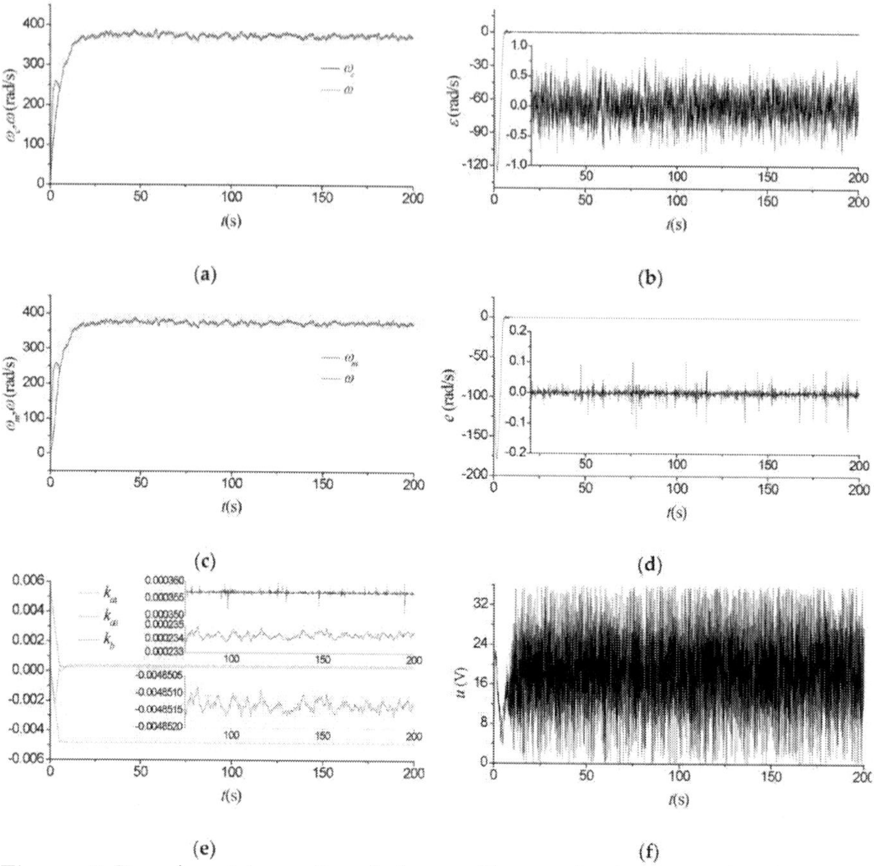

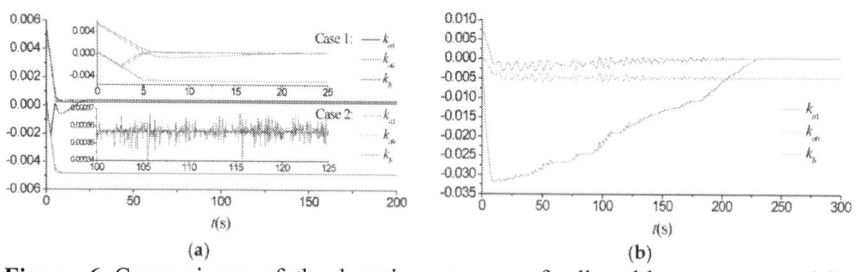

Figure 5. Experimental results of the nonlinear adaptive control scheme. (a) ω_c and ω; (b) ε; (c) ω_m and ω; (d) e; (e) Adjustable parameters; (f) u.

Figure 6. Comparisons of the learning process of adjustable parameters. (a) Adjustable parameters with soft limiting elements: *Case 1:* $\mu=0$, *Case 2:* $\mu=1000(a_{m1}e' + a_{m0}e)$; (b) Adjustable parameters without soft limiting elements.

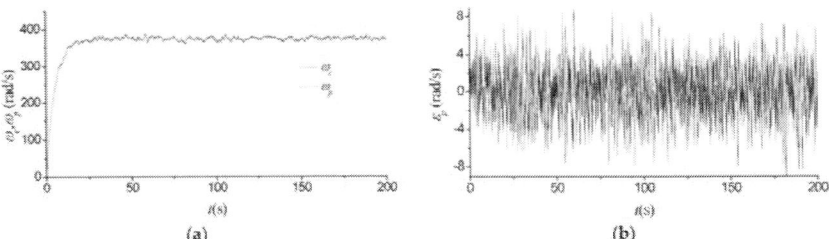

(a)   (b)

Figure 7. Experimental results of the optimized proportional-integral-differential (PID) controller. (**a**) ω_c and ω_p; (**b**) ε_p.

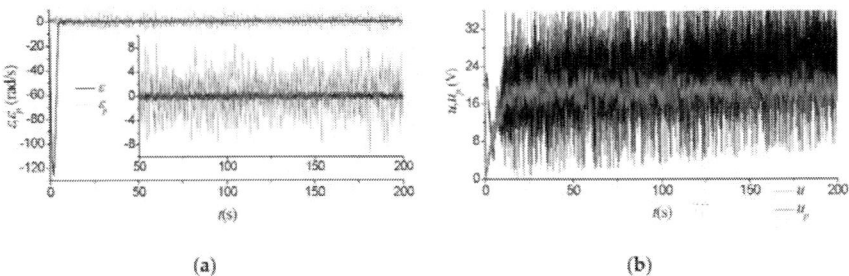

(a)   (b)

Figure 8. Comparisons of some experimental results of two types of control schemes. (**a**) ε and εp; (**b**) u and up.

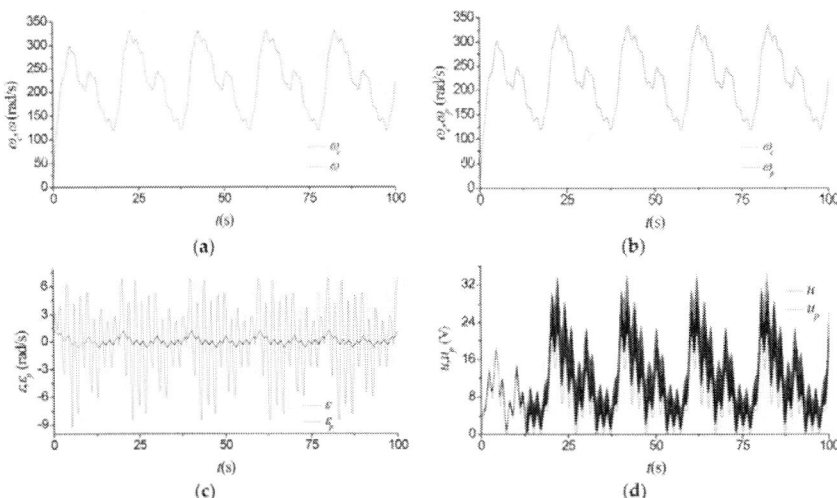

(a)   (b)

(c)   (d)

Figure 9. Comparisons of some experimental results of two types of control schemes in a wide range of rotational speeds: (**a**) ω_c and ω; (**b**) ω_c and ω_p; (**c**) ε and ε_p; (**d**) u and u_p.

5.3. Flight Tests and Discussions

In this subsection, the positive effect of the proposed adaptive rotational speed control strategy on the flight performance of the MAV is comparatively demonstrated by a series of flight tests. The control scheme for the flight tests is shown in Figure 10, where we employ the existing adaptive controllers in [2] or PID controllers in the altitude loops, and the PID controller with or without the closed-loop rotational speed control structure in the altitude loop. For the MAV, the control software runs on the DSP platform, and the sensor units mounted on the top of the fuselage are provided with the measured states of the MAV. In order to ensure the flight test safety, the following steps are conducted: (i) vertical take-off to 20 m above the ground; (ii) holding altitude, as shown in Figure 11. During flight, the serial-based data links provide a link to the ground station computer that allows monitoring the real-time flight information and uploading remote control commands such as the altitude command signals from the manual control unit.

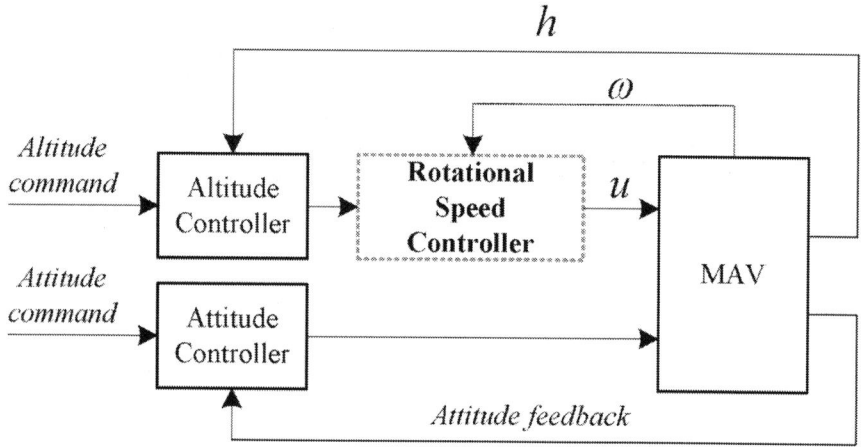

Figure 10. The control scheme.

Figure 11. Flight test.

Some experimental results are comparatively shown in Figure 12 and Figure 13 or Figure 14 and Figure 15, where h is the flight height of the MAV above the ground; θ_c and θ are the pitch command signal and the pitch of the MAV; $e\theta$ is the tracking error between θ_c and θ. The comparison of the experimental results demonstrates that (1) the proposed control scheme can significantly improve the control accuracy of the flight height of the MAV from about 0.8 m to 0.3 m, as comparatively shown in Figure 12b and Figure 13b or Figure 14b and Figure 15b; and (2) that the proposed control scheme can effectively and greatly suppress the oscillation of the rotational speed of the propeller from the approximate range of 100 rad/s to 6 rad/s, as comparatively shown in Figure 12a and Figure 13a or Figure 14a and Figure 15a. The positive effect of the proposed control scheme on the altitude tracking performance is somewhat obvious in different cases, and the altitude tracking error is averagely reduced by more than 50%, as comparatively shown in Figure 12d and Figure 13d orFigure 14d and Figure 15d. We can therefore conclude that the proposed adaptive rotational speed control scheme can contribute to the significant improvement of the flight performance of the MAV.

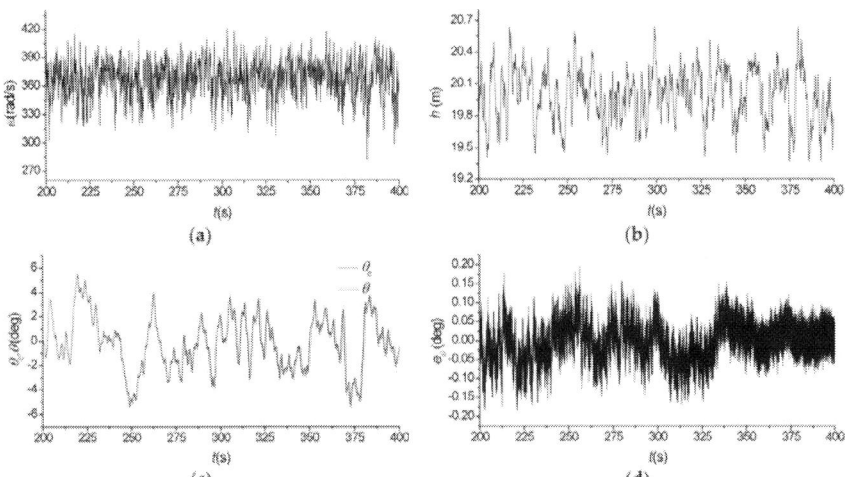

Figure 12. The experimental results using the existing adaptive controllers in the altitude loops, and the PID controller without the closed-loop rotational speed control structure in the altitude loop: (**a**) ω; (**b**) h; (**c**) θ_c and θ; (**d**) eθ.

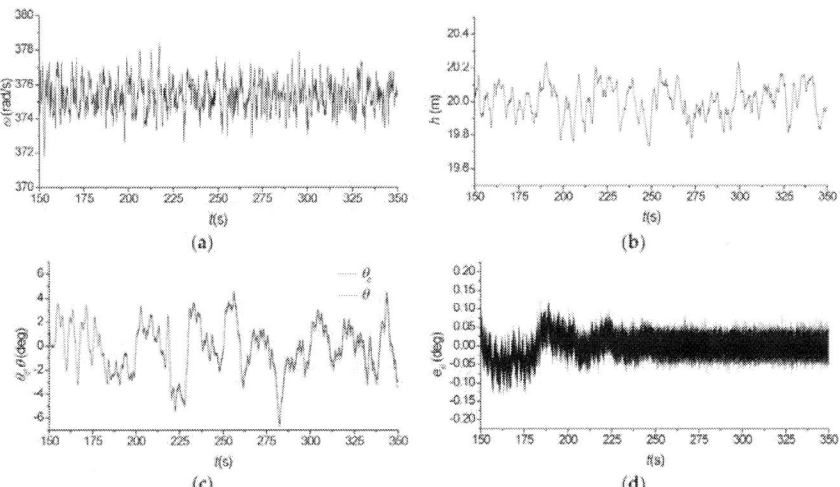

Figure 13. The experimental results using the existing adaptive controllers in the altitude loops, and the PID controller with the closed-loop rotational speed control structure in the altitude loop: (**a**) ω; (**b**) h; (**c**) θ_c and θ; (**d**) eθ.

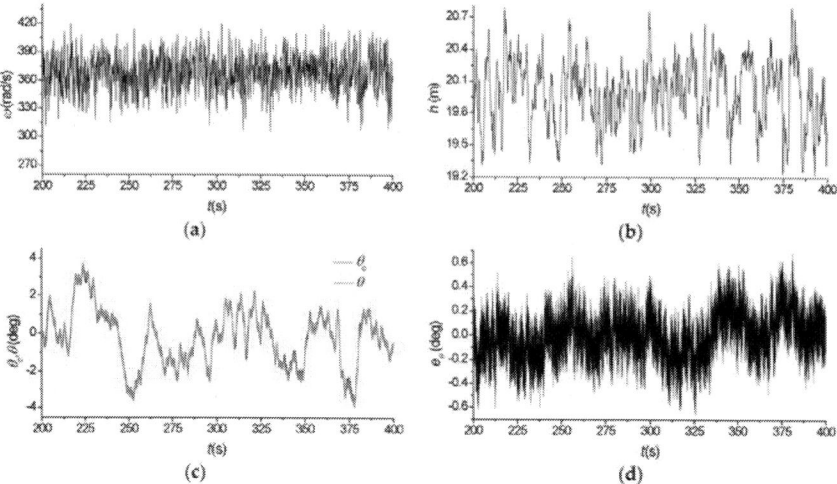

Figure 14. The experimental results using the PID controllers in the altitude loops, and the PID controller without the closed-loop rotational speed control structure in the altitude loop: (**a**) ω; (**b**) h; (**c**) θc and θ; (**d**) eθ.

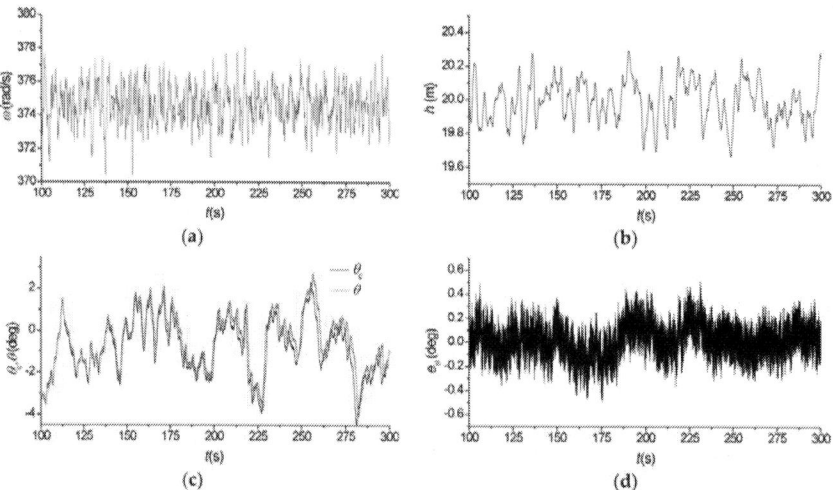

Figure 15. The experimental results using the PID controllers in the altitude loops, and the PID controller with the closed-loop rotational speed control structure in the altitude loop: (**a**) ω; (**b**) h; (**c**) θc and θ; (**d**) eθ.

It is finally noted that from the comparison of the experimental results, the proposed adaptive rotational speed control scheme can directly and significantly improve the control quality of the rotational speed of the propeller, which further results in the improvement of the control qualities of the flight height and altitude. It is therefore demonstrated both theoretically and experimentally that the proposed control scheme is effective and essential to further improving the overall flight performance of such MAVs.

6. CONCLUSIONS

This paper presents the nonlinear adaptive control development for the rotational speed system of the propeller of the prototype electric MAV. The proposed nonlinear adaptive control scheme can contribute to the simplification of the generalized plant, based on which the significant improvement in the dynamic tracking accuracy of the rotational speed can be easily achieved by further using linear control techniques. The experimental test results have demonstrated the performance of the nonlinear adaptive controller. Note that the nonlinear adaptive control scheme can provide quite a satisfactory tracking accuracy from the beginning, given the learned parameters.

The resulting dynamic tracking accuracy of the rotational speed is sufficient to meet the application requirements of a high-performance MAV.

ACKNOWLEDGMENTS

This study was supported in part by National Natural Science Foundation of China (NSFC) (Under Grant No. 61374188), Natural Science Foundation of Jiangsu Province of China (Under Grant No. BK20141412), Applied Basic Research Programs of Natural Science Foundation of Jiangsu Province, China (Under Grant No. BY2015003-10), the Fundamental Research Funds for the Central Universities under Grant No. NP2014601.

AUTHOR CONTRIBUTIONS

All authors discussed the contents of the manuscript. Shouzhao Sheng contributed to the research idea and the framework of this study. Chenwu Sun completed the realization of the test bench and the experimental work.

REFERENCES

1. Sheng, S.; Sun, C. A Near-Hover Adaptive Attitude Control Strategy of a Ducted Fan Micro Aerial Vehicle with Actuator Dynamics. *Appl. Sci.* **2015**, *5*, 666–681.
2. Sheng, S.; Sun, C.; Zhao, H. Indirect Adaptive Attitude Control for a Ducted Fan Vertical Takeoff and Landing Microaerial Vehicle. *Math. Probl. Eng.* **2015**, *2015*, 1–11.
3. Morris, S.; Holden, M. Design of micro air vehicles and flight test validation. In Proceeding of the Coference on Fixed, Flapping and Rotary Wing Vehicles at Very Low Reynolds Numbers, Notre Dame, IN, USA, 5–7 June 2000; pp. 153–176.
4. Jung, D.; Tsiotras, P. Modeling and hardware-in-the-loop simulation for a small unmanned aerial vehicle. In Proceedings of the AIAA Infotech at Aerospace 2007 conference and exhibit, Rohnert Park, CA, USA, 7–10 May 2007.

5. Martin, P.; Boxwell, D. Design, analysis and experiments on a 10-inch ducted rotor VTOL UAV. In Proceedings of the American Helicopter Society (AHS) International Specialists Meeting on Unmanned Rotorcraft: Design, Control and Testing, American Helicopter Society, Phoenix, AZ, USA, 18–20 January 2005; pp. 18–20.

6. Pereira, J.; Chopra, I.; Gessow, A. Effects of shroud design variables on hover performance of a shrouded rotor for micro air vehicle applications. In Proceedings of the AHS International Specialists' Meeting on Unmanned Rotorcraft, American Helicopter Society, San Marcos, CA, USA, 18–20 January 2005.

7. Salluce, D.N. Comprehensive System Identification of Ducted Fan UAVs. Ph.D. Thesis, California Polytechnic State University, San Luis Obispo, CA, USA, January 2004.

8. Ko, A.; Ohanian, O.J.; Gelhausen, P. Ducted fan UAV modeling and simulation in preliminary design. In Proceedings of the AIAA modeling and simulation technologies conference and exhibit, Keystone, CO, USA, 20–23 August 2007.

9. Johnson, E.N.; Turbe, M.A. Modeling, control, and flight testing of a small ducted fan aircraft. *J. Guid. Control Dyn.* **2006**, *29*, 769–779.

10. Metni, N.; Pflimlin, J.M.; Hamel, T.; Soueres, P. Attitude and gyro bias estimation for a VTOL UAV. *Control Eng. Pract.* **2006**, *14*, 1511–1520.

11. Peddle, I.K.; Jones, T.; Treurnicht, J. Practical near hover flight control of a ducted fan (SLADe). *Control Eng. Pract.* **2009**, *17*, 48–58.

12. Sheng, S.; Mian, A.A.; Zhao, C.; Jiang, B. Autonomous take-off and landing control for a prototype unmanned helicopter. *Control Eng. Pract.* **2010**, *18*, 1053–1059. [Google Scholar] [CrossRef]

13. Naldi, R.; Gentili, L.; Marconi, L.; Sala, A. Design and experimental validation of a nonlinear control law for a ducted-fan miniature aerial vehicle. *Control Eng. Pract.* **2010**, *18*, 747–760.

14. Pflimlin, J.M.; Soueres, P.; Hamel, T. Position control of a ducted fan VTOL UAV in crosswind. *Int. J. Control* **2007**, *80*, 666–683.

15. Pflimlin, J.M.; Binetti, P.; Soueres, P.; Hamel, T.; Trouchet, D. Modeling and attitude control analysis of a ducted-fan micro aerial vehicle. *Control Eng. Pract.* **2010**, *18*, 209–218. Gur, O.; Rosen, A. Optimizing Electric Propulsion Systems for Unmanned Aerial Vehicles. *J. Aircr.* **2009**, *46*, 1340–1353.

16. Sinibaldi, G.; Marino, L. Experimental analysis on the noise of propellers for small UAV. *Appl. Acoust.* **2013**, *74*, 79–88.

17. Uragun, B.; Tansel, I.N. The noise reduction techniques for unmanned air vehicles. In Proceedings of the 2014 International Conference on Unmanned Aircraft Systems, Orlando, FL, USA, 27–30 May 2014; pp. 800–807.

18. Ragot, P.; Markovic, M.; Perriard, Y. Optimization of electric motor for a solar airplane application. *IEEE Trans. Ind. Appl.* **2006**, *42*, 1053–1061.

19. .R.; McDonald, R.A. Modeling and Test of the Efficiency of Electronic Speed Controllers for Brushless DC Motors. In Proceedings of the 15th AIAA Aviation Technology, Integration, and Operations Conference, Dallas, TX, USA, 22–26 June 2015; pp. 3191–3200.

20. Beard, R.W. *Quadrotor Dynamics and Control*; Brigham Young University: Provo, UT, USA, 2008.
21. Shirsat, A.R. Modeling and Control of a Quadrotor Aircraft UAV. Ph.D. Thesis, Arizona State University, Tempe, AZ, USA, May 2015.
22. Bouabdallah, S.; Murrieri, P.; Siegwart, R. Design and control of an indoor micro quadrotor. In Proceedings of the 2004 IEEE International Conference on Robotics and Automation, New Orleans, LA, USA, 26 April–1 May 2004; Volume 5, pp. 4393–4398.
23. Hanselman, D.C. *Brushless Permanent Magnet Motor Design*, 2nd ed.; The Writers' Collective: Cranston, RI, USA, 2003; pp. 183–201.
24. Johnson, W. *A Comprehensive Analytical Model of Rotorcraft Aerodynamics and Dynamics. Part 1. Analysis Development*; National Aeronautics and Space Administration Moffett Field CA Ames Research Center: Moffett Field, CA, USA, 1980.
25. PX4 Autopilot. Available online: https://pixhawk.org/choice (accessed on 12 December 2015).

CHAPTER 9

Virtual Sensor for Failure Detection, Identification and Recovery in the Transition Phase of a Morphing Aircraft

Guillermo Heredia 1,* and Aníbal Ollero 1,2

[1] *Robotics, Vision and Control Group, University of Seville, Camino de los Descubrimientos s/n, 41092, Seville, Spain*

[2] *Aerospace Technologies Advanced Center (CATEC), Aeropolis, 41309, Seville, Spain*

ABSTRACT

The Helicopter Adaptive Aircraft (HADA) is a morphing aircraft which is able to take-off as a helicopter and, when in forward flight, unfold the wings that are hidden under the fuselage, and transfer the power from the main rotor to a propeller, thus morphing from a helicopter to an airplane. In this process, the reliable folding and unfolding of the wings is critical, since a failure may determine the ability to perform a mission, and may even be catastrophic. This paper proposes a virtual sensor based Fault Detection, Identification and Recovery (FDIR) system to increase the reliability of the HADA aircraft. The virtual sensor is able to capture the nonlinear interaction between the folding/unfolding wings aerodynamics and the HADA airframe using the navigation sensor measurements. The proposed FDIR system has been validated using a simulation model of the HADA aircraft, which includes real phenomena as sensor noise and sampling characteristics and turbulence and wind perturbations.

KEYWORDS

virtual sensors; UAVs; fault detection; identification and recovery; morphing aircraft

1. INTRODUCTION

Reconfigurable or morphing aircraft are flight vehicles that change their shape to effect both a change in their mission and to perform flight control without the use of conventional control surfaces [1]. The aircraft with morphing capability provides the advantage of being able to fly multiple missions, which is not possible by those based on the conventional aircraft designs.

In the last 60 years there have been many attempts to design a reconfigurable vehicle that could join the VTOL (Vertical Take-Off and Landing) and hovering capabilities of helicopters and the cruise speed and efficiency of airplanes, as the tilt-rotor concept [2], the compound helicopter [3] or the XV1 convertiplane [4].

The "Helicopter ADaptive Aircraft" (HADA) is a project under development since 2007 by a consortium of 26 Spanish companies, universities and research institutions, led by the National Institute for Aerospace Technology (INTA). The objective of the HADA [5,6] project is the development of a reconfigurable Unmanned Aerial Vehicle (UAV) that performs as a helicopter for take-off, landing and hovering flight (Figure 1a), but that "morphs" to a conventional fixed wing configuration (Figure 1b) for cruise flight. The innovative design of the HADA concept relies on that it reconfigures itself in flight by deploying the two half-span wings which are hidden beneath and along the fuselage while in helicopter mode. The morphing process is completed by transferring the engine power to a pusher propeller at the rear end of the fuselage, stopping the rotors and holding backwards the blades of the main rotor. The process is reversed when morphing to helicopter flight mode.

(a)   (b)

Figure 1. HADA conceptual design. (a) Helicopter configuration. (b) Airplane configuration.

The HADA project envisages the development of three prototype UAVs. The first one is the *Colibri* (the Spanish word for hummingbird), a reduced scale model for aerodynamic wind tunnel testing and flight testing. Several instances of the Colibri have been developed since 2007 for different aerodynamic tests in airplane, helicopter and transition modes. The second prototype is the *Alondra* (Spanish word for lark), a 2/3 scale UAV demonstrator with all the relevant mechanisms and functions of the HADA concept, which is in development and will be available in 2010. The third is known as *Libelula* (Spanish word for dragonfly), the full scale HADA prototype which is also in development. The preliminary design for the Libelula UAV system includes a main rotor diameter of 6 m, a wing span of 6 m and a wing area of 4 m². The mass is 380 kg, the required power is 130 kW and the transition speed is approximately 50 m/s.

A typical HADA mission may have the following phases:

1. Take-off in helicopter mode with the wings folded under the fuselage; flight control as a conventional helicopter.
2. When flying forward at the transition speed, gradually unfold the wings while controlling the aircraft with helicopter controls.
3. Once the wings are unfolded, begin power transition between main and tail rotor and the propeller.
4. When full power is transferred to the propeller, fold main rotor and continue flying in aircraft mode with aircraft controls.

Once the targeted area has been reached, the inverse process can be performed and the HADA can operate in helicopter mode to fulfill the mission.

Operational reliability and safety of HADA is extremely important, and a health monitoring and condition-based maintenance system for it is being developed [6]. Being a morphing UAV, new failure modes may appear during the reconfiguration process, as for example sensor and actuator failures in morphing surfaces and failures in power transmission mechanisms. Since one of the main HADA morphing processes is the folding and unfolding of the wings, special efforts have to be devoted to assure reliability of sensors and actuators in this transition phase. This paper concentrates on detection of faults in the wing deployment sensors and actuators.

Reliability has always been a main issue in UAVs [7], where Fault Detection, Identification and Recovery (FDIR) techniques play an important role in the efforts to increase the reliability of the systems. Most FDIR applications to UAVs that appear in the literature use model-based methods, which try to diagnose faults using the redundancy of some mathematical description of the system dynamics. FDIR has been applied to unmanned aircraft, either fixed wing UAVs [8] or helicopter UAVs [9–11]. However, in most cases FDIR has been applied to navigation sensors and actuators, and not to sensors and actuators used in aircraft internal reconfiguration. Furthermore, wing deployment changes significantly the aerodynamics of the aircraft as well as the inertia and mass distribution, being a nonlinear dynamic process.

Virtual Sensors [12] (also known as soft sensors) are software modules which utilize measurable signals (Virtual Sensor inputs) in order to reconstruct a signal of interest (Virtual Sensor output). Virtual Sensors are useful in replacing physical sensors, thus reducing hardware redundancy and acquisition cost, or as part of fault detection methodologies by having their output "compared" to that of a corresponding actual sensor.

Although Virtual Sensors may be developed based upon mathematical models obtained directly from the physics of the system and first principles, in many cases such mathematical models are unavailable, or their exact parameter values are unknown, or they are too complicated to be used. For this reason the development of Virtual Sensors often has to be based upon system identification techniques.

There exist in the literature some works that propose nonlinear Virtual Sensors in aerospace applications [12–14], although they have primarily been applied to navigation sensors and actuators. In this paper, a nonlinear virtual sensor is proposed to estimate the sweeping angle of the wing. The virtual sensor is obtained using Nonlinear AutoRegressive with eXogenous excitation (NARX) system identification.

The paper is organized as follows. Section 2 presents an overview of the model of the HADA aircraft. Section 3 describes the structure of the proposed virtual sensor based FDIR system. Section 4 presents some simulation experiments to illustrate the behavior of the virtual sensor in case of sensor and actuator failures. Finally, the conclusions in Section 5 complete the paper.

2. OVERVIEW OF THE HADA MODEL

At the present development stage of the HADA project it is not possible to make flight tests which include wing folding or unfolding, and therefore the proposed virtual sensor-based FDI system cannot be tested in flight. Thus, a HADA model and simulator have been developed for that purpose, and also for controller development. Furthermore, there are available several wind tunnel tests with variable sweep wings with one of the prototypes, and they have been used to validate HADA aerodynamics modeling.

The HADA aircraft can be modeled as a common airplane or helicopter when it is operating either in airplane or helicopter modes, and the motion equations can be found in many textbooks (see, for example [15,16]). Aerodynamic forces and moments are incorporated in the equations of motion in the form of *aerodynamic stability and control derivatives*. Stability and control derivatives are measures of how particular forces and moments on an aircraft change as other flight variables change, and are based on linear Taylor series expansions to define the aerodynamics of the aircraft [15]. The linear aerodynamic model is based on small perturbation control theory around a single flight condition, which for this work is the transition of HADA between airplane and helicopter modes.

On the other hand, it is not so easy to model HADA aerodynamics in the transition phases when the wings are folding or unfolding. Mode transition in the HADA aircraft will take place at velocities around 40–50 m/s. The velocity of the wing tip in the folding/unfolding movement of the wing is very low compared to the airflow speed around the wing. Thus, it is reasonable to assume that it is a quasi-steady process, and the aerodynamic stability and control derivatives approach can be used.

The Vortex Lattice Method (VLM) is a numerical method built on the theory of potential flow, which has been widely used for aircraft design [17–18]. The method is based on the idea of a vortex singularity as the solution of Laplace's equation. The VLM models the lifting surfaces of an aircraft as an infinitely thin sheet of discrete vortices to compute lift and induced drag. The Tornado program [19] has been used as the VLM code in this paper. Tornado is able to calculate the lift and drag forces and also the stability and control derivatives.

Modelling of the HADA for different wing sweep angles has been done in the following way: for each value of the sweep angle the effective airfoil (the wing section that the main airflow "sees") has been obtained as the result of an oblique section of the wing. Then, the VLM has been used to obtain the stability and control derivatives corresponding to this wing sweep angle. This set of coefficients has been integrated in the HADA simulation model, which has been developed using Matlab/Simulink. A detailed description of the model can be found in [20].

The actuation mechanism for wing folding and unfolding is an electric linear actuator linked to the two halfwings which will rotate around pivot points. The wing actuator has been modelled as a first order system with limited angular deflection and maximum angular rate.

The HADA navigation system provides to other subsystems the position, velocity and attitude information they need to control the aircraft, manage the mission or inform the remote pilot. The navigation sensor pack that will be installed on the HADA prototypes will typically include an Inertial Measurement Unit (IMU), which provides data from three accelerometers, three gyroscopes and a 3-axis magnetometer, a Global Positioning System (GPS) receiver which gives absolute position and also absolute velocity estimations, absolute and differential pressure sensors, which are used to measure aircraft altitude and total velocity with respect to the air mass, and a nose probe providing measurements for the angle of attack and the sideslip angle. The measurements of these navigation sensors are integrated by the navigation system to obtain the aircraft state using a Kalman filter [21].

Although all these sensor measurements are combined to obtain an estimation of the HADA position and orientation angles, the FDIR virtual sensor uses the sensor readings directly to exploit all the available information.

The HADA aircraft will include also a Hall effect rotary position sensor to measure directly the wing sweeping angle, although it may not be present in one of the prototypes. For this reason, the virtual sensor FDIR system has been developed considering both possibilities, that the wing sweeping angle sensor is available or not. All these sensors have been modelled in the simulator, including noise characteristics.

3. VIRTUAL SENSOR FOR FDIR

3.1. FDIR Structure

The monitoring of faults in feedback control system components is known as Fault Detection, Identification and Recovery (FDIR), which is composed of three main functions: the process of determining that a fault has occurred (detection), the localization of the fault within the system (identification) and the process of limiting the fault propagation and enabling the service to be restored to an acceptable state (recovery). The procedure of generating a control action which has a low dependency on the presence of certain faults is known as fault tolerant control.

Figure 2 shows the general schematic arrangement appropriate to many fault tolerant control systems [22] with four main components: the plant itself (including sensors and actuators), the Fault Detection and Identification (FDI) unit, the feedback controller and the supervision system. The plant is considered to have potential faults in sensors, actuators or other components. The FDI unit provides the supervision system with information about the onset, location and severity of any fault. Based on system inputs and outputs together with fault decision information from the FDI unit, the supervision system will reconfigure the sensor set and/or actuators to isolate the faults, and tune or adapt the controller to accommodate the fault effects. The FDI and Supervisor blocks in Figure 2 jointly perform the FDIR functions.

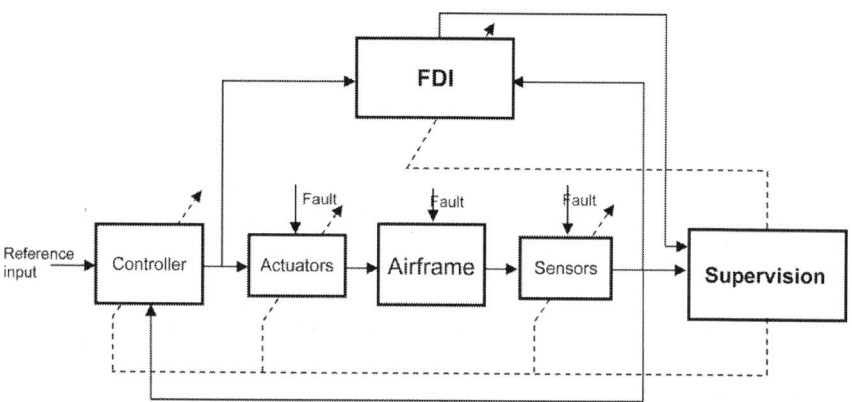

Figure 2. FDIR general structure.

The scheme proposed in this paper for FDIR on sensors and actuators of the wing deployment process is presented in Figure 3, where **u** and **y** are the inputs and outputs vectors of the HADA UAV, respectively. The virtual sensor block implements the wing sweeping angle virtual sensor. The output of this block is an estimation of the sweeping angle, γ_{est}. The FDI block contains the logic for fault detection and identification, and generates a fault signal **F**, which is used by the supervision unit to decide if a reconfiguration of HADA sensors, actuators and controller is necessary to recover from a declared fault.

The virtual sensor takes as inputs the HADA inputs **u** and outputs **y**, and reconstructs the sweeping angle γ from these signals. The wing sweeping angle is an internal variable which does not participate directly in the HADA dynamics. The relationship is indirect: as the wings are unfolding, the lift and drag forces generated by the wing interact with the airframe, and this interaction is reflected in the inputs and outputs of the HADA. This relationship will be nonlinear, and therefore it is not possible to use standard linear input-output models. Thus, a nonlinear input-output model has been used, which is described in the next section.

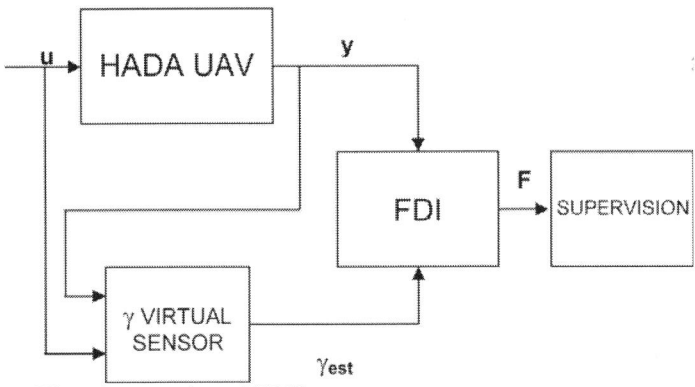

Figure 3. Virtual sensor based FDIR.

3.2. Nonlinear Input-Output Model for Wing Sweeping Angle Virtual Sensor

In order to account for the nonlinear dynamical relationships between the signal under reconstruction (virtual sensor output, the wing sweeping angle) and the measurable signals used for this purpose (virtual sensor inputs, the inputs and outputs of the HADA aircraft), a stochastic Multiple-Input Single-Output (MISO), Nonlinear AutoRegressive with eXogenous excitation (NARX) [23] model structure of polynomial form is adopted for virtual sensor implementation.

Consider a general discrete time nonlinear system with one output y and m inputs $u_1, \ldots, u_m$. The well-known linear AutoRegressive with eXogenous inputs (ARX) [23] model supposes that the current output $y(t)$ is predicted as a weighted sum of past output values and current and past input values (known as regressors). The NARX structure is an extension of the linear ARX structure: instead of the weighted sum that represents a linear mapping, the NARX model has a much more flexible nonlinear mapping function:

$$y(t) = f(y(t-1), \ldots, y(t-n_y), u_1(t-1), \ldots, u_1(t-n_{u1}), \ldots, u_m(t-1), \ldots, u_m(t-n_{um}))$$

where n_y and n_{ui}, $i = 1,2,\ldots,m$, denote the maximum lags in the output and inputs, and $f(\cdot)$ is a suitable nonlinear function, a priori unknown. If $f(\cdot)$ is approximated as a polynomial of degree M, model (1) can be written as:

$$y(t) = \sum_{i=0}^{r} \beta_i x_i(t)$$

where r depends on m, M, n_y, and n_{ui}, $i=1,2,\ldots,m$. The parameters β_i, $i = 1,2,\ldots,r$, are suitable coefficients to be identified, $x_0 = 1$ and $x_i(t)$, $i = 1,2,\ldots,r$, are monomials made up of delayed outputs and/or inputs. For instance, if n_y,

$= m = n_u = 1$ and $M = 2$, the most general model (2) (with positive powers) would be:

$$y(t) = \beta_0 + \beta_1 y(t-1) + \beta_2 y(t-1)^2 + \beta_3 u(t-1) + \beta_4 u(t-1)^2 + \beta_5 y(t-1)u(t-1)$$

In the following, $\beta = [\beta_0 \; \beta_1 \dots \beta_r]$ is the vector of unknown parameters, and:

$$\varepsilon(t) = y(t) - \sum_{i=0}^{r} \beta_i x_i(t)$$

denotes the residual sequence. Assuming that N observations of the input and output variables are available, the least-squares (LS) estimate β^{LS} of β is:

$$\beta^{LS} = \arg\min_{\beta} \sum_{t=1}^{N} \varepsilon(t)^2$$

If the nonlinear model structure (2) is assigned, that is the number and the form of the regressors x_i in (2) are specified, the estimation problem is easily solved by means of standard algorithms. Since this NARX model formulation is linear in the parameters, standard LS algorithms can be used to estimate the parameters β^{LS}. The NARX models are chosen with the structure that achieve the smallest Akaike's Information Theoretic Criterion (AIC) [24], according to a simple search algorithm, in which the first half of data is used for estimation and the second for cross validation.

3.3. FDIR Logic

The FDI block has three signals available for detection and identification of faults in sensors and actuators of the wing folding/unfolding mechanism. The first one is the actual sweeping angle sensor signal (GSENS) if it is available, the second one is the estimation of the sweeping angle reconstructed by the virtual sensor using measurements of other sensors (GVIRT), and the third one is generated from the actuator command, modeling the actuator as a first order system (GACT).

For fault detection, if the sweeping angle sensor is available it is sufficient to compare the signals GSENS and GACT and compute a residual between them. In fact, this is possible also without implementing the virtual sensor. However, if only these two signals are available, it is impossible to isolate the fault to determine if it is located in the actuator or in the sensor. The residuals generated from the comparison of the virtual sensor output GVIRT with GSENS and GACT give independent additional information on the presence of a fault. In case the sweeping angle sensor is not available, only the residual of GVIRT and GACT can be used.

The residuals R_i in this case are defined as the difference of the signals. Ideally, if no fault is present, the residual would be zero. In practice, the residual will take non-zero values due to estimation errors, sensor noise, perturbations, *etc.* Usually, the residual for a specific sensor will be bounded, and therefore a "threshold level" can be defined so that the absolute value of the residual is always below it in the absence of failures. The first time the residual goes above the threshold level, the fault is assumed to be present.

The fault detection procedure is designed to decide if the observed changes in the residual signal R_i can be justified in terms of the disturbance (measurement noise) and/or modelling uncertainty as opposed to failures. It is critical to minimize the detection delay associated with a 'true' fault; furthermore, the false alarm rate should be minimized while, at the same time, no 'true' faults should remain undetected.

A well-known filter for the detection of moderate persistent shift in the mean value of the residual is the cumulative sum (CUSUM) filter [25]. This filter is used to detect both positive and negative changes in the mean value of the residual R_i caused by the occurrence of a fault. The CUSUM filter with forgetting factor will be used to detect changes in the mean of the residuals of the wing sweeping angle.

Once a fault has been detected, it is very important to isolate the location of the fault. If there is no sweeping angle sensor installed it is obvious that the fault will be located at the wing actuator. But in case there is an angle sensor, it is critical to know if the fault is located in the sensor or in the actuator, because the fault location will determine the actions taken by the supervisor. This is impossible to do if the virtual sensor is not implemented. But if the three signals explained above are available, a simple fault identification logic can be defined. In brief, if GSENS is similar to GVIRT, and GACT differs from the other two signals, it means that the angle sensor is working properly and the fault is in the actuator, which is receiving a command that it is not able to execute. On the other hand, if GVIRT is different from GSENS, and it is similar to GACT, it means that the actuator is working properly, and it is the angle sensor that is faulty.

The fault identification that is provided by the virtual sensor is very important for the supervision unit. If the fault is in the sensor, the supervisor may decide to continue with the mission (maybe with degraded performance), substituting the faulty angle sensor signal with the angle virtual sensor signal. However, if it is the actuator that is faulty, it will probably decide to abort the mission, return to base and perform an emergency landing (depending on the wing sweeping angle that the actuator has stuck).

4. SIMULATION EXPERIMENTS

Many simulation experiments have been performed with the HADA simulator presented in Section 2 to test the virtual sensor based FDIR system. The simulations have been done trying to reconstruct real flight conditions, and therefore sensor noise and perturbations have been taken into account. Sensor characteristics have been considered modeling their sampling frequency and additive Gaussian white noise. The main sources of perturbations in aircraft flight are air turbulence and wind gusts. Moreover, it is important to note that the mechanism which lets the virtual sensor act works in the following way: as the wings are unfolding, lift and drag forces are produced in the wing, and these forces and moments are transferred to the aircraft. As a consequence, flight variables (velocities, orientation angles, *etc.*) change, and the control system acts changing the actuator commands to maintain level flight. The virtual sensor described in this paper models the relationship between the flight variables and actuators (which are available to the FDIR system through the sensors) and the wing sweeping angle. It is clear that wind turbulence will modify these relationships, and therefore it will introduce errors in the virtual sensor estimate.

In this section, several simulation experiments are presented. In them, the HADA aircraft is flying in level flight at a constant velocity, in helicopter mode with the wings folded under the fuselage (sweeping angle of 0 degrees). At a given time instant, wing unfolding is initiated at constant angular velocity, and it lasts 10 seconds. Sensor noise is considered as explained above and light to moderate atmospheric turbulence is also considered using the Dryden turbulence model [26,27].

The residuals are obtained as the difference between the signals, and a CUSUM filter with forgetting factor is used to analyze them. Figure 4 shows the residual of the comparison between the virtual sensor angle estimate (GVIRT) and the wing angle derived from the actuator command (GACT), in fault-free conditions. The dashed vertical lines mark the initial and final times of the unfolding process. It can be seen that the turbulence and sensor noise cause variations in the residual, and that these variations are larger when the wings are unfolding and when they are unfolded than when the wings are under the fuselage. The red dashed horizontal lines denote the fault detection threshold level.

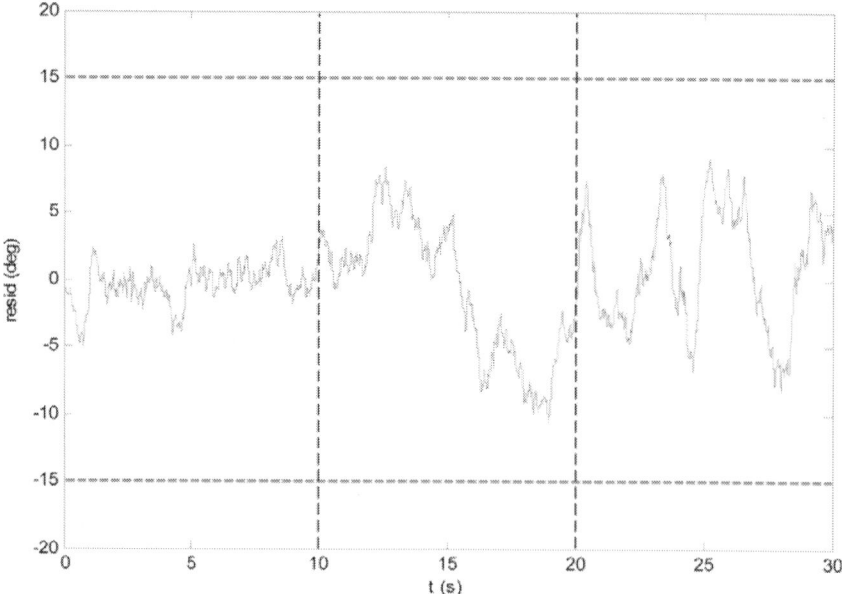

Figure 4. CUSUM filtered residual of GVIRT and GACT.

4.1. Sensor and Actuator Failure with Angle Sensor

This subsection presents an experiment with a failure in the wing angle sensor. The experiment is similar to the one described above: the HADA is in level flight in helicopter mode, and the wings are unfolded between $t = 10$ s and $t = 20$ s. The aircraft has an angle sensor connected to the wing unfolding mechanism. The sensor has had a hard failure, and it gives the same output regardless of the real wing sweeping angle. In this case it is straightforward to detect that there is a failure just comparing the angle sensor (GSENS) with the GACT signal (it is not shown here). From this comparison it will be detected that there has been a fault, but there is no information about where it is located.

Figure 5 shows the results of the comparison between GSENS and the virtual sensor GVIRT signal. It can be seen that as the wings are unfolding, the residual begins to grow and clearly goes above the threshold level at $t = t_d$, thus detecting that the fault is located at the angle sensor.

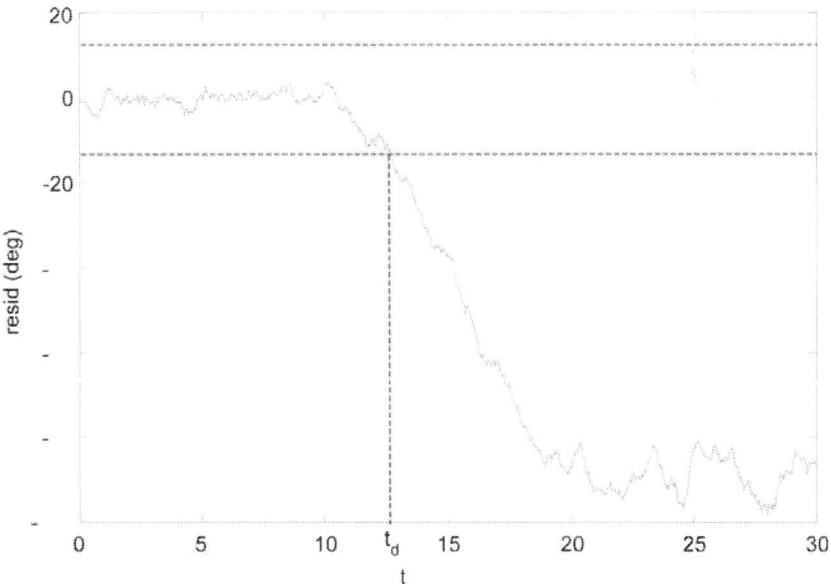

Figure 5. Angle sensor failure: CUSUM filtered residual of GVIRT and GSENS.

Moreover, it is possible to reconfigure the control system to use the angle virtual sensor once the fault has been detected and isolated. Figure 6 shows the "best" estimation of the wing sweeping angle: it is the angle sensor output when no fault is declared. At t_0 a failure is detected (comparing GACT and GSENS), but the fault location is still unknown; at t_d the fault is isolated and identified, and the virtual sensor reading is taken as the best estimate of the angle. Clearly the estimation of the virtual sensor is much noisier and less accurate, but it can give important information to the control system as to what action perform next.

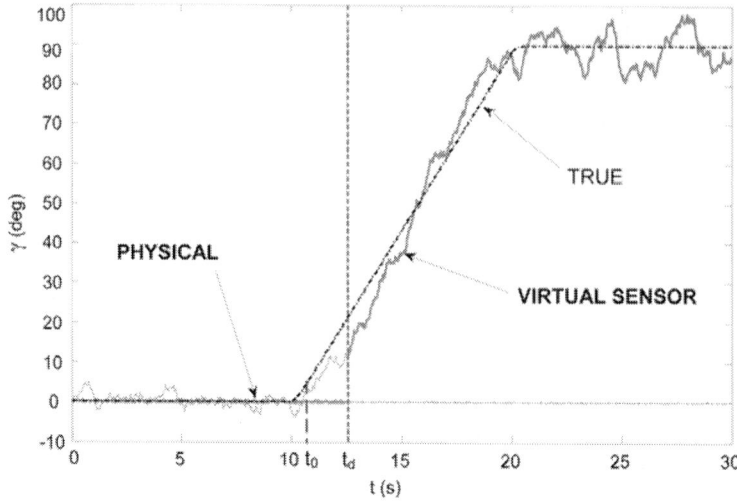

Figure 6. Wing sweeping angle estimation in case of sensor failure.

In the same way, if the residual of GVIRT and GSENS do not grow larger that the threshold value it can be considered that the fault is located at the actuator and not the angle sensor.

4.2. Actuator Failure with No Angle Sensor Installed

If there is no wing angle sensor installed, the virtual sensor can still be used to detect faults in the wing actuator. Figure 7shows the residual generated by the comparison of the GACT and GVIRT signals when there is a fault in the wing actuator.

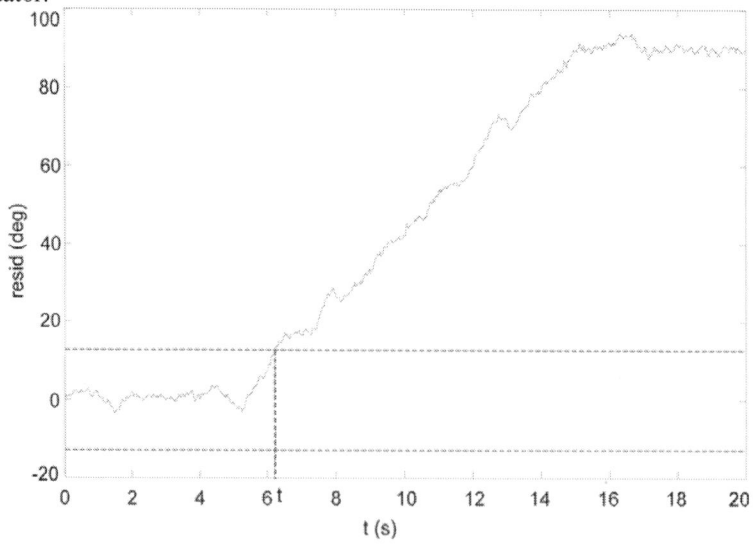

Figure 7. Actuator fault detection.

The actuator is instructed to unfold the wings at constant angular velocity between $t = 5$ s. and $t = 15$ s. It can be seen that the residual error begins to grow after the command, and goes above the threshold level, since the virtual sensor GVIRT detects that the actuator has not moved the wings but the GACT signal (reconstructed from the actuator command) "thinks" it has done so. In this case, it is also possible to use the virtual sensor as an anytime estimator of the wing sweeping angle.

5. CONCLUSIONS

Flight control in the transition phase of the HADA morphing aircraft is very important for mission success. When HADA is flying in helicopter mode and the wings are folding or unfolding, it is critical to know the real wing sweeping angle. This paper has shown how a virtual sensor can be used for FDIR of wing angle sensor and actuator failures. A nonlinear system identification technique has been used, since wing folding and unfolding is a complex nonlinear aerodynamic problem. The proposed method has been validated with experiments using a HADA simulator which includes real phenomena as sensor noise and sampling characteristics and turbulence and wind perturbations. The simulation experiments include detection of faults in the angle sensor and the angle actuator. In case of an angle sensor failure or if it is not available, the virtual sensor output can be used as a rough estimate of the wing sweeping angle, which gives valuable information to the HADA controller.

ACKNOWLEDGMENTS

This work has been partially supported by the HADA-PLATINO program (Spanish Ministry of Innovation and Science, MICIN, PSE-370000-2009-64) and the ROBAIR (DPI2008-03847) Spanish National Research project.

REFERENCES AND NOTES

1. Seigler, T.M.; Neal, D.A.; Bae, J.S.; Inman, D.J. Modeling and Flight Control of Large-Scale Morphing Aircraft. *J. Aircr* **2007**, *44*, 1077–1087.
2. Yeo, H.; Johnson, W. Performance and Design Investigation of Heavy Lift Tilt-Rotor with Aerodynamic Interference Effects. *J. Aircr* **2009**, *46*, 1231–1239.
3. Yeo, H.; Johnson, W. Optimum Design of a Compound Helicopter. *J. Aircr* **2009**, *46*, 1210–1221.
4. Markman, S.; Holder, B. *Straight Up—A History of Vertical Flight*; Schiffer Publishing: Atglen, PA, USA, 2000.
5. Mulero, M. PLATINO-HADA Programme: a Step Forward. Proceedings of Innovation in Unmanned Air Systems Conference (UAS07), Madrid, Spain, November, 2007.
6. Perez, J.; Gorostiza, A.; Mulero, M. Helicopter Adaptive Aircraft (HADA)—Condition Based Maintenance. Proceedings of RTO Applied

Vehicle Technology Panel (AVT) Symposium, Evora, Portugal, April, 2009.

7. Ollero, A.; Merino, L. Control and Perception Techniques for Aerial Robotics. *Annu. Rev. Control* **2004**, *28*, 167–178.

8. Napolitano, M.; Windon, D.; Casanova, J.; Innocenti, M.; Silvestri, G. Kalman Filters and Neural-Network Schemes for Sensor Validation in Flight Control Systems. *IEEE Trans. Control Syst. Technol* **1998**, *6*, 596–611.

9. Drozeski, G.; Saha, B.; Vachtsevanos, G. A Fault Detection and Reconfigurable Control Architecture for Unmanned Aerial Vehicles. Proceedings of the IEEE Aerospace Conference, Big-Sky, MT, USA, March, 2005.

10. Heredia, G.; Ollero, A.; Bejar, M.; Mahtani, R. Sensor and Actuator Fault Detection in Small Autonomous Helicopters.*Mechatronics* **2008**, *18*, 90–99.

11. Heredia, G.; Ollero, A. Sensor Fault Detection in Small Autonomous Helicopters using Observer/Kalman Filter Identification. Proceedings of the 5th IEEE International Conference on Mechatronics, Málaga, Spain, April, 2009.

12. Samara, P.; Fouskitakis, G.; Sakellariou, J.; Fassois, S. Aircraft angle-of-attack virtual sensor design *via* a functional pooling NARX methodology. Proceedings of the European Control Conference, University of Cambridge, UK, September, 2003; p. 309.

13. Oosterom, M.; Babuska, R. Virtual Sensor for Fault Detection and Isolation in Flight Control Systems—Fuzzy Modeling Approach. Proceedings of the 39th IEEE Conference on Decision and Control, Sydney, Australia, December, 2000; pp. 2645–2650.

14. Tomczyk, A. Simple Virtual Attitude Sensors for General Aviation Aircraft. *Aircr. Eng. Aerosp. Technol* **2006**, *78*, 310–314.

15. Cook, M. *Flight Dynamics Principles*, 2nd ed; Elsevier: Burlington, NC, USA, 2007.

16. Padfield, G. *Helicopter Flight Dynamics: the Theory and Application of Flying Qualities and Simulation Modeling*, 2nd ed; AIAA Educational series; Blackwell Science: Malden, MA, USA, 2007.

17. Katz, J.; Plotkin, A. *Low-Speed Aerodynamics*; Cambridge University Press: Cambridge, UK, 2001.

18. Cardenas, E.M.; Boschetti, P.J.; Amerio, A. Stability and Flying Qualities of an Unmanned Airplane Using the Vortex-Lattice Method. *J. Aircr* **2009**, *46*, 1461–1464.

19. Melin, T. *User's Guide and Reference Manual for Tornado*; Department of Aeronautics, Royal Institute of Technology (KTH): Stockholm, Sweden, 2000.

20. Duran, A.; Heredia, G.; Ollero, A. Aerodynamic Modelling of the HADA Reconfigurable Aircraft. *J. Aircr*. submitted for publication..

21. Esposito, F.; Accardo, D.; Moccia, A.; Ciniglio, U.; Corraro, F.; Garbarino, L. Real-Time Simulation and Data Fusion of Navigation Sensors for Autonomous Aerial Vehicles. In *Advances and Innovations in Systems, Computing Sciences and Software Engineering*; Elleithy, K., Ed.; Springer: Dordrecht, The Netherlands, 2007; pp. 127–136.

22. Patton, R.J.; Chen, J. Observer-Based Fault Detection and Isolation: Robustness and Applications. *Contr. Eng. Pract* **1997**, *5*, 671–682.
23. Ljung, L. *System Identification—Theory for the User*, 2nd ed; Prentice-Hall: Upper Saddle River, NJ, USA, 1999.
24. Soderstrom, T.; Stoica, P. *System Identification*; Prentice-Hall: Upper Saddle River, NJ, USA, 1989.
25. Gustafsson, F. *Adaptive Filtering and Change Detection*; John Wiley & Sons: West Sussex, UK, 2000.
26. Hoblit, F.M. *Gust Loads on Aircraft: Concepts and Applications*; AIAA Education Series; AIAA: Washington, DC, USA, 1988.
27. Dogan, A.; Lewis, T.A.; Blake, W. Flight Data Analysis and Simulation of Wind Effects during Aerial Refueling. *J. Aircr* **2008**, *45*, 2036–2048.

CHAPTER 10

Panel/Full-Span Free-Wake Coupled Method for Unsteady Aerodynamics of Helicopter Rotor Blade

Jianfeng Tan , Haowen Wang,

School of Aerospace, Tsinghua University, Beijing 100084, China

ABSTRACT

A full-span free-wake method is coupled with an unsteady panel method to accurately predict the unsteady aerodynamics of helicopter rotor blades in hover and forward flight. The unsteady potential-based panel method is used to consider aerodynamics of finite thickness multi-bladed rotors, and the full-span free-wake method is applied to simulating dynamics of rotor wake. These methods are tightly coupled through trailing-edge Kutta condition and by converting doublet-wake panels to full-span vortex filaments. A velocity-field integration technique is also adopted to overcome singularity problem during the interaction between the rotor wake and blades. Helicopter rotors including Caradonna–Tung, UH-60A, and AH-1G rotors, are simulated in hover and forward flight to validate the accuracy of this approach. The predicted aerodynamic loads of rotor blades agree well with available measured data and computational fluid dynamics (CFD) results, and the unsteady dynamics of rotor wake is also well simulated. Compared to CFD, the present method obtains accurate results more efficiently and is suitable to rotorcraft aeroelastic analysis.

KEYWORDS

Free wake method; Helicopter rotor; Rotor wake; Unsteady aerodynamics; Unsteady panel method

1. INTRODUCTION

The flow field around a rotorcraft is extremely complex because it is dominated by vortices shedding from rotor blades. In hover and low-speed forward flight, the rotor wake remains in the vicinity of the operating rotor blades. The close interaction of the rotor wake with rotor blades has a strong effect on rotor blade airloads and rotor wake dynamics. Due to the complexity of rotor wake, determining pressure and load distribution is of critical importance to accurately predict rotorcraft performance, stability, loads, and aeroacoustic.

Base on Navier–Stokes equations for primitive variables, computational fluid dynamics (CFD) has been used to simulate the rotor wake [1,2,3 and 4] and has great potential to provide insights into complex aerodynamic processes. This approach can adequately predict unsteady aerodynamic loads of blades, and has been used in comprehensive rotorcraft analysis in recent years. [5 and 6] However, CFD methods are susceptible to excessive numerical dissipation of vorticity and are rather computationally expensive.

The free-wake method, which represents the rotor wake with vortex filaments, is a powerful approach to simulating complex rotor wake. [7,8,9 and 10] In general, aerodynamics of rotor blade in the free-wake method is represented by lifting-line or lifting-surface method; however, these methods cannot consider the effect of airfoils which has significant influence on aerodynamics.

For their simple and accurate computations, unsteady panel methods [11,12,13 and 14] have been widely used as an alternative to CFD, lifting-line, and lifting-surface methods to predict aerodynamic loads in rotorcraft applications. However, simulating distorted wake behavior for a panel is difficult because a singularity problem arises when wake panels are close to one another. Therefore, the panel method is usually coupled with other wake dynamic methods. A panel method was coupled with a free-wake method by replacing doublet-wake panels with vortex-ring wake filaments, [15] however, variation of doublet strength distribution was prescribed over blade camber line, and it was difficult to account for the effect of airfoils. The doublet panels were then distributed on blade surface in unsteady time-marching free-wake method to predict blade vortex interaction (BVI) noise. [16] Moreover, based on a novel boundary integral formulation, a direct panel method was coupled with a free-wake method to analyze the BVI. [17]

Although the free-wake method coupled with a panel method can predict blade airloads and rotor wake dynamics, an accurate and efficient prediction has yet to be fully achieved. Moreover, there is a numerical singularity problem when the rotor wake approaches blades in forward flight. Therefore, a full-span free-wake method is coupled with an unsteady panel method to accurately predict the unsteady aerodynamics of rotor blades in this study. The method is achieved through trailing-edge Kutta condition and by converting doublet-wake panels to full-span vortex filaments. The doublet panels are distributed on the blade and shed wake surface. The shed wake is connected to the blade panels through trialing-edge Kutta condition. The rotor wake is described by full-span vortex filaments shedding from the shed wake. A second-order time-stepping scheme is applied to predict wake convection. A viscous vortex core model is adopted to avoid numerical singularity problem in velocity calculation, and a velocity-field integration technique is also used to reduce numerical instability of wake and panels when the rotor wake approaches rotor blades in forward flight.

Numerical results are compared with experimental data and CFD results for helicopter rotors, such as Caradonna–Tung, UH-60A, and AH-1G rotors, in hover and forward flight to validate the accuracy of the present method.

2. COMPUTATIONAL METHOD

2.1. Unsteady panel method

The fluid surrounding rotorcraft is assumed to be inviscid, irrotational, and incompressible.12 and 14 Therefore, a velocity potential ϕ can be defined and the continuity equation in terms of ϕ in the inertial frame of reference becomes Equation (1)

$$\nabla^2 \phi = 0$$

Eq. (1) does not include time derivatives, and the time dependency is introduced through boundary conditions. The boundary condition for body surface, which requires the velocity component normal to the body surface S_B be zero, can be expressed as Equation (2)

$$\frac{\partial \phi}{\partial n} - v_B \cdot n = 0$$

where v_B is the velocity of a point on body surface S_B and n denotes the outward unit normal vector at this point. In general, v_B includes all the contributions of blade's rigid motion, blade's structural deformation, and fuselage's motion in maneuvering flight for rotorcraft. An infinity boundary condition requires that the flow disturbance should diminish far away from the body owing to the body's motion through the fluid. Equation (3)

$$\lim \nabla \phi_{r \to \infty} = 0$$

where r is the position vector (x, y, z). This condition is automatically fulfilled through Green's function.

Following Green's identity, we can construct a general solution to Eq. (1) by summing up the source σ and doublet μ distributions placed on the body surface S_B and wake surface S_w. equation (4)

$$\phi(x, y, z, t) = \frac{1}{4\pi} \int_{S_B} \mu n \cdot \nabla \left(\frac{1}{r}\right) dS - \frac{1}{4\pi} \int_{S_B} \sigma \left(\frac{1}{r}\right) dS + \frac{1}{4\pi} \int_{S_w} \mu n \cdot \nabla \left(\frac{1}{r}\right) dS$$

2.2. Boundary condition

According to the Neumann boundary condition, the velocity potential jump on S_B is $-\sigma$, and the source strength is equation (5)

$$-\sigma = \boldsymbol{n} \cdot \boldsymbol{v}_B$$

The potential inside the body (without internal singularities) will not change for an enclosed boundary (e.g., S_B); therefore, the internal potential is set to $\phi_{int} = 0$. By dividing the blade surface into N panels and wake surface into N_w panels, integration on the surfaces in Eq. (4) can be equivalently written as the superposition of integrations on panels that constitute those surfaces. A constant-strength rectilinear panel is adopted in current study. Thus, the boundary condition for body surface can be rewritten as equation (6)

$$\sum_{k=1}^{N} \mu_k C_k + \sum_{l=1}^{N_w} \mu_l C_l + \sum_{k=1}^{N} \sigma_k B_k = 0$$

Where equation (7)

$$C = \frac{1}{4\pi} \int_{\text{body or wake panel}} \boldsymbol{n} \cdot \nabla\left(\frac{1}{r}\right) dS$$

$$B = -\frac{1}{4\pi} \int_{\text{body panel}} \left(\frac{1}{r}\right) dS$$

To define this problem uniquely, the wake doublets should be known or related to the unknown doublets on S_B. The wake doublets can be expressed in terms of the unknown surface doublet through the Kutta condition. Defining upper and lower trailing edge (TE) doublets μ_u and μ_l, respectively, the TE wake doublet μ_{tw} is given as equation (8)

$$\mu_{tw} = \mu_u - \mu_l$$

Eq. (6) can then be expressed as equation (9)

$$\sum_{k=1}^{N} \mu_k A_k = -\sum_{l=1}^{N_w} \mu_l C_l - \sum_{k=1}^{N} \sigma_k B_k$$

where A_k includes contributions of the body surface as well as wake surface, and $A_k, B_k,$ and C_l can be solved by the analytic formulations of Hess and

Smith [18] for a constant strength of potential distribution on each panel. If the kth panel is not at the TE, then $A_k = C_k$; if it is, then $A_k = C_k \pm C_{tw}$.

2.3. Panel pressure and force

Once the flowfield is determined, the pressure on the body surfaces can be computed by using the velocity potential and flow velocity through the unsteady Bernoulli equation: Equation (10)

$$\frac{\partial \phi}{\partial t} + \frac{1}{2} v^2 + \frac{1}{\rho} p = \frac{1}{\rho} p_{ref}$$

where v, p, and p_{ref} are the local fluid velocity, local pressure, far-field reference pressure, respectively.

The non-dimensionalized form of the pressure is given as equation (11)

$$C_p = \frac{p - p_{ref}}{(1/2)\rho v_{ref}^2} = 1 - \frac{v^2}{v_{ref}^2} - \frac{2}{v_{ref}^2} \frac{\partial \phi}{\partial t}$$

where v_{ref} is the reference velocity. The reference velocity v_{ref} includes the flight velocity, induced velocity by fuselage's motion, blade's rigid motion, and blade's deformation.

The aerodynamic force on the panels can be computed as equation(12)

$$\Delta F_k = -C_{pk}\left(\rho v_{ref}^2 / 2\right)_k \Delta S_k n_k$$

where ΔF_k is the aerodynamic load on the panel, ΔS_k the panel area, and n_k the normal vector.

2.4. Time-step free-wake method

The solution of the blade-doublet distribution is related to rotor wake, which is simulated by the free-wake method8, 9 and 10 in the present study.

The free-wake method is based on the assumption of an incompressible flow, with all of the vorticity in the rotor wake discretizated into straight-line vortex filaments. A description of the vorticity field in the rotor wake is governed by the three-dimensional, incompressible, viscous flow Navier–Stokes equations, which can be expressed with velocity–vorticity (u,ω) as Equation (13)

$$\frac{\partial \omega}{\partial t} + u \cdot \nabla \omega = \nabla u \cdot \omega + v \nabla^2 \omega$$

where v is kinematic viscosity. The first term on right hand side is strain term, and the second term is diffusion term. In many practical problems, it can be justified that the viscous term has little effect on rotor flow field, and the vortex filaments will then move freely as material lines with constant circulation. In this case, Eq. (13) can be a simpler convection form equation(14)

$$\frac{dr}{dt} = V(r, t), \quad r(t = 0) = r_0$$

where r_0 is the initial position vector, and V is the local velocity field. Vector r is a function of the age of the vortex filament ζ and blade azimuth ψ where filament was trailed into rotor wake. Therefore, Eq. (14) can be written as the partial differential equation. equation(15)

$$\frac{\partial r(\psi, \zeta)}{\partial \psi} + \frac{\partial r(\psi, \zeta)}{\partial \zeta} = \frac{1}{\Omega} V(\psi, \zeta)$$

where Ω is the rotor speed.

The right hand side of Eq. (15) is the total convected velocity of the filament, which includes all the contributions of freestream, blades' sources and doublets, shed-wake panels, and full-span vortex filaments. equation(16)

$$V = V_\infty + V_{\text{body panel}} + V_{\text{shedwake panel}} + V_{\text{fullspan wake}}$$

where V_∞ is freestream velocity, $V_{\text{body panel}}$, $V_{\text{shedwake panel}}$, $V_{\text{fullspan wake}}$ are velocity induced by blade panels, shed-wake panels, and full-span wake filaments, respectively.

The induced velocity of vortex filaments is calculated through the Biot–Savart law. Vatistas vortex model[19] is adopted in velocity calculation to avoid the singularity problem when $r \to 0$. Viscous effect can be integrated into vortex model and the growth of viscous core radius for vortex filaments can be modeled by[20] equation(17)

$$r_c(\zeta) = \sqrt{r_{c0}^2 + 4\alpha\delta v\zeta/\Omega}$$

where r_{c0} is the initial vortex core radius, α the Oseen constant equaling to 1.25643, δ the eddy viscosity parameter.

Eq. (15) is solved through finite differences to approximate the space and time derivatives. The displacements of the filaments are then found in a time-stepping manner through numerical integration of the governing equations. The second-order-accurate predictor–corrector scheme (PC2B scheme)8 and 10 is employed in the present approach.

2.5. Blade motion

The rotor wake and blade airloads are closely related to blade-flapping response. Therefore, the individual blade-flapping dynamics is solved simultaneously with the rotor wake dynamics. The flap equation is given by the equilibrium of moments about the flapping hinge, which include all the contributions of blade aerodynamics, fuselage's motion, and blade's rigid motion. The blade-flapping behavior can be expressed as a set of ordinary differential equations, and the blade-flapping response is determined through the fourth-order Runge–Kutta scheme in current work.[10]

2.6. Coupled method

In general, the shed wake and trailing wake can be represented by wake panels. However, the wake panels can be replaced with a vortex wake to overcome the singularity problem when wake panels are close to one another.[15, 16 and 17] Therefore, the doublet-wake surface is replaced by full-span vortex filaments to describe complex wake behavior, which provides a connection between the unsteady panel method and the free-wake method. The blade panels, shed wake panels, and full-span vortex filaments are shown in Fig. 1. Rotor blades are represented by panels, the shed wake is described by doublet-shed wake panels, and the rotor wake is indicated by the full-span vortex filaments which connect to the shed wake.

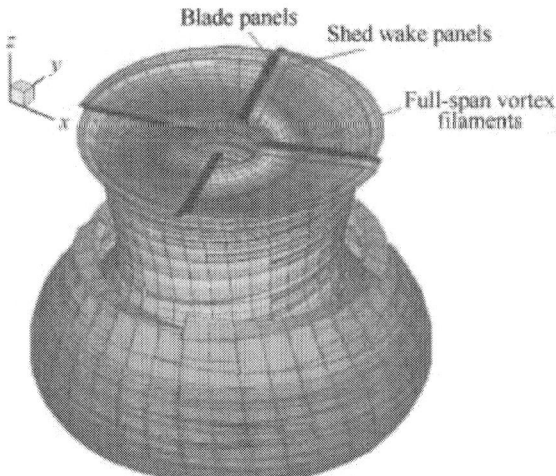

Figure. 1. Rotor wake system.

The doublet-wake panels can be replaced by vortex filaments, and the equivalence of doublet panel surface and vortex filament can be expressed as[21] equation (18)

$$V_{\text{wake}} = \nabla\phi = -\frac{\mu}{4\pi} \int_{S_w} \nabla\frac{z}{r^3}dS = \frac{\Gamma}{4\pi} \int_C \frac{d\mathbf{l} \times \mathbf{r}}{r^3}$$

where C denotes the curve bounding of the wake panel, and $\Gamma = \mu$ is the circulation strength of the vortex filament along C. The order of vortex filament is one less than that of the doublet panel, and the vortex filament is then less singular behavior than the doublet panel.

The strength of full-span vortex filaments is derived from the shed wake at each time step base on Eq. (18), which is shown in Fig. 2. The strength of the shed wake is obtained through the Kutta condition in Eq. (8). The strength of I th shed wake panel is μ_i^t, and the next time step is $\mu_i^{t+\Delta t}$, the strength of vortex filament parallel and perpendicular to the blade are then $\mu_i^{t+\Delta t} - \mu_i^t$, $\mu_i^t - \mu_{i+1}^t$, respectively.

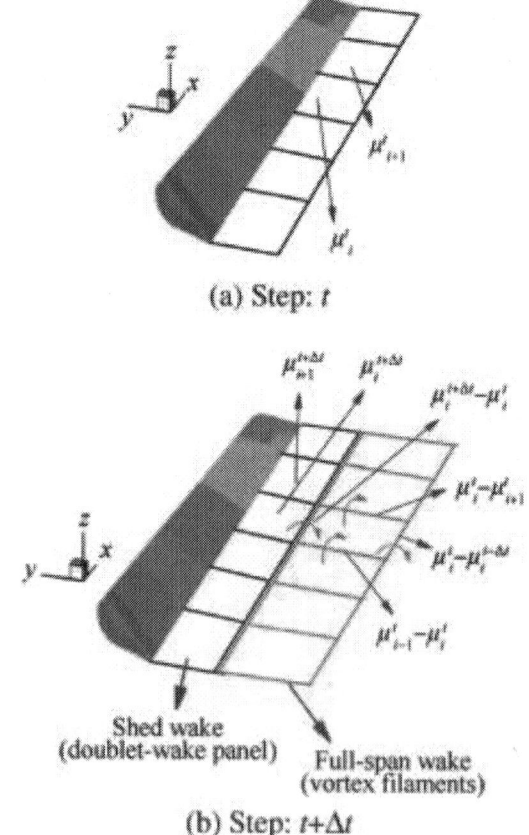

(a) Step: t

(b) Step: $t+\Delta t$

Figure. 2. Generation of full-span vortex filaments.

The strength of the full-span vortex filaments is determined by blade panels, and the strength of doublet on blade panel is affected by the full-span vortex filaments at each time step, therefore, the full-span free-wake method is tightly coupled with the unsteady panel method.

2.7. Singularity problem

It is necessary to determine wake geometry before solving Eq. (9), and singularity problem in the numerical integration in Eq. (9) will occur when the rotor wake approaches blades on the advancing and retreating sides in forward flight (see Fig. 3).

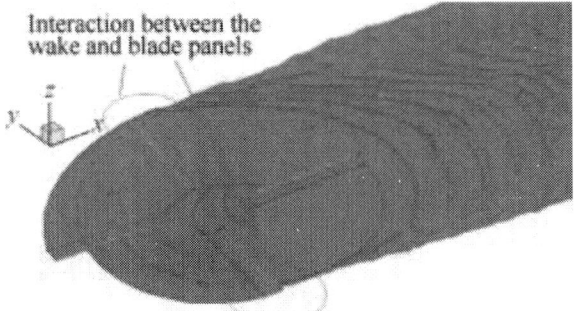

Figure. 3. Interaction between the wake and blade panels.

A velocity-field integration technique 17 and 22 is employed to overcome the singularity problem in the present study. The singularity problem is similar to that of velocity calculation; therefore, this problem can be avoided through that the wake potential is obtained indirectly by integrating the velocity field.

Velocity induced by wake vortex filaments can be expressed in Eq. (18), and the singularity can be overcome through Vatistas vortex model. The wake potential can be determined by integrating the velocity along an arbitrary integral path. Equation (19)

$$\phi_{\text{wake}} = \int_1^2 \left[\frac{\Gamma}{4\pi} \int_C G(r) \frac{d\boldsymbol{l} \times \boldsymbol{r}}{r^3} \right] \cdot \boldsymbol{t}_{\text{p}} ds$$

where $\boldsymbol{t}_{\text{p}}$ is the tangential vector of the integral path; the upper and down index 1, 2 are the initial point and ending point of integration, and G is Vatistas vortex model. This approach prevents the numerical singularity problem which is very favorable to accurately predict pressure distribution when a BVI occurs in forward flight.

3. NUMERICAL RESULTS AND DISCUSSION

The blade pressure distribution and unsteady blade loads of the Caradonna–Tung, UH-60A, and AH-1G rotors are predicted and compared with available experimental data and CFD results to validate the proposed method.

3.1. Caradonna–Tung rotor in hover

Caradonna and Tung's[23] two-bladed rotor experiment is analyzed for a hovering rotor. This rotor has two untwisted, untapered, rectangular blades with NACA0012 airfoil. The rotor radius is 1.143 m and the aspect ratio is 6. This case involved a rotor with a collective pitch angle of 8°, rotating at 1250, 1500, or 1750 rpm, corresponding to a tip Mach number of 0.439, 0.526, and 0.612, respectively.

The computational model was composed of 60 panels in the chordwise direction and 20 panels in the spanwise direction on each blade. The results for the current case were obtained with an azimuthal step of 10° and about 11 h with one CPU for 31 revolutions.

Fig. 4, Fig. 5 and Fig. 6 show the pressure coefficient C_p at two blade-span sections ($r/R = 0.68$, 0.96, where r is radial distance, R is blade radius) with different tip Mach numbers. The experimental data, [23] CFD results,[24] and Hybrid CFD results[25] are shown for comparison. The CFD method[24] used multi-bock structured mesh to account for blade's rigid motion through sliding-plane approach, and employed mesh-deformation technique to account for blade deflections through trans-finite interpolation (TFI) method. The Hybrid CFD method[25] included a compressible Navier–Stokes solver, a compressible potential flow solver, and a free wake model to accurately and efficiently predict unsteady blade pressure. The present numerical results are in fairly good agreement with the experimental results and the two different CFD results for all regions. However, the suction peak at the leading edge near the blade tip region is underestimated by the present and hybrid CFD results. Compared with CFD and hybrid CFD, the present method predicts aerodynamics of rotor blades in hover more efficiently.

3.2. UH-60A rotor in hover

The present method is also applied to the typical four-bladed UH-60A rotor.[26] This rotor has four complex blades with an aspect ratio of 15.3 and a maximum nonlinear twist of −13°. The blade has a rearward sweep of 20° starting from a rotor radius of 93% and is made up of two airfoils section, SC1095 as the main airfoil in inside and outside, and SC1095R8 section in the mid-span. The rotor is operated at a tip Mach number of 0.628 and with a collective pitch angle of 9°. The computational model is composed of 60 panels in the chordwise direction and 20 panels in the spanwise direction. An azimuthal step is 10°.

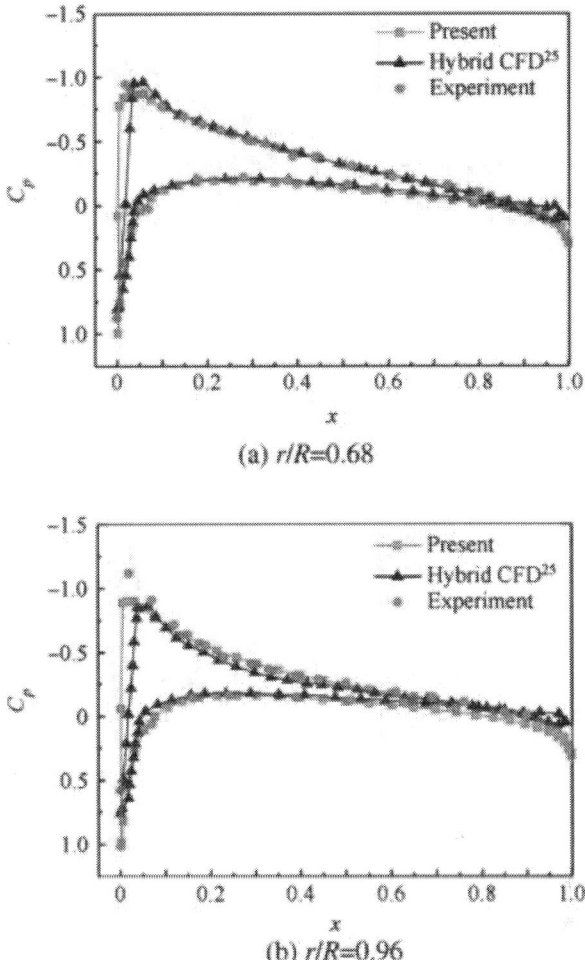

Figure. 4. Pressure coefficient over blade chord ($\Omega = 1250$ rpm).

The rotor wake geometry is shown in Fig. 1. Wake contraction near the rotor plane and vortex bundle far below the plane is well captured. A comparison of the predicted pressure distribution at three radial stations ($r/R = 0.675$, 0.865, 0.990) with the corresponding experiment data and the hybrid CFD results[25] is shown in Fig. 7. The predicted results agree well with the experimental data and the hybrid CFD results. At the station $r/R = 0.990$, the suction peak is underestimated by the present and hybrid CFD results. The pressure at trailing edge is overestimated by the present result, and one of the reasons is that the roll-up of shed wake described by doublet-wake panels is neglected in the present method, and this behavior will affect panels near trailing edge in tip region.

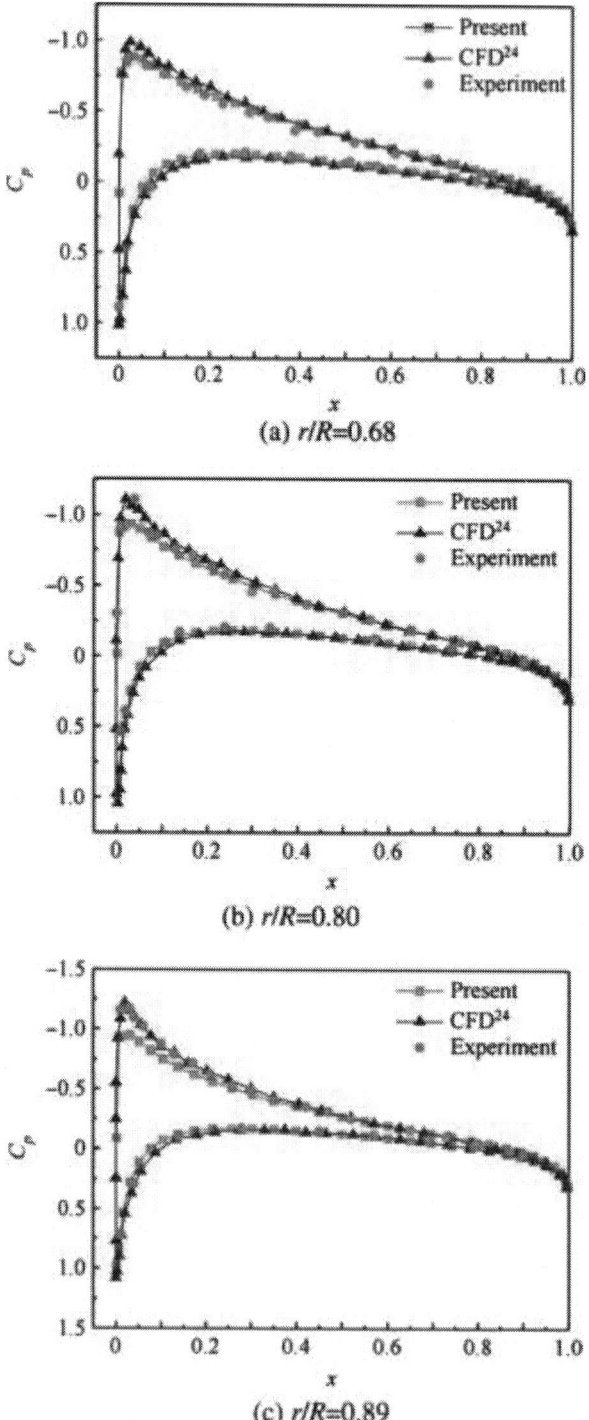

Figure. 5. Pressure coefficient over blade chord ($\Omega = 1500$ rpm).

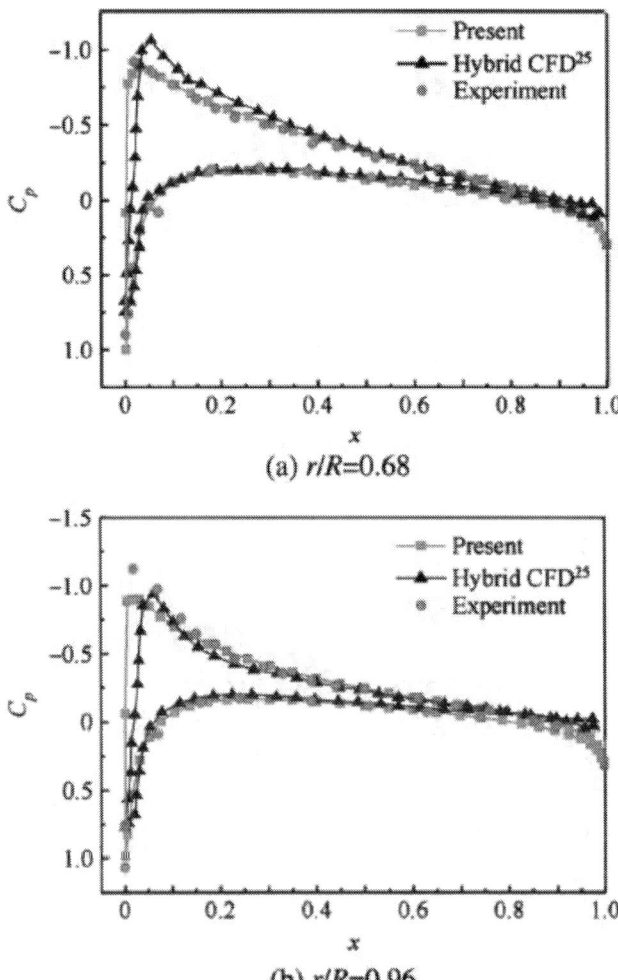

(a) r/R=0.68

(b) r/R=0.96

Figure. 6. Pressure coefficient over blade chord (Ω = 1750 rpm).

3.3. AH-1G rotor in forward flight

The AH-1G model27 and 28 in forward flight is also simulated. This rotor has two untapered, rectangular blades with OLS/TAAT airfoil. The rotor radius is 6.7 m and the aspect ratio is 9.2. The linear twist is $-10°$. The rotor is operated at a tip Mach number of 0.68 and an advance ratio of 0.19. The computational blade model was composed of 60 panels in the chordwise direction and 20 panels in the spanwise direction. The results were obtained with an azimuthal step of 5° and about 24 h with one CPU for 10 revolutions.

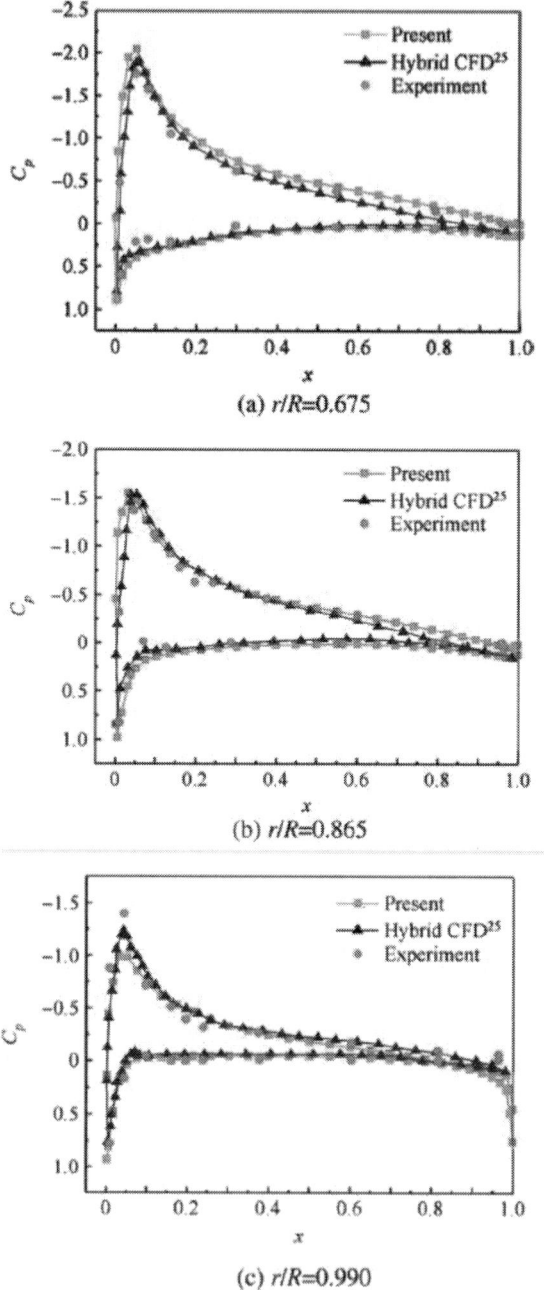

(a) $r/R=0.675$

(b) $r/R=0.865$

(c) $r/R=0.990$

Figure. 7. Pressure coefficient over blade chord (UH-60A rotor).

The wake structure is shown in Fig. 8. The roll-up wake on the advancing and retreating side is captured well and wake convection is also observed. The wake in the forward part of the rotor is raised up towards the rotor disc, and the blade tip vortex approaches to the other blades, resulting in an interaction between the wake and blades. The singularity problem on the advancing side and retreating side is overcome.

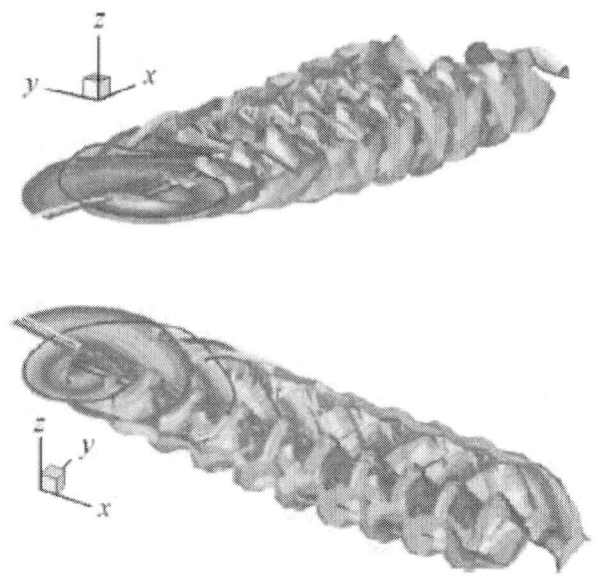

Figure. 8. AH-1G rotor wake geometry in forward flight.

Fig. 9 and Fig. 10 show the pressure coefficients at specific span positions (r/R = 0.60, 0.91) at azimuth angles of 30°, 90°, 180°, 270°, 285°, and 300°. The CFD results[29] are also shown for comparison. The predicted results agree well with measured data and CFD results at different azimuths. The suction peaks on the advancing side and retreating side are captured fairly well with the CFD results; however, the suction peaks at azimuths of 270° are overestimated by both methods.

Fig. 11 shows the variation of the sectional thrust coefficient at the three blade radial stations over one revolution. The results are compared with measured data,[27 and 28] CFD,[29 and 30] and conventional free-wake results.[31] The predicted unsteady thrust coefficients agreement well with the flight test data and CFD results in different spans, and are more accurate than conventional free-wake results. The variations of the thrust coefficient correlate well with the measured data and CFD results near the azimuths of 90° and 270°. The influence of the interaction between the wake and blades on sectional thrust distributions is observed on the advancing side at azimuths of around 60°–100° and on the retreating side at around 280°. Because the blade elastic deformation has impact on blade geometry, there is a discrepancy among the present, both CFD results, and the measured data from azimuthal angles of 180°–280°.

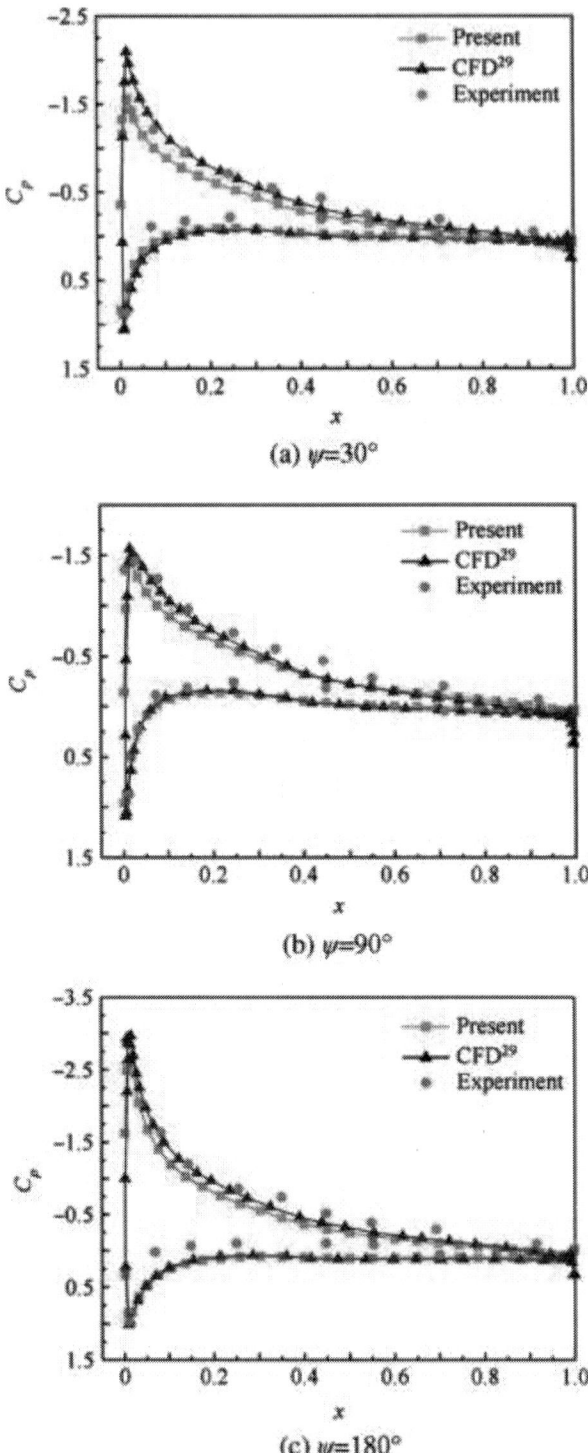

Figure. 9. Pressure coefficient at $r/R = 0.60$ (AH-1G).

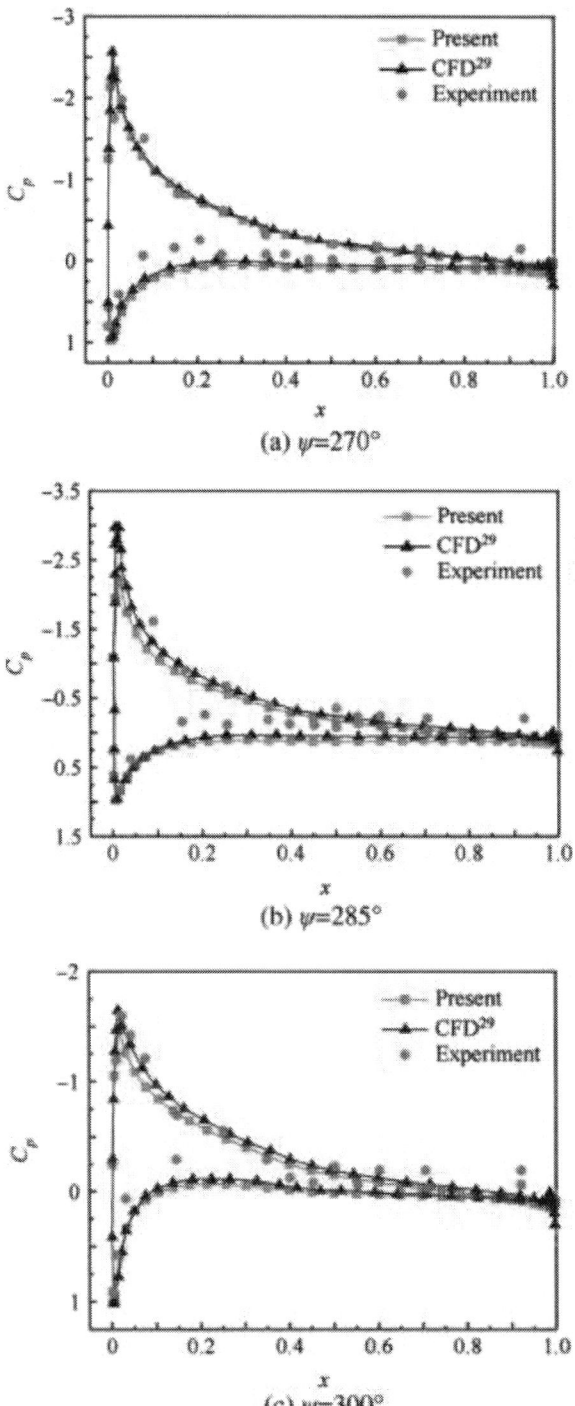

Figure. 10. Pressure coefficient at $r/R = 0.91$ (AH-1G).

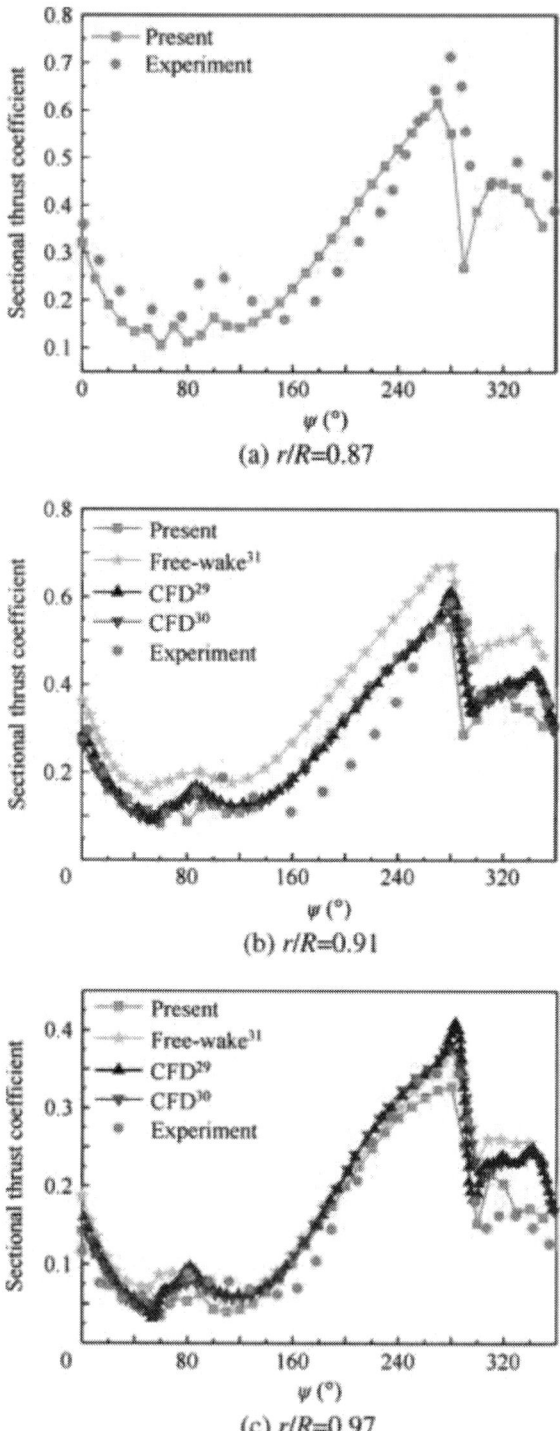

(a) $r/R=0.87$

(b) $r/R=0.91$

(c) $r/R=0.97$

Figure. 11. Sectional thrust coefficients history (AH-1G).

The section oscillatory pitching moment coefficient (the mean value have been removed from the measured data and calculations) is shown in Fig. 12. The predicted results are compared with the measured data and CFD results.[29] The overall variation of oscillatory pitching moment agrees well with the measured data and CFD results; however, there are still discrepancies. One of the reasons is that there is discrepancy of pressure at leading edge and trailing edge, and the pressure distribution in chordwise direction has significant impact on section pitching moment. Another reason is the influence of aeroelasticity which is neglected in present simulation.

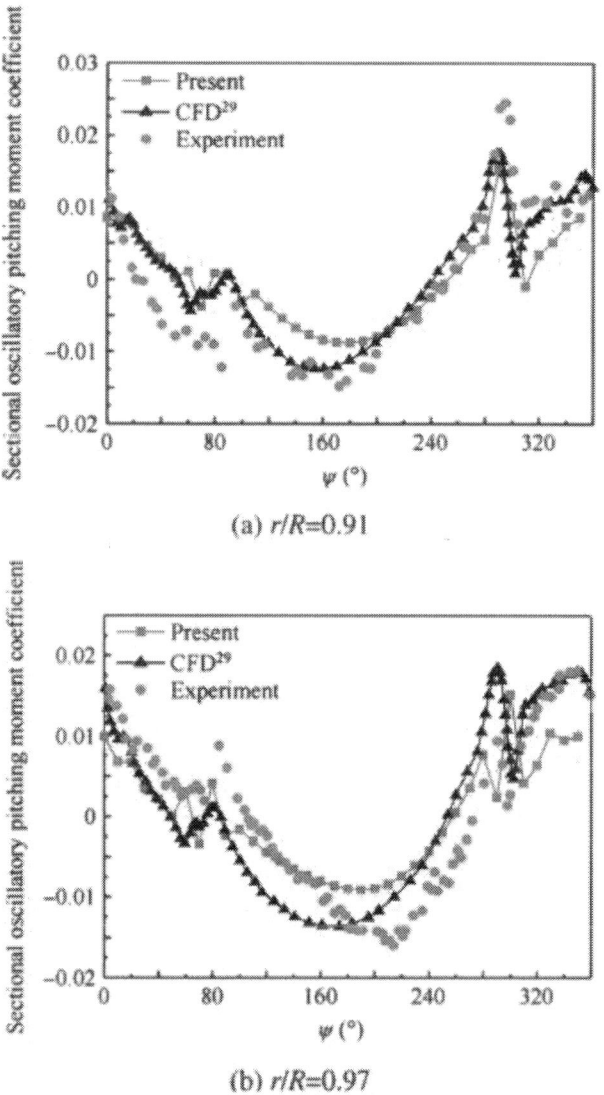

(a) $r/R=0.91$

(b) $r/R=0.97$

Figure. 12. Oscillatory pitching moment history (AH-1G).

The present method can accurately predict unsteady aerodynamic loads of rotor blades. Compared with CFD, this method is more efficient and suitable to rotorcraft aeroelastic analysis.

4. CONCLUSIONS

1. The blade pressure distribution predicted by the present unsteady panel/full-span free-wake coupled method is in a good agreement with measured data and CFD results in hover and forward flight for this article. The unsteady aerodynamic loads of blade agree fairly well with measured data and CFD results, and are more accurate than conventional free-wake results.
2. Wake contraction, convection, and roll-up behavior are captured well in hover and forward flight. The singularity problem of the interaction between the wake and blades is overcome.
3. The present method predicts unsteady aerodynamics of rotor blade more efficiently than CFD, and is suitable to rotorcraft aeroelastic analysis.

REFERENCES

1. Markus D, Manuel K, Ewald K, Siegfried W. Tip vortex conservation on a helicopter main rotor using vortex-adapted chimera grids. AIAA J 2007;45(8):2062–74.
2. Steijl R, Barakos GN. Computational study of helicopter rotorfuselage aerodynamic interactions. AIAA J 2009;47(9):2143–57.
3. Shi YJ, Zhao QJ, Fan F, Xu GH. A new single-blade based hybrid CFD method for hovering and forward-flight rotor computation. Chin J Aeronaut 2011;24(2):127–35.
4. Cao YH, Yu ZQ, Su Y, Kang K. Combined free wake/CFD methodology for prediction transonic rotor flow in hove. Chin J Aeronaut 2002;15(2):65–71.
5. Sitaraman J, Roget J. Prediction of helicopter maneuver loads using a fluid–structure analysis. J Aircr 2009;46(5):1770–84.
6. Mishra A, Ananthan S, Baeder JD. Coupled CFD/CSD prediction of the effects of trailing edge flaps on rotorcraft dynamic stall alleviation. In: 47th AIAA aerospace sciences meeting including the new horizons forum and aerospace exposition, 2009. p. 1–17.
7. Bagai A, Leishman JG. Rotor free-wake modeling using a pseudoimplicit algorithm. J Aircr 1995;32(6):1276–85.
8. Bhagwat MJ, Leishman JG. Stability consistency and convergence of time-marching free-vortex rotor wake algorithms. J Am Helicopter Soc 2001;46(1):59–71.
9. Li CH, Xu GH. The rotor free-wake analytical method for tiltrotor aircraft in hover and forward flight. ACTA Aerodyn Sin 2005;23(2):152–156 [Chinese].
10. Tan JF. Analysis of rotor aerodynamic response under manoeuvring conditions dissertation. Nanjing: Nanjing University of Aeronautics and Astronautics; 2009 [Chinese].

11. Morino L, Kuo C. Subsonic potential aerodynamics for complex configuration. AIAA J 1974;12(2):191–7.
12. Katz J, Maskew B. Unsteady low-speed aerodynamic model for complete aircraft configurations. J Aircr 1988;25(4):302–10.
13. Morino L, Bharadvaj BK, Freedman MI, Tseng K. Boundary integral equation for wave equation with moving boundary and applications to compressible potential aerodynamics of airplanes Fig. 12 Oscillatory pitching moment history (AH-1G). and helicopter. Comput Mech 1989;4(4):231–43. 542 J. Tan, H. Wang
14. Richason TF, Katz J. Unsteady panel method for flows with multiple bodies moving along various paths. AIAA J 1994;32(1):62–8.
15. Ahmed S, Vidjaja VT. Unsteady panel method calculation of pressure distribution on BO105 model rotor blades. J Am Helicopter Soc 1998;43(1):47–56.
16. Chung KH, Hwang CJ, Park YM, Jeon WJ, Lee DJ. Numerical predictions of rotorcraft unsteady air-loadings and BVI noise by using a time-marching free-wake and acoustic analogy. In: 31th European rotorcraft, forum; 2005. p. 1–8.
17. Gennaretti M, Bernardini G. Novel boundary integral formulation for blade-vortex interaction aerodynamics of helicopter rotors. AIAA J 2007;45(6):1169–76.
18. Hess J, Smith AMO. Calculation of non-lifting potential flow about arbitrary three-dimensional bodies. J Ship Res 1964;8(2):22–44.
19. Vatistas GH, Kozel V, Mih WC. A simpler model for concentrated vortices. Exp Fluids 1991;11(1):73–6.
20. Bhagwat MJ, Leishman JG. Generalized viscous vortex model for application to free-vortex wake and aeroacoustic calculations. In: The international 58th annual forum and technology display of the American Helicopter Society; 2002. p. 11 3.
21. Morino L. Boundary integral equations in aerodynamics. Appl Mech Rev 1993;46(8):445–66.
22. Seong YW, Seongkyu L, Duck JL. Potential panel and timemarching free-wake coupling analysis for helicopter rotor. J Aircr 2009;46(3):1030–41.
23. Caradonna FX, Tung C. Experimental and analytical studies of a model helicopter rotor in hover. NASA TM-81232; 1982.
24. Steijl R, Barakos GN, Badcock KJ. A CFD framework for analysis of helicopter rotors. In: 17th AIAA computational fluid dynamics conference; 2005. p. 1–14.
25. Berkman ME, Lakshmi NS, Charles RB, Michale ST. A Navier–Stokes/full potential/free wake method for rotor flows. AIAA- 1997-401; 1997. p. 1–8.
26. Lorber PF, Stauter RC, Landgrebe AJ. A comprehensive hover test of the airloads and airflow of an extensively instrumented model helicopter rotor. In: Proceedings of 45th annual forum of the AHS, Boston; 1989.
27. Cross JL, Watts ME. Tip aerodynamics and acoustics test: a report and data survey. NASA-RP-1179; 1988.
28. Cross JL, Wilson T. Tabulation of data from the tip aerodynamics and acoustics test. NASA TM-102280; 1990.

29. Jee WK, Soo HP, Yung HY. Euler and Navier–Stokes simulations of helicopter rotor blade in forward flight using an overlapped grid solver. In: 19th AIAA computational, fluid dynamics; 2009. p. 1–13.
30. Park YM, Kwon OJ. Simulation of unsteady rotor flow using unstructured adaptive sliding meshes. J Am Helicopter Soc 2004;49(4):391–400.
31. Joonbae L, Kwanjung Y, Sejong O. Development of an unsteady aerodynamic analysis module for rotor comprehensive analysis code. J Aeronaut Space Sci 2009;10(2):23–33

CHAPTER 11

Ornithopter Type Flapping Wings for Autonomous Micro Air Vehicles

Sutthiphong Srigrarom [1],* and Woei-Leong Chan [2]

[1] *Aerospace Systems, University of Glasgow Singapore, 500, Dover Rd., #T1A-02-24, Singapore 139651*

[2] *Temasek Laboratories, National University of Singapore, #09-02, 5A Engineering Drive 1, Singapore 117411*

ABSTRACT

In this paper, an ornithopter prototype that mimics the flapping motion of bird flight is developed, and the lift and thrust generation characteristics of different wing designs are evaluated. This project focused on the spar arrangement and material used for the wings that could achieves improved performance. Various lift and thrust measurement techniques are explored and evaluated. Various wings of insects and birds were evaluated to understand how these natural flyers with flapping wings are able to produce sufficient lift to fly. The differences in the flapping aerodynamics were also detailed. Experiments on different wing designs and materials were conducted and a paramount wing was built for a test flight. The first prototype has a length of 46.5 cm, wing span of 88 cm, and weighs 161 g. A mechanism which produced a flapping motion was fabricated and designed to create flapping flight. The flapping flight was produced by using a single motor and a flexible and light wing structure. A force balance made of load cell was then designed to measure the thrust and lift force of the ornithopter. Three sets of wings varying flexibility were fabricated, therefore lift and thrust measurements were acquired from each different set of wings. The lift will be measured in ten cycles computing the average lift and frequency in three different speeds or frequencies (slow, medium and fast). The thrust measurement was measure likewise but in two cycles only. Several observations were made regarding the behavior of flexible flapping wings that should aid in the design of future flexible flapping wing vehicles. The wings angle or phase characteristic were analyze too and studied. The final ornithopter prototype weighs only 160 g, has a wing span of 88.5 cm, that could flap at a maximum flapping frequency of 3.869 Hz, and produce a maximum thrust and lift of about 0.719 and 0.264 N respectively. Next, we proposed resonance type flapping wing utilizes the near resonance phenomenon of a two-degree of freedom elastic system, that is, the wing is supported by the springs for flapping and feathering motions. Being oscillated close to the resonance frequency of the system, only by the torque in flapping motion, the amplitude gained is a few times higher than that of normal case. The first prototype was made from acrylic using a laser cutting machine. The wings were made up of carbon rods and kite material Ripstop. First test showed that the wings were too heavy for the mechanism to work. The third prototype was a smaller single gear crank design which was fabricated using a 3D printer. Initial test proved that the second prototype could withstand the high frequency flapping and near resonance amplitude as designed. With remote control, the third prototype was able to take off, climb, cruise and land in flapping mode successfully.

KEYWORDS:

ornithopter-like flapping wing; albatross-like wing; micro air vehicle (MAV)

1. INTRODUCTION

There is a growing recognized need for miniature flight vehicles with multifunctional capabilities, such as micro air vehicles (MAVs) for both military and civilian surveillance [1,2,3,4]. Flapping wing flight of birds provides us with a sophisticated example of utilizing unsteady aerodynamics to mechanize the miniature flight structures at low Reynolds numbers (10^3–10^5) [5,6]. We attempt to mimic both the long-distance birds, due to their natural long-endurance manner, and their high lift production during take-off (start-up). The albatross, as shown in Figure 1 and Figure 2, is chosen to represent long distance migratory birds. It is about the size of our intended flapping wing based surveillance MAV. We hope to mimic albatross's flapping flight to achieve this long-distance characteristic. It is used for investigating flow characteristic aiming at better design of flapping MAV. The wingspan of the albatross is 60 cm. The flapping pattern of the albatross is of the avian type, *i.e.*, vertical motion as shown in Figure 3. Our model simulated the complete, three-dimensional, unsteady flow fields around this type of wing with large-scale vortices.

At University of Glasgow Singapore, we are developing the resonance type flapping wing models, as shown in Figure 4and Figure 5 below. The proposed resonance type flapping wing will utilize the resonance phenomenon of a two-degree of freedom elastic system, that is, the wing is supported by the springs for flapping and feathering motions, being oscillated, at the resonance frequency of the system, as shown in Figure 6 [7]. The amplitudes of flapping and feathering motions and the phase angle between them are controlled by changing the amount of the damping.

Figure 1. Albatross in flight.

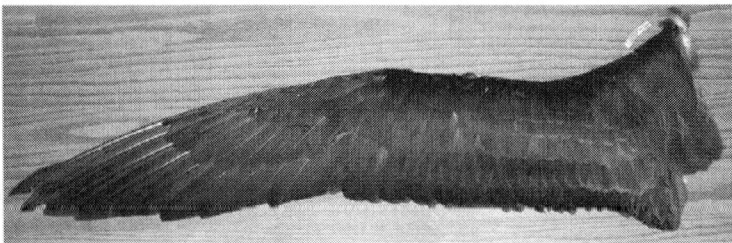

Figure 2. Albatross wing model.

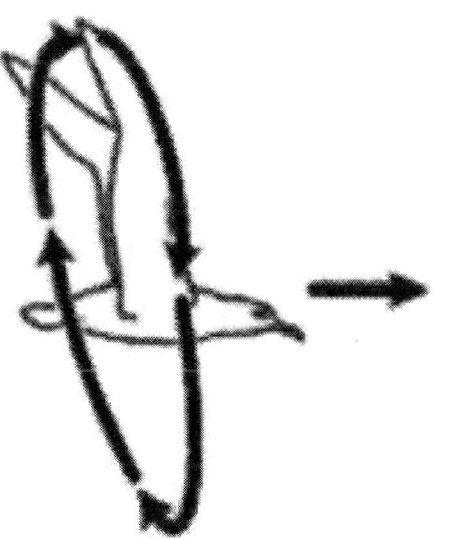

Figure 3. The flapping pattern of albatross is avian type, *i.e.*, more up/down vertically.

Figure 4. Near-resonance ornithopter-like flapping wing model.

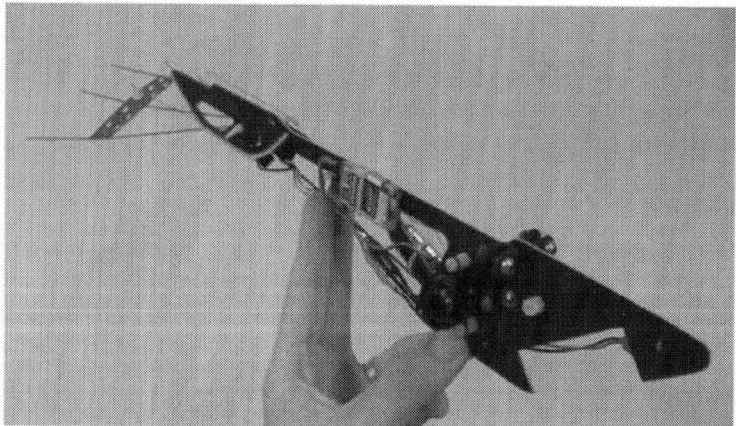

Figure 5. Near-resonance ornithopter-like flapping wing mechanism.

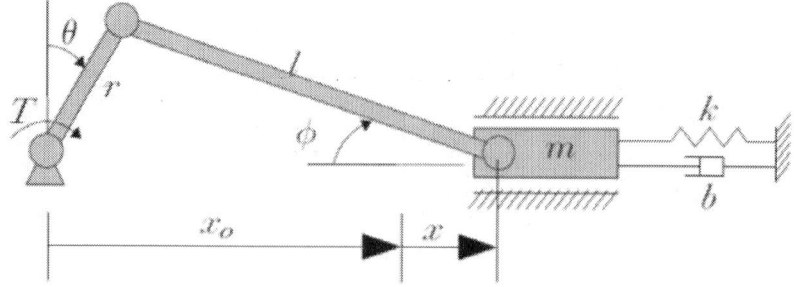

Figure 6. Concept of resonance-type flapping mechanism [7].

2. COMPUTATIONAL FLUID DYNAMICS

Firstly, we examined the flight dynamics as well as unsteady flow characteristics of the flapping wings for long-distance bird (e.g., albatross-like) by Computational Fluid Dynamics (CFD). These birds mostly flap their wings about their body axis (chordwise) with little change in twist (spanwise), as depicted in Figure 7 [8]. Therefore, for preliminary analyses, we apply only chordwise flapping to our model.

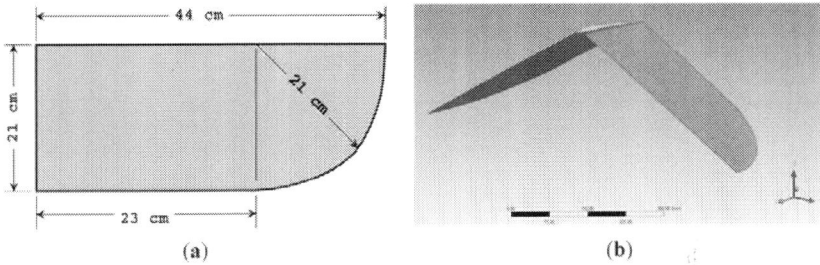

(a)   (b)

Figure 7. (a) Simplified albatross flapping model with detailed geometry; (b) The wing is shown at −15 flapping position about its body (X) axis.

The flow around the flapping wing was simulated by using ANSYS Fluent® (ANSYS Inc., Canonsburg, PA, USA) unsteady three-dimensional compressible Navier–Stokes equations. The machine that ran these problems was a 64 bit computer, Intel® Core i7-2600 CPU at 3.4 GHz. It had eight processors at 16 GByte RAM. The geometric model of the albatross wing was the idealization of an albatross. The model wing had a wing span of 30 cm, mean wing chord length of 5 cm, a thickness of 2.5% of the mean wing chord length, as shown in Figure 7. The computing domain extended to 50-chord lengths in all directions around the full model wing, and that, there were about 10^7 meshes of the tetrahedral type. The flow condition was unsteady transient flow, with the built-in Large–Eddy Simulation (LES) turbulence model.

The chosen wing's geometry is shown in Figure 7 below, of which Wingspan = 0.88, Wing Surface Area = $1.66 \times 10^{-1} \, \text{m}^2$. This wing was originally set at rest. From past studies [8,9,10,11], the starting flapping frequency of the albatross wing was at 0.5 rad/s, corresponding to the reduced frequency k (defined as $\omega c/2U_\infty$) of 0.0025. The take-off speed of the albatross (*i.e.*, the freestream velocity in body-fitted coordinates), U_∞ was approximately 5 m/s. This gave the Reynolds number, based on chord Re of 25,000. The wing was set to flap down and up about its body axis (chordwise), mimicking the start-up of the flapping motion (taking-off) and return to its initial position. The flapping (dihedral) angle changed from +30° to −30° in downstroke motion, and from −30° to +30° in upstroke motion. The input angular velocity about its chord (ω_x) is shown in Figure 8. This downwards stroke in the first half of the cycle is the lift generation stroke, whereas the upwards stroke in the latter half of the cycle is the recovery stroke.

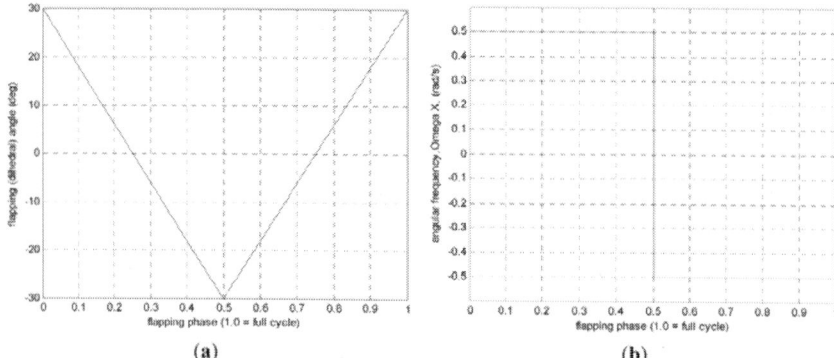

(a)   (b)

Figure 8. (a) Flapping patterns for simplified albatross flapping model about its body axis (chordwise, ω_x); and (b) the corresponding angular velocity, in each flapping cycle.

Initially, the wing was positioned at +30 degree dihedral angle. When the flapping motion started, the wing flapped downwards about its chord. At the outer half of the wing the flow separated from the wing, forming the start-up vortex which more visible at the top. There was flow separation at the trailing edge, closer to the root, at the area near the scapula (inboard of the wing). This leads to the creation of strong leading edge vortex [8] as shown in Figure 9, resulting in high lift (Figure 10) at the beginning of this downstroke motion (at +30 degree dihedral angle).

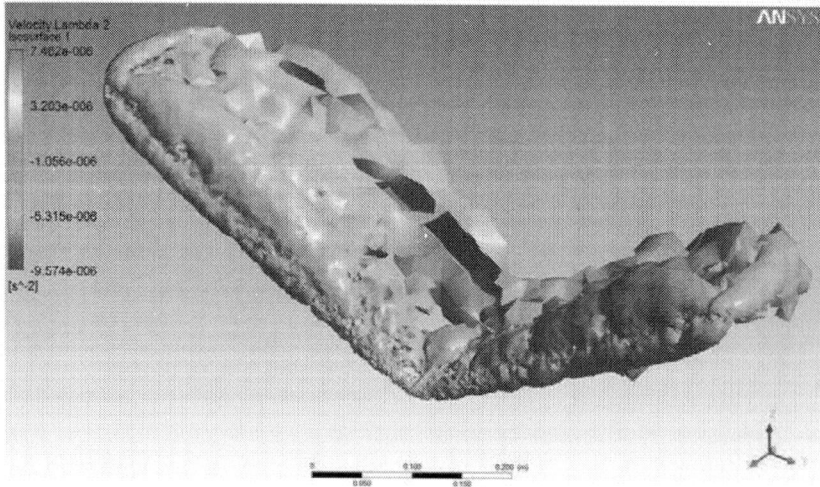

Figure 9. Vortex structure on the wing, shortly after start-up. The strong leading edge vortex on the upper part is clearly shown.

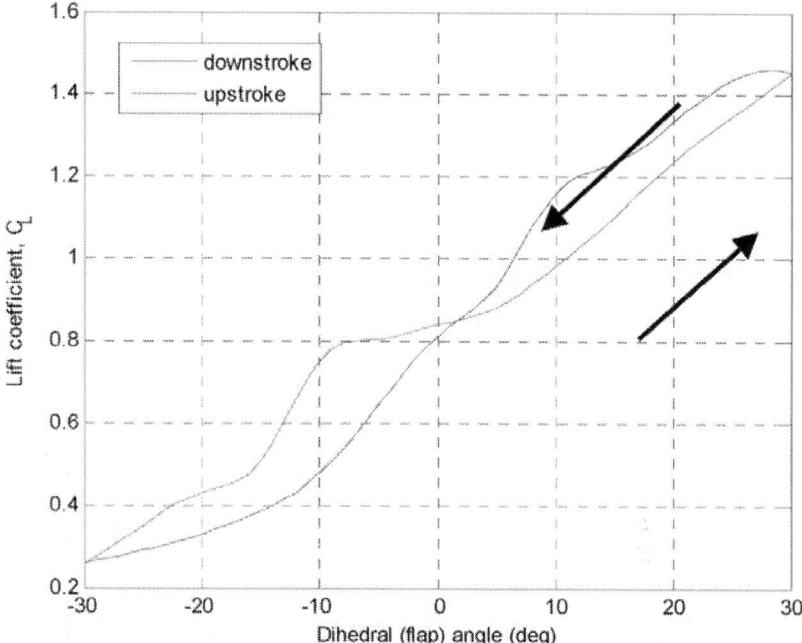

Figure 10. Lift coefficient during downstroke and upstroke motion of the wing.

As time goes on the flow separates across the wing, forming the strong leading edge vortex over the wing. In the inboard area near the trailing edge and scapula, the wake (shear layers) becomes more obvious. The lift is generated more by to this push-down motion of the wing. After mid-point (dihedral angle = 0°), the start-up vortex becomes bigger, and the flow separates from the wing entirely. At the leading edge near the alula (about half way between the root and the tip). The wake at the trailing edge of the wing also breaks up into outer and inner parts.

Figure 10 shows the lift coefficient in both downstroke and upstroke motion of the wing. The C_L values are obtained from directly from ANSYS Fluent® software. The locus of the lift coefficient in both downstroke and upstroke motion resembles the inclined Figure 8, as reported in many other literatures (e.g., [4,6,8,10,12,13,14,15]). Figure 11 shows the lift coefficients derivatives with respect to flapping angle ($dC_L/d\alpha$) this is for flight dynamics and control consideration.

Figure 12 shows the vortex structure (invariant-Q) at dihedral angle = −15°. The leading edge vortex is dominant, covering the entire the wing. At the wing tip, the wingtip vortices is clear together with the wake behind the wing.

Figure 13 shows the surface pressure on the albatross wing also at dihedral angle = −15°. At the tip, there is region of lower pressure, shown in blue color. This is the area where the force is less generated. So, we deduce the albatross does maneuvering by flapping its entire wingspan at different amplitudes than using (flapping or twisting) its wing tip only.

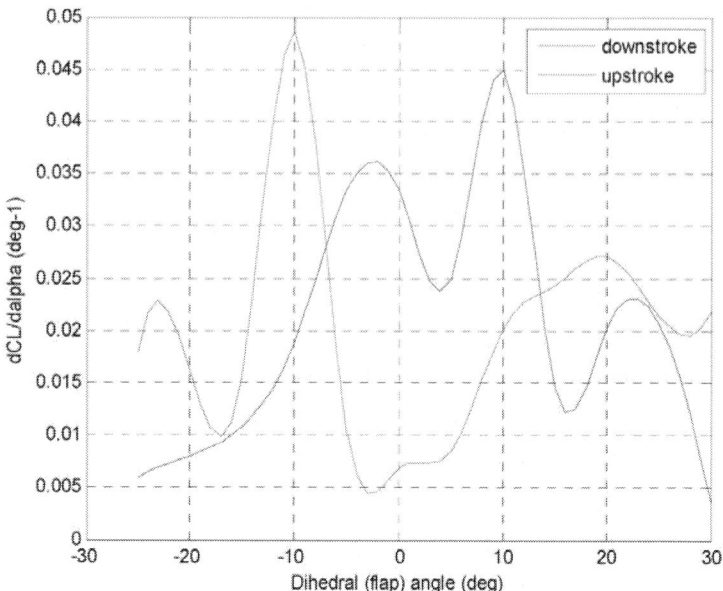

Figure 11. The lift coefficients derivatives with respect to flapping angle ($dC_L/d\alpha$), from C_L.

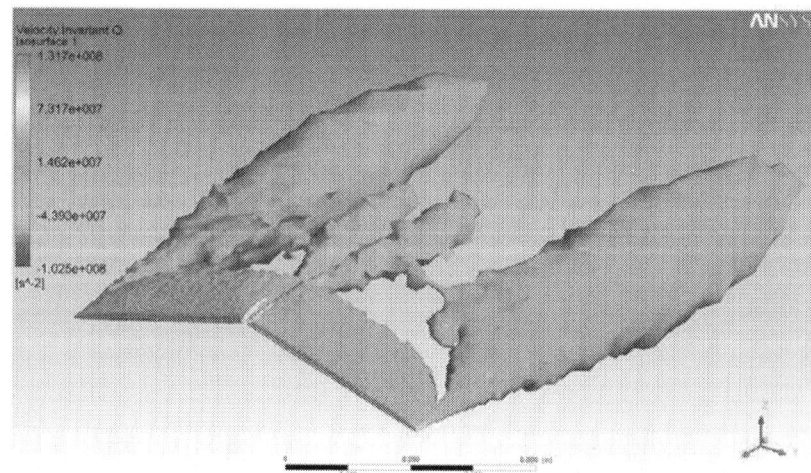

Figure 12. Isosurface of the vorticity on the upper part of the albatross wing, at dihedral angle $= -15°$. The leading edge vortex covers the whole wing. The trailing edge vortices on both ends are outstanding.

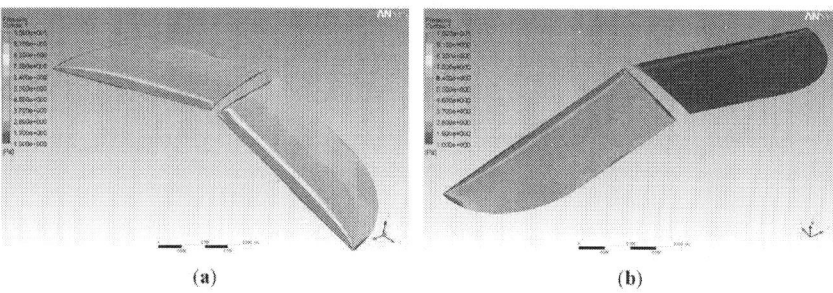

(a)   (b)

Figure 13. Surface pressure on the albatross wing at dihedral angle = −15°, upper surface (**a**) and lower surface (**b**).

These unsteady flow results show that the albatross generates lift on its wing mainly by vortex lift mechanism, as reported in other literatures ([15,16,17,18,19,20,21,22,23,24,25,26,27,28]). The lift is generated by flapping its entire left and right wings at different amplitudes than using (flapping or twisting) its wing tip only. Any additional spanwise flapping (pitching) will fine-tune its position inflight. Therefore, for the sake of developing flapping-wing like MAV (next section), we could introduce only the chordwise flapping motion and neglect the spanwise flapping motion.

3. UGS FLAPPING WING MAV PROTOTYPES 1

3.1. Mechanism

A flapping wing mechanism capable of producing flapping motion is designed and built. An overall view of this mechanism with a length of 46.5 cm can be seen in Figure 14. This figure also depicts the relative scale between the flapping mechanism and the tail. This mechanism is powered by a small brushless motor and has a maximum output of 55 W that weight only 15 g. A Lithium–Polymer battery that has a nominal voltage of 7.4 V is required to drive the motor.

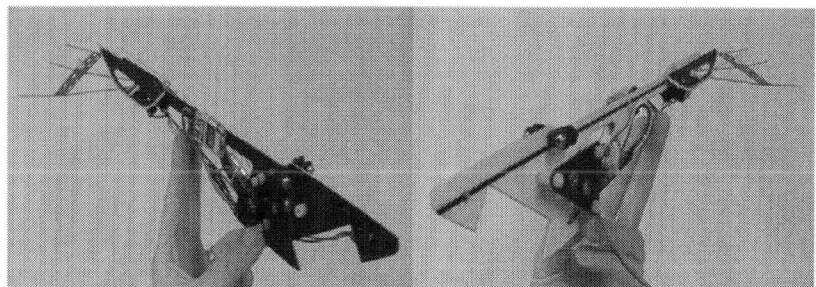

Figure 14. Overall view of entire flapping mechanism.

A complete set of computer aided design (CAD) models are developed to ensure an accurate fit between parts. The software used to generate CAD models is done using CATIA (Computer Aided Three-dimensional Interactive Application (Dassault Systèmes, Cedex, France). The various parts of the mechanism can be seen in Figure 15 and Figure 16.

Figure 15. Left side view of flapping mechanism (computer aided design (CAD) model).

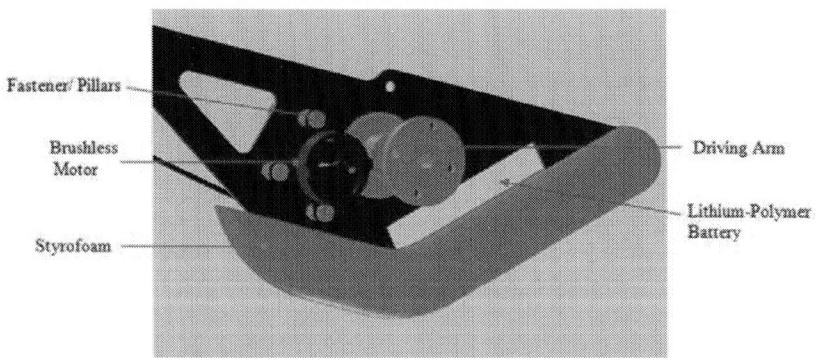

Figure 16. Right side view of flapping mechanism (CAD model).

From Figure 16 the brushless DC motor is mounted onto the fuselage of the ornithopter using three nylon pillars. The motor is powered directly by Lithium–Polymer battery. In order for both wings to flap in sync, a single shaft is installed and attached onto both sides of the driving arms. The driving arms are then attached along with the wings. The Styrofoam at the front side of the fuselage is used to reduce the impact if the ornithopter would to fall. Carbon fiber sticks are used to strengthen the fuselage to prevent twisting and bending stress when it flaps. The Lithium–Polymer battery is placed in the front fuselage, so that the center of gravity for the prototype is achievable.

It is important to make the mechanism as light as possible, therefore the lightest, and most cost effective materials available are considered. Initially, carbon fiber which is strong and light-weight is one of the considerations. However, there are limited resources to manufacture and cut into the desired shape. As a result, the next option is to use cardboard, which is equivalent light-weight and easy to cut but not as durable as carbon fiber. Each component and sub-assembly for this ornithopter is weighted as shown in Table 1.

Table 1. Weight of each Component and Sub-assembly.

Component and Sub-assembly	Weight (g)
Kypom Lithium–Polymer Battery	29
Hacker A10-A12s Brushless Motor	15
Futaba R6004FF Receiver and Castle Creation Thunderbird 9 Electronic Speed Controller	13
2× Tahmazo TS-1002 Servo Motor	8
2× Driving Arm and Linkage	8
Tail Structure	11
Main Shaft	6
Main Gear	7
Middle Gear	5
Pinon Gear	1
Fuselage	15
Fasteners, Screws, Nuts and Wheel-Locks	7
Wings	35
Total Weight	160

3.2. Wing Construction

In the design and construction of any ornithopter, one of the most crucial components governing overall performance is the wings. Especially in a flapping-wing air vehicle, the wings are vital to flight performance aspects including endurance, speed, manoeuvrability, and many other useful behaviors. An effective ornithopter must have wings capable of generating both thrust, the force that propels the craft forward, and lift, the force, perpendicular to the direction of flight, that keeps the ornithopter airborne. These forces must be strong enough to counter the effects of drag and the weight of the ornithopter. The wings produce lift and thrust primarily due to their ability to change shape during the flapping motion. As the wing is accelerated, aerodynamic loading causes the wing to deform. The result is a fairly large camber change of the wing that results in the normally flat plate shape changing to an airfoil shape. The aerodynamic loading also produces large angles of attack that create thrust. When these two effects are combined, the airfoil wings are placed in a moving airstream, thus creating a flight sustaining lift force.

In order to achieve the desired flexibility and minimum weight, carbon fiber rods are used in the construction of the wings to provide a lightweight and stiff structure of spars, similar to the skeletal structure of a flying animal as shown inFigure 17. Ornithopters do not necessarily act like flying animals in flight. Typically flying animals have thin and cambered wings to produce lift and thrust. Ornithopters with thinner wings have a limited angle of attack but provide optimum minimum-drag performance for a single lift coefficient. The Pigeon Hawk has the closest characteristic as our ornithopter prototype, with the weight of 181.3 g and therefore the dimension of the constructed wing is as shown in Figure 18 [9].

There are two different materials used for the wings. The first material used is a polyester film made from stretched Polyethylene Terephthalate (PET). It has high tensile strength, and is commonly used in ornithopter flights. The second material used is Orcon, which is a better material than PET film. The Orcon is a strong and ultra-lightweight material. Unlike other plastic and mylar-like films, Orcon will not continue to tear even a small slit is detected.

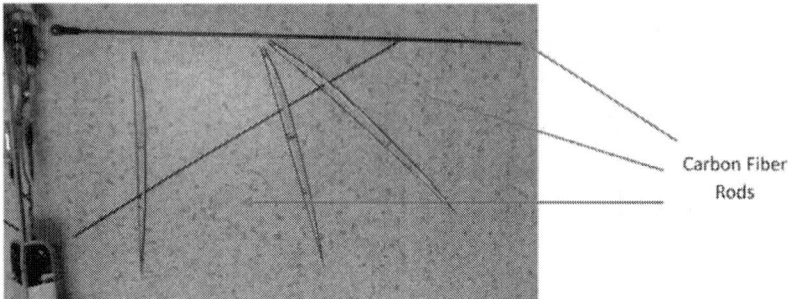

Carbon Fiber
Rods

Figure 17. View of wing skeletal structure—one half.

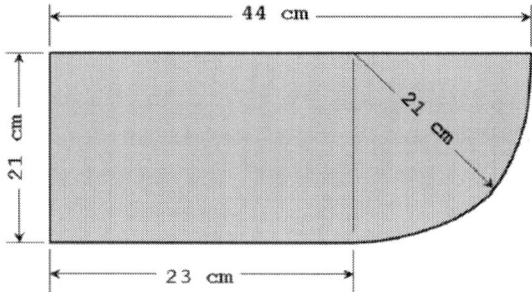

Figure 18. Dimension of the constructed wing [9].

Three different types of wings designs with various material and skeletal structure are fabricated. The first type is called PET cambered thin wing with a skeletal structure as shown in Figure 19. Figure 20 shows the PET cambered thin wing that is fabricated using the PET material.

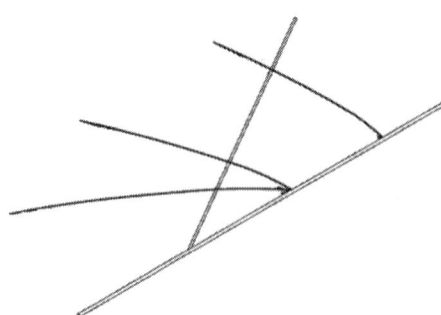

Figure 19. Skeletal structure for polyethylene terephthalate (PET) cambered thin wing (CAD Model).

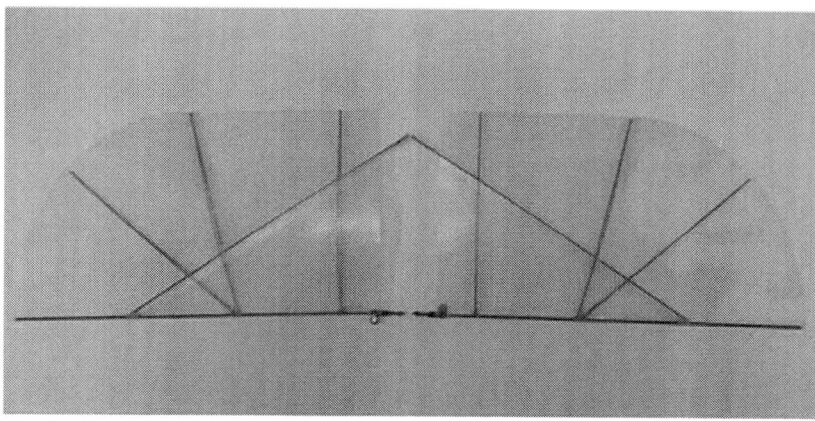

Figure 20. PET cambered thin wing.

The second type of wing is called Orcon cambered thick wing with a skeletal structure as shown in Figure 21.

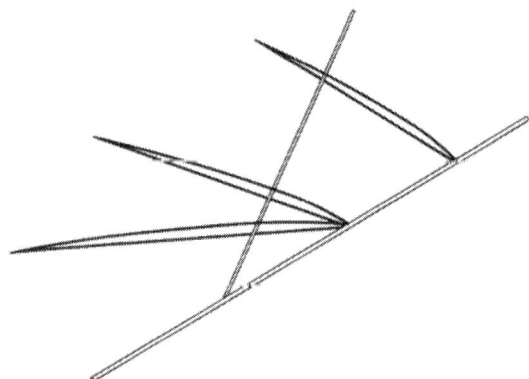

Figure 21. Skeletal structure for Orcon cambered thick wing (CAD model).

Figure 22 shows the Orcon cambered thick wing that is fabricated using the orcon material.

Figure 22. Orcon cambered thick wing.

The first two types of wings that are mentioned above are designed to have cambered due to the advantages of the increased lift-drag ratios and more desirable stall characteristics. It has a higher lift coefficient than the symmetrical airfoil. The top edge of the airfoil is shaped differently than the bottom edge, which changes the way air flows over it. This causes the air to move faster, which creates more lift.

The third type of wing is designed with a thin airfoil, called the Orcon flat wing with a skeletal structure as shown in Figure 23.

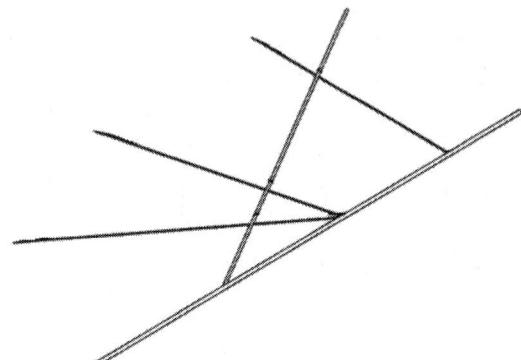

Figure 23. Skeletal structure for Orcon flat wing (CAD model).

The lift coefficient is sacrificed for a lighter weight. Lighter weight can produce smaller inertia of the wing and thus faster flapping speed, which will in turn increase air velocities. Figure 24 shows the Orcon flat wing which is fabricated using the orcon material.

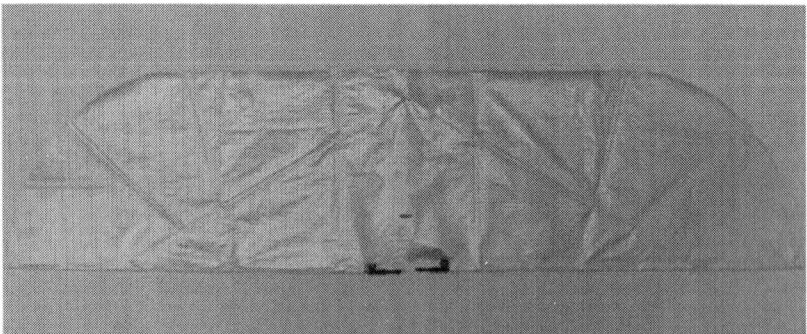

Figure 24. Orcon Flat Wing.

3.3. Center of Gravity (CG)

The center of gravity may be defined as the average location of the mass distribution. It is a point in space where, for the purpose of various calculations, the entire mass of a body may be assumed to be concentrated. The center of gravity is an important point on an aircraft, which significantly affects the stability of the aircraft to fly safely. Therefore the center of gravity must fall within specified limits to ensure the ornithopter is stable. A simple method is adopted to determine the CG of the prototype ornithopter and that is to hang the prototype ornithopter from the ceiling using a kevlar thread as shown inFigure 25, the CG is obtained when the prototype ornithopter is equilibrium and balanced horizontally with the ground level.

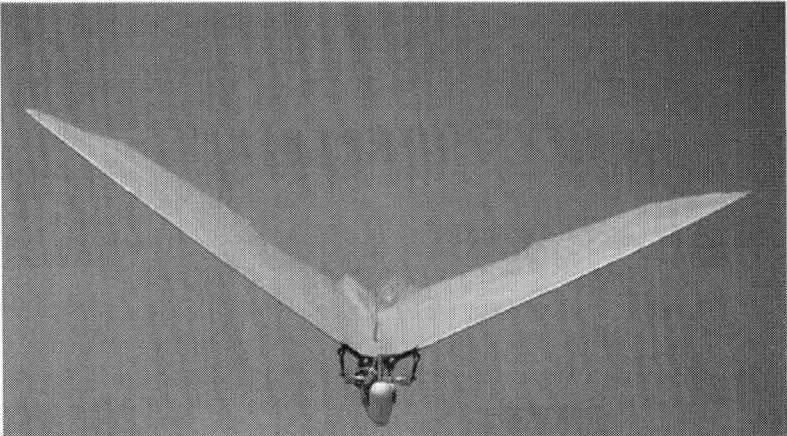

Figure 25. Equilibrium position of the prototype ornithopter.

3.4. Lift and Thrust Force Measurement

The experimental apparatus utilized by the authors include a force measuring load cell, a test assembly, and video recording is also taken of the mechanism with the various wing designs flapped. This section also presents the calibration of the load cell to ensure that the data measured is accurate. After the calibration is done, the data unit of the force measurement is converted from electrical signal into gram or Newton unit.

3.4.1. Force Measuring Load Cell

The first and most important component of the force measurement system is the transducer. A load cell is a transducer that is used to convert the loads generated by the flapping wing into voltage signal for subsequent recording and processing. Through a mechanical arrangement, the force being sensed deforms a strain gauge. In most cases, four strain gages are used to obtain maximum sensitivity and temperature compensation. Two of the gauges are usually in tension, and two in compression. A strain gauge is a series of thin wire filaments wound in a serpentine fashion and placed in a Wheatstone Bridge configuration.

Voltage is supplied to the strain gauge and as a load is applied to the wire filaments, they will either elongate or shrink, changing the resistance in the wires. This variation in resistance results in different input and output voltages from the strain gauge. The difference in voltage is then used to calculate the strain.

It should be noted that the expected measured thrust and lift are very small, *i.e.*, around 0.1–1.0 N. The force measuring load cell utilized in this project is an Omega Engineering, LCMFD-20N load cell which is shown in Figure 26. This LCMFD-20N load cell is small in size and capable of providing highly accurate readings of a 2041 gram capacity. According to Omega Engineering's website (www.omega.co.uk) on technical specification, this LCMFD-20N load cell has its accuracy (linearity and hysteresis combined) and uncertainty of 0.15% FSO (Full-Scale Output), *i.e.*, 0.03 N (0.15% of 20 N). Since, the accuracy and uncertainty numbers are about 0.03 N (order of 10^{-2}), *i.e.*, about 1 order of magnitude less of the expected minimum measured thrust and lift values around 0.1–1.0 N (order of 10^{-1} to 1), this is acceptable. In addition, this model is selected because of its high frequency resonant characteristics, minimal contamination from off-axis loads, and robust overloading tolerances.

Figure 26. Omega engineering, LCMFD-20N load cell.

The load cell measures the forces when the prototype ornithopter flaps its wing on a force balance by converting these forces into electric signals, which will be displayed on the Omega DP41 digital panel.

3.4.2. Test Assembly

One of the challenges in this project is to design and build a test assembly made of pinewood in which the mechanism could flap and also measure both lift and thrust forces with the use of the load cell. A flapping mechanism in the test assembly can be seen in Figure 27. The position of the load cell is mounted to a fixed point on the test assembly, and by rotating the test assembly vertically together with the respective kevlar thread connected to the flapping mechanism, this testing assembly could measure the lift and thrust forces generated by the flapping mechanism. In addition, this only allow the flapping mechanism to move in either $\pm X$ axis, which measures the thrust force, or $\pm Y$ axis, which measures the lift force.

G-clamps are used to clamp the test assembly to a metal stand. This allows the wing tests to be done at a distance above ground so that we do not take into account the ground effects when the wings are flapping.

Figure 27. Test assembly and flapping mechanism.

3.4.3. Lift and Thrust Forces Measurement Methods

As the objective of this project is to characterize the different lift and thrust generation performance from three different wing designs, namely PET cambered thin wing, Orcon cambered thick wing and Orcon flat wing, different thrust measurement techniques will be investigated. This section presents the various methods for testing the lift and thrust generation of the prototype ornithopter. Kevlar thread is used extensively in this project as this innovative thread is light and its strength to weight ratio makes it five times as strong as steel. Kevlar thread does not break instantly but progressively, providing a non-catastrophic failure mode allowing a margin of safety.

Thrust Measurement by Swing Method
In this method, the prototype ornithopter is tightly attached to one end of the connecting kevlar thread and the other end is attached to the ceiling, making the body axis of the prototype ornithopter perpendicular to the connecting kevlar thread, as shown in Figure 28. The three different pairs of wings are tested using this method to measure the thrust produced.

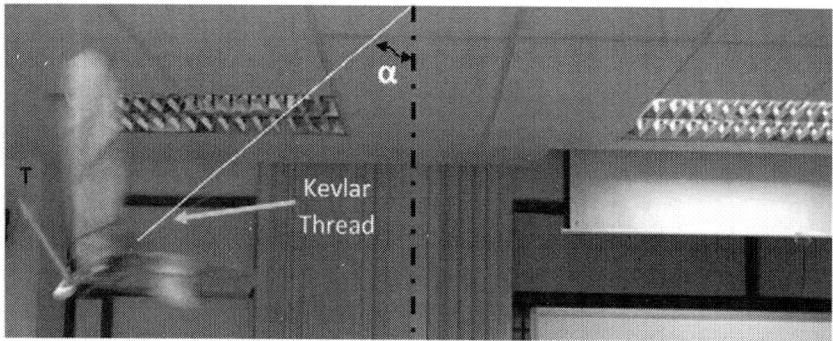

Figure 28. Thrust measurement of swing method.

When the prototype ornithopter is excited, it produced thrust and moved to an equilibrium position. Assuming that all forces act at the center of gravity of the prototype ornithopter as shown in Figure 1, the equilibrium equation of the ornithopter after applying voltage is

$$T + F + L + W = 0$$

where T, F, L, W are the thrust, tension of the connecting kevlar thread, side force, and weight of the prototype ornithopter, respectively. T and L are two components of the resultant aerodynamic force projected on the ornithopter's body axis and the direction perpendicular to the ornithopter's axis, respectively. The equilibrium condition in vertical and horizontal directions can be modified as follows:

$$\sin(\alpha + \beta) + F\cos\alpha + L\cos(\alpha + \beta) - W = 0$$
$$T\cos(\alpha + \beta) - F\sin\alpha - L\sin(\alpha + \beta) = 0$$

where α and β are the swing angle of the connecting kevlar thread and the body angle of the ornithopter, respectively, as shown in the Figure 29.

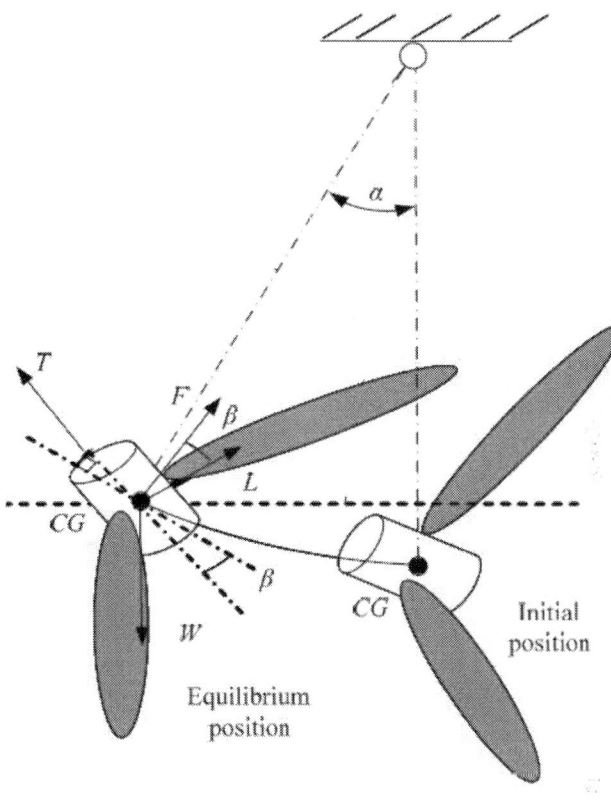

Figure 29. Force diagram of swing test.

The body angle β is the angle between the ornithopter's body axis and the line perpendicular to the connecting kevlar thread. By eliminating the tension F in Equation (2), the thrust is calculated as follows:

$$T\cos\beta - L\sin\beta = W\sin\alpha$$

Hence, if the ornithopter's body axis is zero (β = 0°; *i.e.*, the body axis of the ornithopter is perpendicular to the connecting kevlar thread), the thrust at equilibrium can be calculated by:

$$T = W\sin\alpha$$

Load Cell Thrust Measurement Experiment Setup
In this thrust measurement setup, the mounting orientation of the load cell is shown in Figure 30. The load cell is fixed onto the test assembly in a stationary

position. The front portion of the flapping mechanism is attached to the load cell while the top and bottom segments of the flapping mechanism are tied vertically to the test assembly with the use of kevlar thread. Thus, after mounting the pair of wings onto the flapping mechanism and turn on the motor, the flapping mechanism is constrained by the kevlar thread so that it is only capable of motion in the X-direction.

Motion in the $+X$ direction represented thrust, while motion in the $-X$ direction represented drag. As the flapping mechanism is moving in the $+X$ direction, the load cell is under compression and therefore the readings displayed on the digital panel are negative values as discussed in Section 2 Load Cell Calibration. In this configuration, the thrust produced for the three different wing designs at three different flapping frequencies will be measured and then recorded in an Excel spreadsheet.

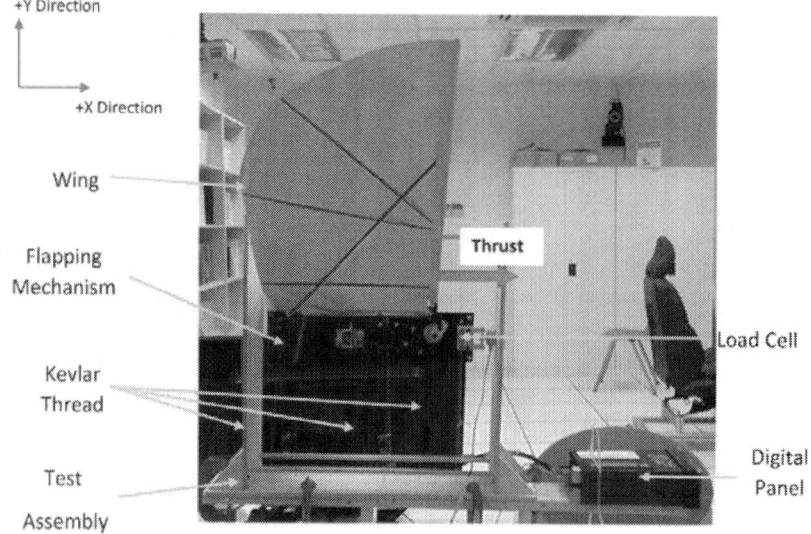

Figure 30. Side view of the load cell thrust measurement setup.

Load Cell Lift Measurement Experiment Setup
In the lift measurement setup, the test assembly is vertically rotated where the bottom part of the flapping mechanism is attached to the load cell as shown in Figure 31. The kevlar thread is tied horizontally at the front and back segments of the flapping mechanism, so that it is only capable of motion in the Y direction which the lift force is measured.

As the flapping mechanism is moving in the Y direction, the load cell is under both compression and tension, which resulting in negative and positive values respectively displayed on the digital panel. This is due to the matter of fact that a flapping wing is an aerodynamic machine with two strokes, the upstroke and the downstroke. In this experiment, the lift produced for the three different wing designs at three different flapping frequency will be measured and recorded in an Excel spreadsheet. Before recording the lift values during the flapping period, the weight of the flapping mechanism with the wings attached is measured at static condition. By subtracting the weight measured at static condition from all the values obtained during the flapping phase, the true lift force produced by the prototype is then recorded for analysis.

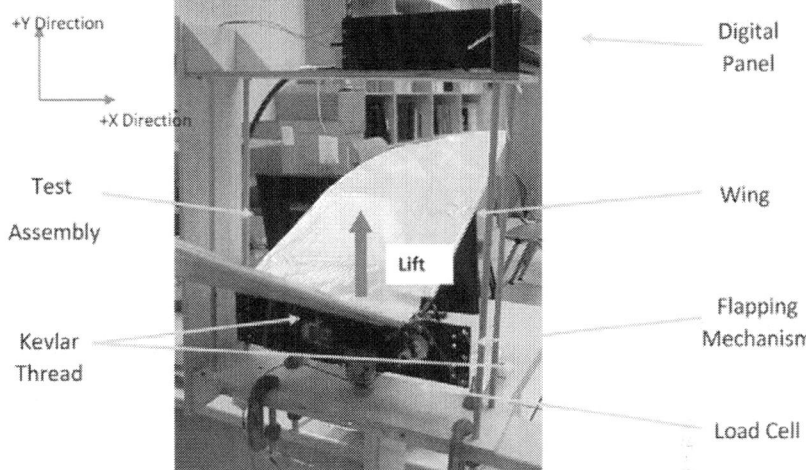

Figure 31. Load cell lift measurement setup.

3.4.4. Determination of Flapping Frequency

The flapping frequency of the wings when different amount of power is supplied to the motor is to be determined so as to make a comparison between the three different wing designs, as shown below (Figure 32), at the various flapping frequency.

Figure 32. Three different wing designs (bottom view).

The measurement of the flapping frequency of the prototype is achieved through the use of a digital camera. The digital camera is placed in front of the test assembly which allows the author to analyze how the wings are flapped in up- and down-stroke manner. The videos are then loaded into the Ulead software to process the captured video into frames. There are 25 frames for each second of the video recorded. Hence, by calculating the number of frames taken for the prototype to produce a full cycle consisting of upstroke and downstroke as shown in Figure 33, the period (T) can be obtained using the following equation:

$$T = (1/25) \times \text{number of frames per full cycle}$$

Once the period is obtained, the flapping frequency (f) is calculated as follows:

$$f = (1/T)$$

Hence, the flapping frequency can then be determined. The figure below shows the motion of the flapping wings in one full cycle.

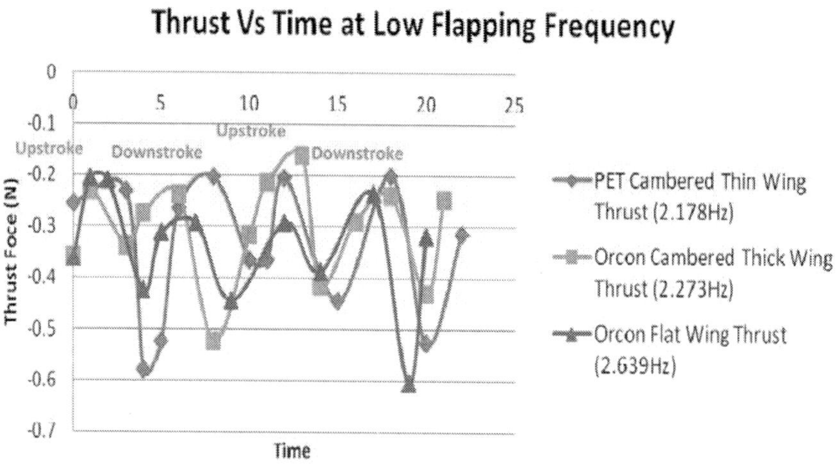

Figure 33. Thrust against low flapping frequency (2 full cycles).

3.5. Lift and Thrust Force Results and Analysis

This section presents the findings of various experiments that the authors have conducted. The main element of this section is the results and analysis of evaluation of different wing designs. The average thrust and lift output are measured and compared to different wing designs to understand what styles of wings are more effective. The time-varying force profile can be examined to gain understanding into why particular wing designs prove more effective than others. The results of thrust and lift testing will be analyzed to establish any trends that explain performance differences and this section will be split into five different parts:

- Comparison of thrust generation performance between various flapping frequencies at a particular wing design.
- Comparison of thrust generation performance between wings of various designs against various flapping frequencies.
- Comparison of thrust performance between the load cell measuring setup and swing method.
- Comparison of lift generation performance between wings of various designs at different flapping frequencies.
- Investigation of lift generation between wings of different designs at various phase angles.

Thrust generation tests are done using the load cell setup method on the three different wing designs. For each wing design, three different flapping frequency staring from low, then medium and lastly to high are evaluated.

It is observed from Figure 33 that the thrust tends to produce four peaks. This is due to the motion of flapping wings. For each test, thrust performance is recorded in terms of two flapping cycles that consist of upstroke, downstroke, upstroke and downstroke motion, which explains the four peaks that are captured in the graph.

As mentioned in earlier, negative values are obtained when the load cell is under compression. Since the load cell is experiencing compression throughout the flapping period, it means that thrust is always being generated throughout the flapping cycle. More thrust is generated on the downstroke as compared to upstroke.

Apart from that, the slight periodic nature of the thrust is due to the inertial effects of the flapping mechanism. In the graphs above, the inertial effects have not been removed. When flapping with the test set up as described earlier, some inertial effects were present in the thrust direction. This is because the clearances between the flapping mechanism and the test assembly did not purely constrain the flapping in the vertical direction.

3.6. Thrust Comparison between Wings of Various Design

It can be observed from the results as shown in Figure 34 below that Orcon Flat wing has a better thrust generation performance as compared to the rest of the wing designs. It is interesting to note that the thrust for the Orcon flat wing resulted in a linear relationship when compared against the flapping frequency.

The flapping wing mechanism constantly appears to produce thrust. The reason for this is that the wing is constantly changing its shape and angle of attack dynamically. On the downstroke the wing is pitched down, forcing air in the $-Y$ direction, while in the upstroke the wing sweeps forward and up, minimizing the movement of air in the direction causing negative thrust. For both cases the thrust is relatively constant throughout the cycle. The magnitude of the thrust appears to slightly increase for the higher frequency.

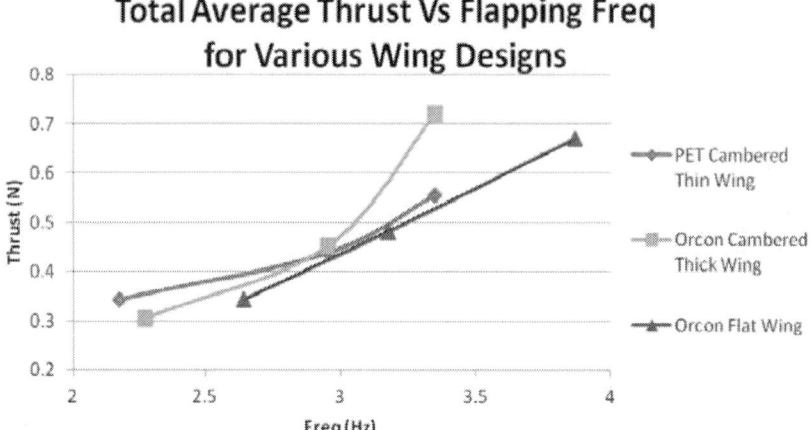

Figure 34. Total average thrust against flapping frequency for various wings.

Orcon Flat wing is much flexible than the other wing designs. Flexible wings can attain efficiency as more elastic the wing is, the more thrust produced and Orcon material is used to achieve a minimum weight. It is observed that PET cambered thin wing has a slightly slower flapping rate. It is because the weight of the flapping mechanism with that pair of wings on is 1.6 N, which weights the highest and tends to have more inertia. Hence, the motor will require more torque to drive the wings. With the same torque, heavier wing will have slightly slower flapping rate which results in a lesser thrust as compared to the Orcon Flat wing.

3.7. Thrust Comparison between Load Cell Setup and Swing Method

Swing angle α is the angle between the initial position and equilibrium position of the prototype ornithopter, and is determined from digital camera images. As discussed, the thrust produced by the ornithopter can be calculated using Equation (5). Table 2 below shows the comparison of thrust results obtained from the various wing designs at three different flapping frequency.

Lift generation tests are done using the load cell setup method as mentioned earlier on three different wing designs. Three sets of wings are tested at three different flapping frequency staring from low, then medium and lastly to high, and the lift force is compared at each speed. As the flapping mechanism is moving in the ±Y direction, the load cell is under both compression and tension, which resulting in negative and positive values respectively displayed on the digital panel. This is due to the matter of fact that a flapping wing is an aerodynamic machine with two strokes, the upstroke and the downstroke. During the upstroke, the flapping mechanism will be "pushing" at the load cell, which results in a compression at the load cell and negative lift values will be displayed. During the downstroke, lift is generated and the flapping mechanism will be "pulling" the load cell, which results a tension at the load cell and positive lift values will be shown.

The first wing design to test is the PET cambered thin wing and the plots in Figure 20 shows the lift against a period of time of 10 full flapping cycle for three different flapping frequency.

Table 2. Comparison of thrust between load cell and swing method.

Parameter	Flapping Mechanism with		
	PET Cambered Thin Wing	Orcon Cambered Thick Wing	Orcon Flat Wing
At Low Flapping Frequency			
Swing Angle (α)	14°	12°	15°
Weight (N)	1.6	1.53	1.47
Swing Method Thrust $T = W\sin\alpha$	0.387	0.318	0.38
Load Cell Measurement Thrust (N)	0.343	0.305	0.344
At Medium Flapping Frequency			
Swing Angle (α)	16°	18°	20°
Weight (N)	1.6	1.53	1.47
Swing Method Thrust $T = W\sin\alpha$	0.441	0.472	0.503
Load Cell Measurement Thrust (N)	0.438	0.452	0.481
At High Flapping Frequency			
Swing Angle (α)	22°	28°	28°
Weight (N)	1.6	1.53	1.47
Swing Method Thrust $T = W\sin\alpha$	0.599	0.718	0.69
Load Cell Measurement Thrust (N)	0.555	0.719	0.67

It is observed from Figure 35 that as the flapping frequency goes higher, the lift is slightly increased. The average lift force generated during the low, medium and high flapping cycle are −0.196, −0.14 and −0.218 N respectively. Figure 36 shows the amount of lift force generated at different angles for low flapping frequency.

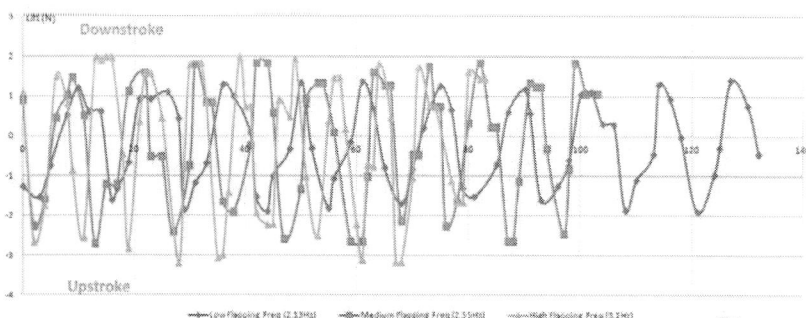

Figure 35. PET cambered thin wing lift against time (10 full cycles).

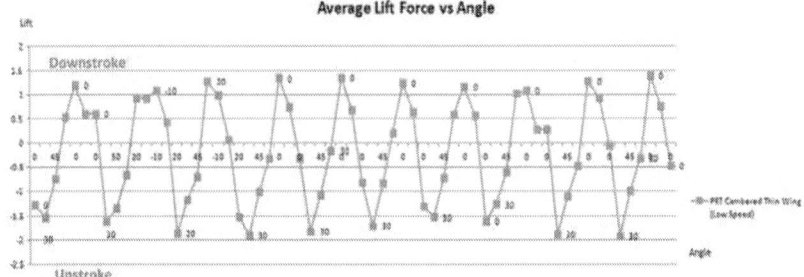

Figure 36. Average lift Force *vs.* angle (PET cambered thin wing).

From the figure above, the results are consistent with the research made as discussed in earlier section that lift is generated at the down stroke, which reaches the maximum lift of 1.38 N at about 0° to −10° angle. However, the minimum lift of −1.92 N is reached when the flapping wing is on its upstroke. Note that the peak negative lift is greater than the peak positive lift, thus indicating that no net lift is generated.

The second wing design to test is the Orcon cambered thick wing and the plots in Figure 37 shows the lift against a period of time of 10 full flapping cycle for three different flapping frequency.

It is observed from the figure above that the faster the flapping frequency, the higher the lift is generated. The average lift force generated during the low, medium and high flapping cycle are −0.23, −0.26 and −0.3 N, respectively. Therefore, it is also observed that faster the flapping frequency will result a higher negative lift force value. Figure 38 shows the amount of lift force generated at different angles for low flapping frequency.

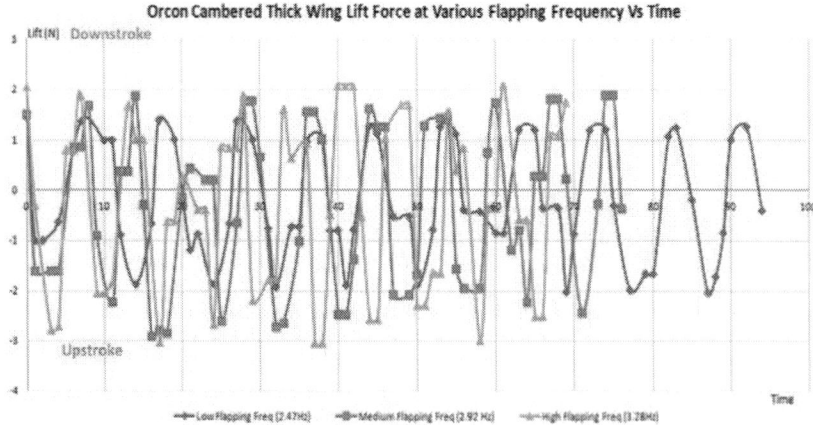

Figure 37. Orcon cambered thick wing lift against time (10 full cycles).

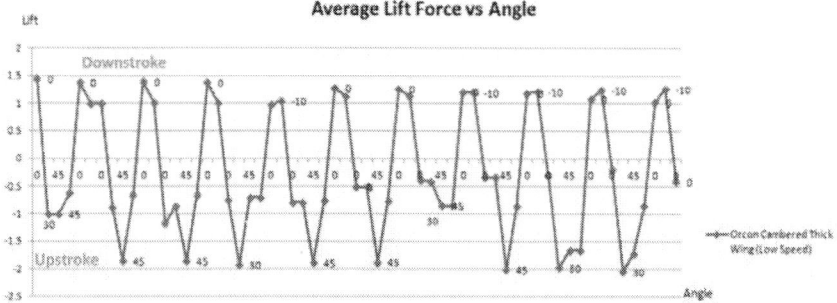

Figure 38. Average lift force *vs.* angle (Orcon cambered thick wing).

From the figure above, the result is similar to the PET cambered thin wing that lift is generated at the down stroke, which reaches the maximum lift of 1.46 N at about 0° to −10° angle. However, the minimum lift of −2.04 N is reached when the flapping wing is on its upstroke. Note that the peak negative lift is greater than the peak positive lift, thus indicating that no net lift is generated.

The third wing design to test is the Orcon flat wing and the plots in Figure 39 shows the lift against a period of time of 10 full flapping cycle for three different flapping frequency. From Figure 39, it is observed that faster the flapping frequency, the higher the lift is generated. The average lift force generated during the low, medium and high flapping cycle are −0.07, 0.26 and −0.14 N, respectively. It is interesting to find that a small amount of lift is generated at medium flapping frequency.Figure 40 shows the amount of lift force generated at different angles for low flapping frequency.

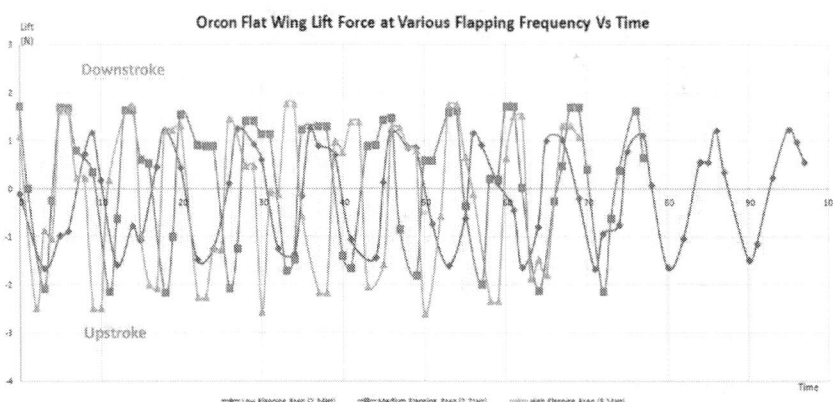

Figure 39. Orcon flat wing lift against time (10 full cycles).

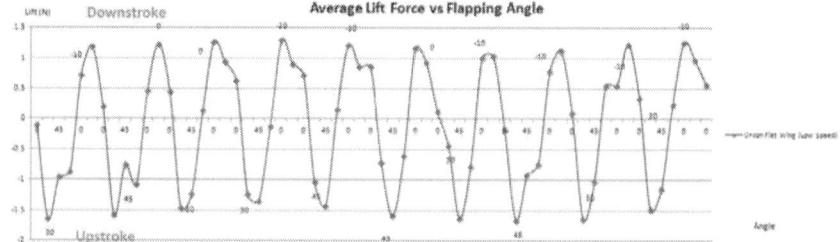

Figure 40. Average lift force *vs*. angle (Orcon flat wing).

From these figures, the result is similar to the previous two tests that lift is generated at the down stroke, which reaches the maximum lift of 1.25 N at about 0° to −10° angle. However, the minimum lift of −1.66 N is reached when the flapping wing is on its upstroke. Note that the peak negative lift is greater than the peak positive lift, thus indicating that no net lift is generated.

3.8. Comparison of Lift between Wings of Various Design

Table 3 shows that the only wing design that produced lift is the Orcon flat wing flapping at medium speed. However, most of the data obtained in this lift test has showed that lift forces produced are generally negative. This means that the prototype ornithopter will not be able to sustain a flight in the air. The Orcon flat wing has shown potential in generating lift with the most lift produced as compared to the rest of the wings. This may be due to the advantage of its light weight and flexibility of wing design. Having a lighter weight will actually allow the flapping mechanism to flap the wings faster. Therefore, improvements and modification are to be made to the wing design in order to produce the sufficient lift to sustain a flight.

Table 3. Comparison of lift forces.

Type	Low Speed (N)	Medium Speed (N)	High Speed (N)
PET Cambered Thin Wing	−0.196	−0.140	−0.218
Orcon Cambered Thick Wing	−0.225	−0.256	−0.305
Orcon Flat Wing	−0.075	0.264	−0.142

Figure 41 below shows the cross section view of the three various wing designs.

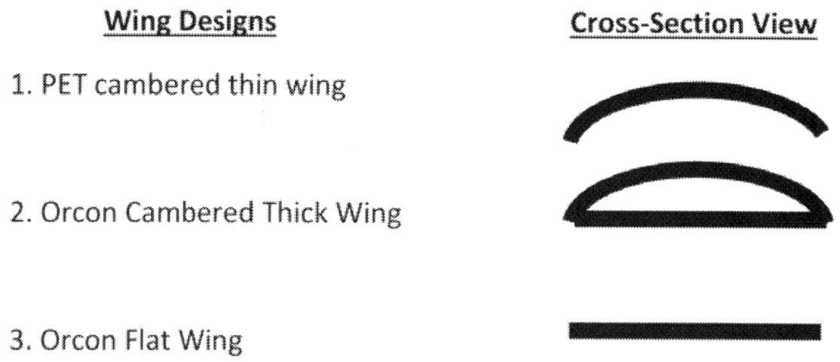

Wing Designs	Cross-Section View
1. PET cambered thin wing	
2. Orcon Cambered Thick Wing	
3. Orcon Flat Wing	

Figure 41. Cross-section view of the three different wing designs.

The next few figures (Figure 42, Figure 43, Figure 44, Figure 45 and Figure 46) will show the investigation of phase angles against the lift generation for the three different wing designs at various flapping frequency. The first wing design to be discussed is the PET cambered thin wing, followed by the Orcon cambered thick wing and lastly the Orcon flat wing.

It is observed from the previous figures that as the flapping frequency increases, there is a transition from a "smooth circular" graph to a "figure-of-eight" graph. An example of the lift generation from the Orcon flat wing flapping at the medium frequency graph as shown below, during the initial downstroke, there is a sudden rise in lift until it reaches its peak. When the lift reaches its peak somewhere around 30°, it is noticed that constant lift is produced until it reaches at an angle of 0°. This could be the leading edge vortex (LEV) that causes it. It appears that LEV can enhance lift by attaching the bounded vortex core to the leading edge during wing translation. The vortex, formed roughly parallel to the leading edge of the wing, is trapped by the airflow and remains fixed to the upper surface of the wing. As air flows around the leading edge, it flows over the trapped vortex and is pulled in and down to generate the lift.

There are two routes that can be seen from the graphs, the first route which is the positive angle transit to negative angle (Downstroke) and the second route is the reverse of the first route (Upstroke). From Figure 46, the net lift can be easily seen by looking at the difference between the two routes.

4. UGS FLAPPING WING MAV PROTOTYPES 2 AND 3

With the lesson learnt on the materials from the first prototype obtained in the previous section, and from other literatures ([29,30,31,32,33]), we designed, built and flew another two flapping wing MAVs using fabrication method such as laser cutting and Rapid Prototyping.

4.1. Prototype 2

Our flapping wing MAV would be based on an albatross-like design. In addition to the results shown above with inspiration from other ornithopter-like MAVs (e.g., Delfly [34,35]). We introduced additional design criteria, e.g., it has to be lightweight, simple and yet strong enough to withstand the stress of the flapping motion and the crash landings during test flight. Simplicity is the key here as most of the components that would be used would be from hobby shops.

4.1.1. Flapping Wing Mechanism

The flapping wing mechanism function is to convert the motor's rotary motion into flapping motion. It is the most important component of the MAV thus much research was done to assess the many different designs available. Generally the mechanism design is about the same to each other with only slight modifications.

Staggered Crank Design
The staggered crank design in Figure 47 is the most basic of the flapping wing design [36]). The connector rods are staggered in a measured distance and angle to ensure that the left and right wing are flapping symmetrically. This design is favoured by a hobbyist who wants to attempt to make their own Ornithopter using household items. Modifications have to be made so that the motor can be used instead of a rubber band as its power source.

Single Gear Crank Design
The single gear crank design in Figure 48 taken from University of California Biomimetic Millisystems Lab [37], looks simple however it is more complicated than it seems. Figure 48 shows the wings at the same level. The center point where the connector rod and the wing hinges are connected to each other has to expand and contract as the mechanism flaps. Contracting and expanding at a very high frequency could result in component failure.

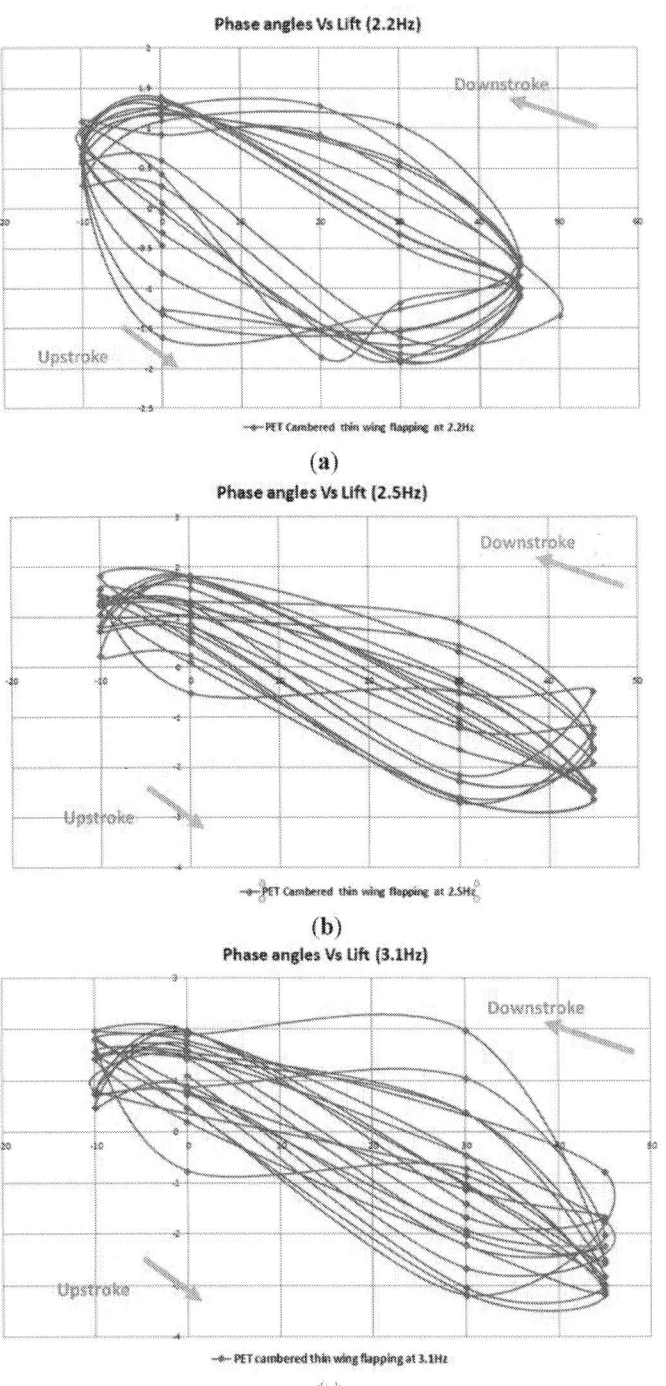

Figure 42. Phase angle *vs*. lift for PET cambered thin wing. (**a**) Low speed; (**b**) medium speed; (**c**) high speed.

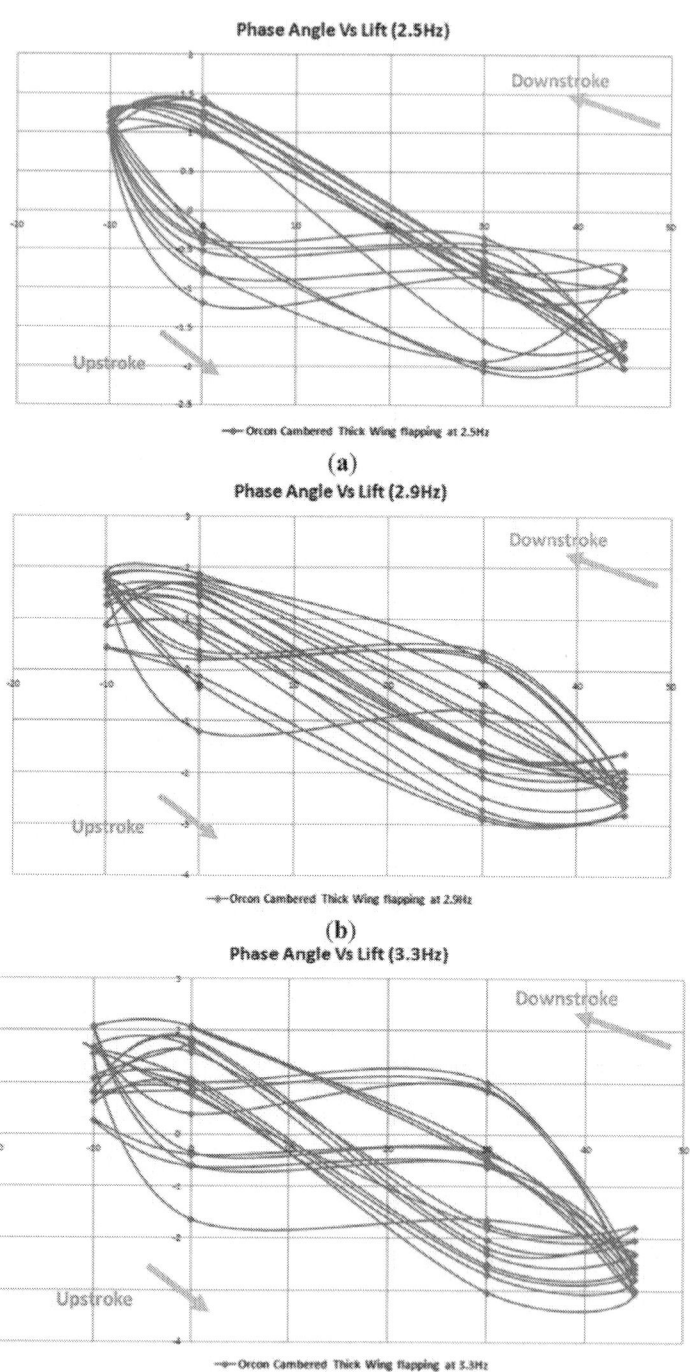

Figure 43. Phase angle *vs.* lift for Orcon cambered thick wing. (**a**) Low speed; (**b**) medium speed; (**c**) high speed.

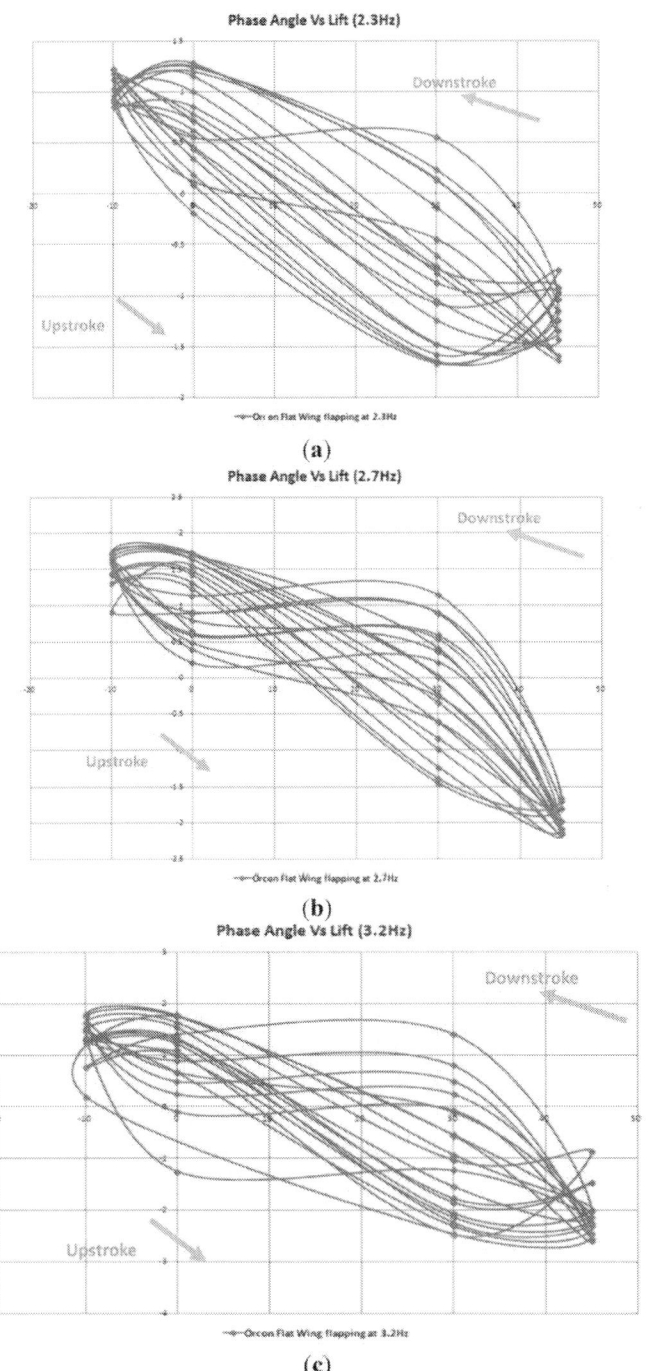

Figure 44. Phase angle *vs.* lift for Orcon flat wing. (**a**) Low speed; (**b**) medium speed; (**c**) high speed.

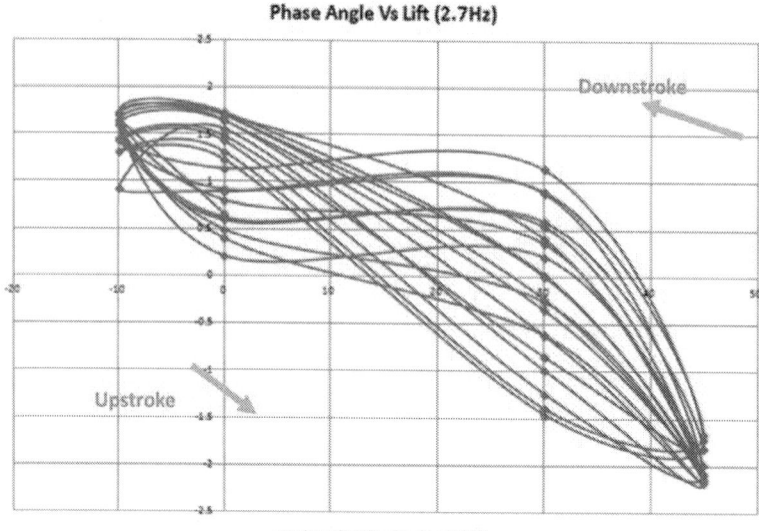

Figure 45. Phase angle *vs.* lift for Orcon flat wing flapping at 2.7 Hz.

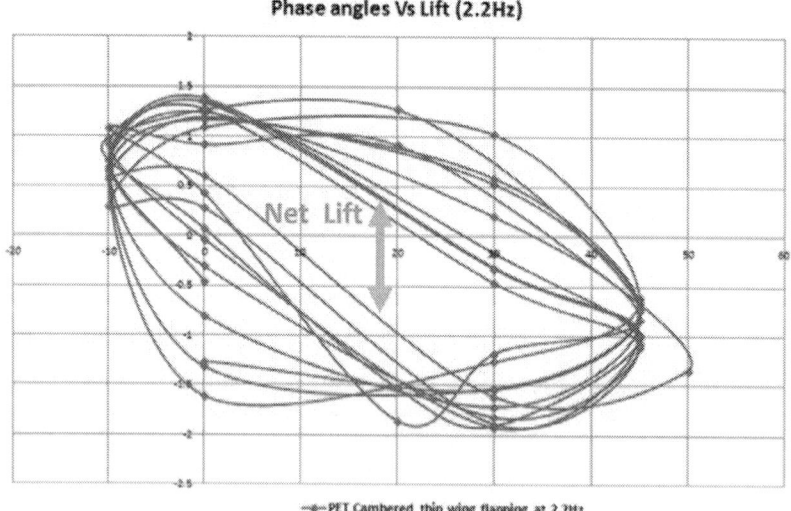

Figure 46. Net lift generation.

Dual Gear Crank Design

Figure 49 shows the dual gear crank design from similarly used in the Festo's SmartBird [38]. It features two gears that controls each wing hinges separately. There are different variation to the drivetrain design. The one shown in Figure 49, uses the pinion wheel to drive both the secondary gears. The secondary gears will rotate in the same direction with each other. In the other design, the pinion gear rotates the secondary gear and this secondary gear rotates another secondary gear. The secondary gears would rotate counter clockwise to each other. This design is much simpler to implement and reduce the wing symmetry misalignment.

Figure 47. Staggered crank.

Figure 48. Single gear crank.

Figure 49. Dual gear crank.

Transverse Shaft

The transverse shaft design shown in Figure 50 is the other variation of flapping mechanism from [39] which allows for the most symmetrical flap, however, it is the heaviest and the most complicated design. The rotating gears and the flapping wings are not in the same plane thus the connector rod has to be able to rotate. The connector rod has a ball bearing inside and this adds weight to just the component itself. The number of gears used in this design is more than any other design. The transverse shaft design is usually used for a bigger MAV design where weight could be overcome by large wings.

4.1.2. Tail

The tail design varies with its intended use. Some of the design uses it only for stability but in most cases they are used for control as well. For stability, the tail is tilted upwards so that it the downward force of the tail would force the nose to pitch up. The angle is typically around 15° or less. For control the more common designs implemented are the swinging tail (Figure 51) and the tilting tail (Figure 52) due to their simplicity. The swinging tail works by causing a rolling moment to when it swings to either side. The tilting tail works like a rudder, when it tilts to the right it causes the MAV to yaw to the right. A horizontal stabilizer tail design unlike the other two designs could provide additional control. It can act as an elevon, providing pitch and roll control. However, this design requires two servos to be used and a more complicated design.

Figure 50. Transverse Shaft.

Figure 51. Swinging tail.

4.1.3. Body

The body is the part where the components like the electronic speed controller, the receiver and the battery is located. The body also has to hold all the components from moving around too much. This is to prevent the shifting of the center of gravity of the MAV. The components would each be taped separately and then hooked to the body by Velcro tape. As the design would not require much space the body design could be hollowed. Figure 53 shows the body design with holes in them. This significantly reduces the total weight of the body. The body design had to be glued to the flapping mechanism at a 90° angle. Small triangles were added in between them as a support structure to prevent the body and the flapping wing mechanism from snapping off.

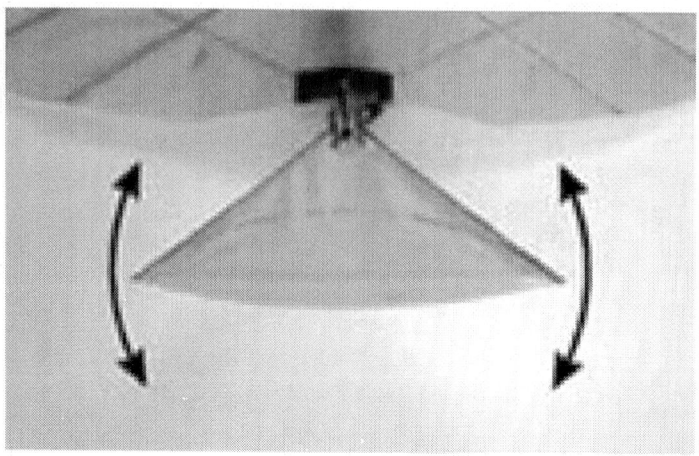

Figure 52. Tilting tail.

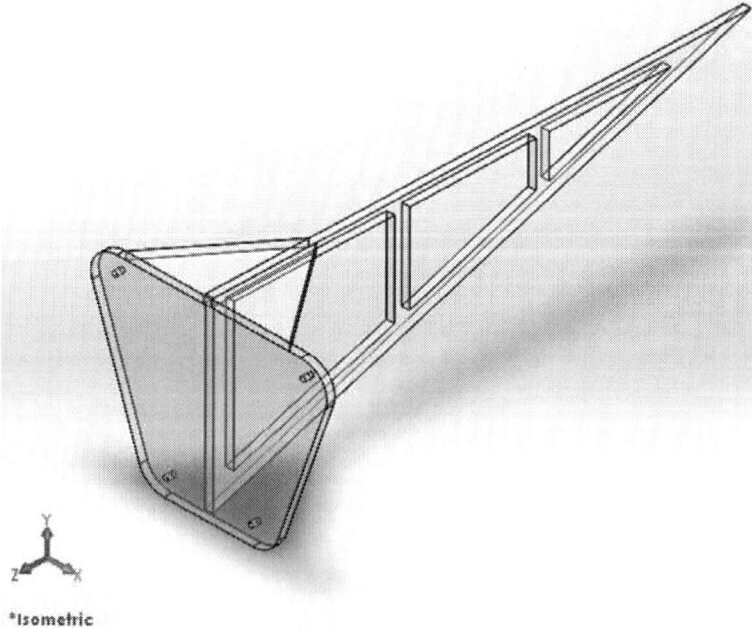

Figure 53. Body design.

4.1.4. Gear and Motor Selection

The gear design was dependent on the motor that is going to be used. The motor rating affects the gear ratio which then affects the flapping frequency. The motor that is used is a brushless outrunner motor. Outrunner motors have lower KV ratings meaning they have more torque but less speed. More torque is needed than speed for this project as the motors have to turn the gears to flap. The motor also needed a front mount so that it could be mounted easily to the flapping mechanism frame instead of a separate mount just for the motor. This narrows down to two motors as shown on the Table 4.

Table 4. Motor specification comparison.

Specification/Motor	Motor 1	Motor 2
Motor Rating (KV)	1200	2800
Load Speed (rpm)	5800	8350
Voltage (V)	11.1	4
Weight (g)	38	25

Motor 2 was chosen as it was lighter and requires lesser voltage. Voltage is linked to the number of cells that the Lithium–Polymer (Li–Po) batteries has and the rating of Electronic Speed Controller (ESC). Each cell on a battery is 3.7 V so the higher the voltage the heavier the battery. It is the same for ESCs, higher ratings means bigger and heavier ESCs.

4.1.5. Fabrication and Material

There were three materials being considered initially: Carbon fiber, balsa wood and acrylic. The first material of choice was to use carbon fiber due to it being strong and light. As it turns out, CO_2 laser cutting a carbon fiber sheet would burn the material. Balsa wood is very light and easy to cut; however, due to complex design of the MAV it was decided that it was not a suitable material. Hence, acrylic was selected. Acrylic is not as light and strong as carbon fiber however it can use CO_2 laser cutting machine to do precision cutting.

4.1.6. CAD Design Dimensions

In order to find out the total dimensions and the weight that is allowed for flight, a lift equation was used. Certain assumptions made before using this equation are as follows:

1. The resulting lift would be higher in reality due to neglecting other flapping wing effects that contribute to lift when flapping.
2. The coefficient of lift is independent of the location on the wing and time.

From the assumptions made the equation for a rectangular wing shaped (of the same area) lift could be expanded to [38,40]:

$$L = \varphi_0^2 \cdot \pi^2 \cdot f^2 \cdot C_L \cdot \rho \cdot c_0 \cdot l^3 \cdot \frac{1}{3}$$

where φ_0 is flapping angle, f is the flapping frequency, c_0 is the chord length and l is the wing span length. C_L is obtained from the CFD results in the previous section (Section 2). This equation is to be used as a rough estimate so that the dimensions and weight of the MAV could be measured. Table 5 shows the results from using the equation.

Table 5. Approximated lift generated.

Parameters	Values	Unit
Flapping Amplitude	70	deg
Flapping Frequency	6.5	Hz
Lift Coefficient	0.8	
Air Density	1.225	kg/m³
Chord Length	0.13	m
Wing Span	0.3	m
Lift	1.684	N

CAD Design
Using the dimensions above and the design criteria, a CAD design using SolidWorks was modelled. It would incorporate a dual gear crank and a horizontal stabilizer tail design. The dual gear crank was the simplest design with not much wing symmetry misalignment. The horizontal stabilizer tail design was chosen as it could provide both pitch and roll control. The initial design showed in Figure 54 featured an articulated wing. This design was not used as there were too many moving parts in the design and may complicate things. Therefore the chosen design is the one shown in Figure 55.

The total weight of the MAV was measured using one of the features in SolidWorks. Now the total weight of the MAV plus the components could be compared to the lift equation result. Table 6 shows the sum of all the component weights. The two measurements show that the weight of the MAV is below the total lift generated. An image of the assembled MAV is shown in Figure 56.

4.2. Prototype 3

Learning from the prototype 2, some design considerations were made, *i.e.*, (1) the flapping mechanism needs to be more simplified; (2) the number of moving parts need to be reduced; (3) the overall design has to be much smaller to reduce weight; and (4) changing the tail design to either a tilting or swinging tail would reduce the number of servos used which would reduce weight.

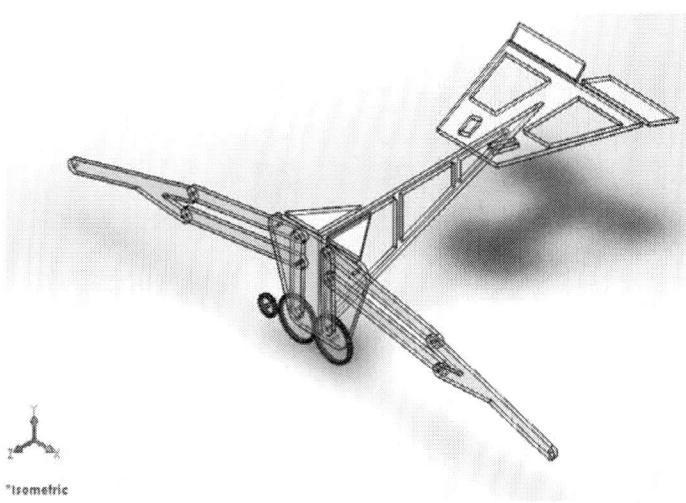

Figure 54. Articulated flapping wing.

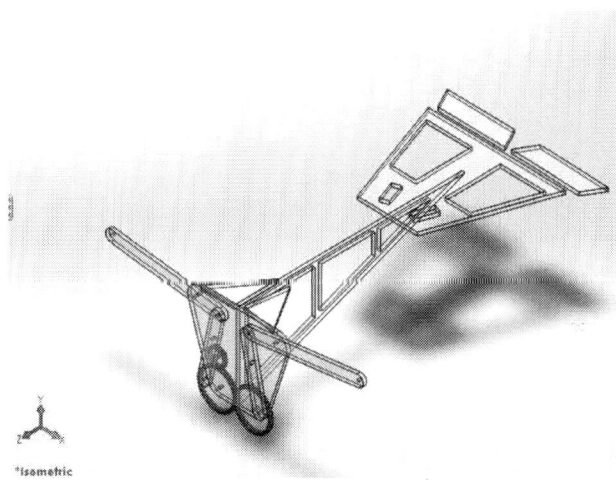

Figure 55. Single flapping wing.

4.2.1. Flapping Wing Mechanism

The flapping wing mechanism for prototype 2 had too many moving parts and was not simplified enough. A simpler design was needed and thus another look at the single gear crank was taken. The design idea was to shift its fixed pivot point from being at the center of the wing to it being at the end of the two wing joints. Figure 57 shows this design. The changes made to its pivot point made the flapping mechanism worked properly. A simulation test was done using the software and it showed that it could hold at high frequency flapping and the flapping movement is synchronized.

Table 6. Total weight of micro air vehicles (MAV).

Components	Values	Unit
Brushless Outrunner Motor	25	g
Radio Receiver	11.5	g
Servos	9	g
Li–Po Battery	15	g
Electronic Speed Controller	10	g
MAV Design	80.24	g
Total Weight	150.74	g

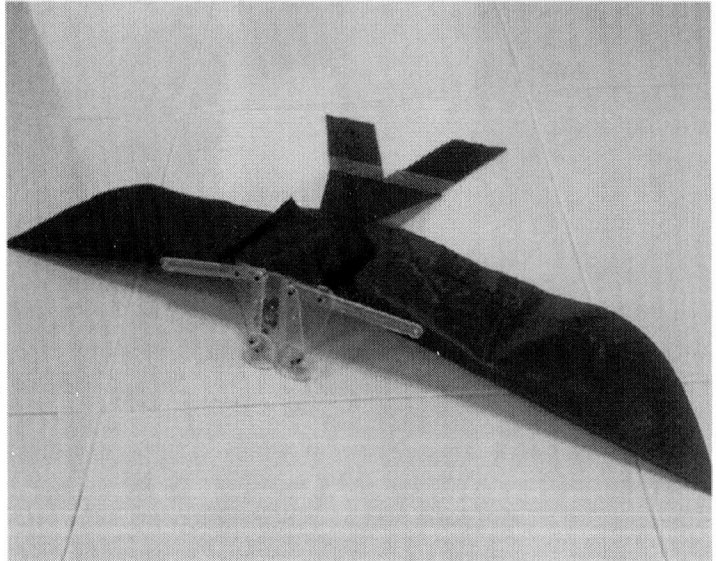

Figure 56. Assembled flapping wing MAV.

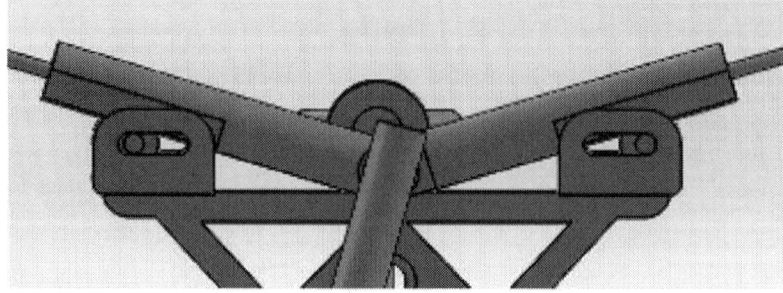

Figure 57. Prototype 2 flapping mechanism.

4.2.2. Tail

The previous prototype was using an elevon tail design which could provide pitch and roll control however it requires two servos to be used. For weight reduction and simplicity sake, a simple tilting tail would be used instead. The tail frame would be made up of carbon rods which would be fixed to the tail piece and covered with Ripstop. The tail piece has a ball bearing inside it so that the tail could tilt easily. Figure 58 shows the tilting tail design.

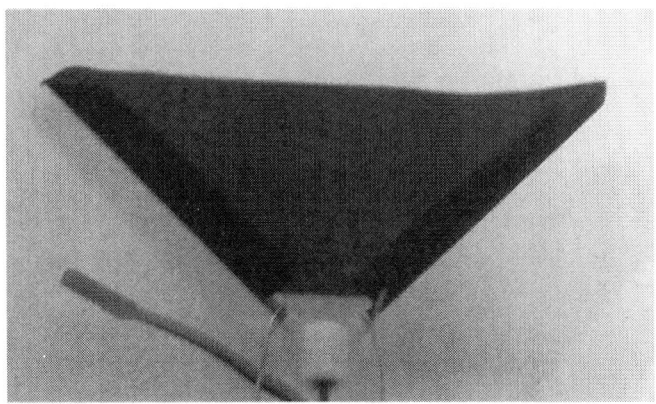

Figure 58. Prototype 2 tail.

4.2.3. Body

In previous design the body was made out of acrylic and had to be solvent weld together. The design was simpler to implement however it was bearing a lot of weight. In order to reduce more weight, carbon rods would be connected to the front piece and the tail piece to form a rigid triangle frame. The frame would then be covered with Ripstop and Velcro tape to secure the components to the platform. Figure 59 shows the CAD design of the body.

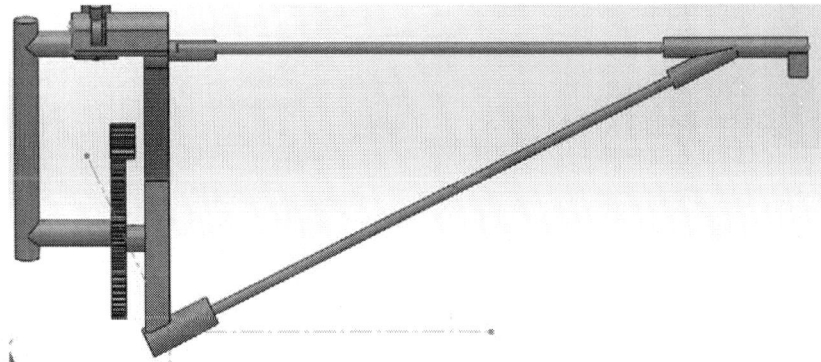

Figure 59. Prototype 2 body.

4.2.4. Gear and Motor Selection

The new flapping mechanism uses only two gears (Table 7). This allows more fine tuning to the gears which allowed a gear ration of 5.5:1. This was acceptable as the newer motor has a slower load speed but higher torque. The new gear is specially hollowed at the center for a ball bearing to be inserted so that it can spin freely around the connecting part of the front piece.

Table 7. Motor specification comparison.

Specification/Motor	Motor 1	Motor 2
Motor Rating (KV)	1700	2800
Load Speed (rpm)	7800	8350
Voltage (V)	7	4
Weight (g)	20	25

4.2.5. Fabrication and Material

In the first design, a laser cutting machine was used. For this second prototype, a Rapid Prototyping Machine or also known as a 3D printer would be used. A 3D printer allows for more freedom of design. An extruded part could be combined during the design process easily, compared to assembling the parts after it has been fabricated. The chosen material was PLA as the design such as the gears needed the material to be strong and durable.

4.2.6. CAD Design Dimensions

Figure 60 shows the completed CAD design of prototype 3. Table 8 below shows that the lift is more than the weight thus the prototype fabrication can proceed.

4.2.7. Flight Test

Figure 61 shows the assembled prototype. Similarly a dry run test was done for the flapping mechanism. Everything was working normally. Next it was the tethered flight. The MAV was also able to move in a circular motion. Finally the free flight test was carried out via remote control. The MAV was held until it flapped at high frequency after which it was hand thrown in the forward direction. After it was thrown, the MAV continued to fly forward while slowly pitching upwards. The left and right controls were tested and the MAV showed that it could maneuver left and right. The last test was the pitching control. Increasing the rpm of the motor pitches the MAV upwards and decreasing the rpm pitches the MAV downwards. The test was a success, the MAV showed that it can fly and was able to be controlled remotely. The video clip can be seen at Youtube website: http://youtu.be/hp-Kpw6sll0.

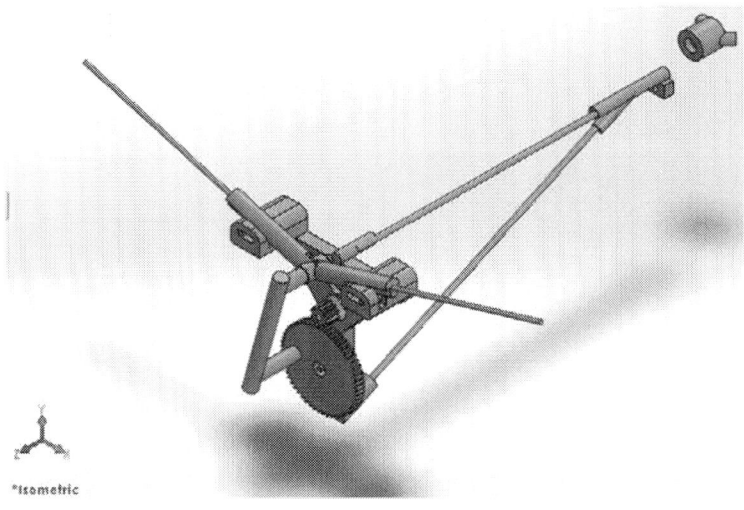

*Isometric

Figure 60. CAD design prototype 2.

Table 8. Lift and weight comparison.

Component	Weight (g)	Parameter	Value
Brushless Outrunner Motor	20	Flapping Amplitude (degree)	50
Radio Receiver	11.5	Flapping Frequency (Hz)	10
Servos	4.5	Lift Coefficient	0.8
Li–Po Battery	4	Air Density (kg/m^3)	1.225
Electronic Speed Controller	10	Chord Length (m)	0.1
MAV Design	14.77	WingSpan (m)	0.15
Total Weight	64.77	Lift (g)	977

Figure 62 shows the flight trajectory captured using Optitrack in our lab. The left figure shows our intended flight path. The right figure shows the actual trajectory of its body. The up/down motion due to the upstroke and downstroke of the wings at 10 Hz flapping frequency has been filtered out using notch filter, hence, the trajectory appears smooth as shown in the right figure. The gradual increase during $1.0 < t$ (s) < 2.0 is due to initial throw to gain altitude and speed by the pilot. Subsequently, $(2.5 < t$ (s) $< 7)$ the flapping wing flies at steady altitude and land $(t > 7$ s). It is clear that the second prototype was able to take off, climb, cruise and land in flapping mode successfully.

Besides unstable flight, certain segments of the flight test were captured well by the Optitrack system. It was then imported into MATLAB. In Figure 62 below the plots on the right segment was captured from three different test flights while the plots on left segment was obtained from the Simulation performing similar outcomes of the captured data.

Figure 61. Assembled prototype 2.

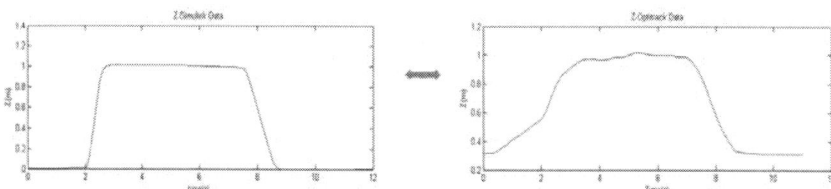

Figure 62. Flight trajectory of the second flapping wing prototype using Optitrack. Left: Intended flight path simulation; Right: Actual flight path captured by Optitack.

5. CONCLUSIONS

This paper reports the research and development of our in-house near-resonance type albatross-like flapping wing models for MAV. The flapping wing models mimic the long-distance migratory bird, similar to albatross. CFD results show that the albatross generates lift on its wing mainly by vortex lift mechanism. They do maneuvering by flapping its entire left and right wings at different amplitudes than using (flapping or twisting) its wing tip only. During forward motion, the wings produce a largely tilted leading edge vortex ring. The flight dynamic parameters is estimated, and used as guidance to predict flying characteristics of this type of ornithopter-like flapping wing MAV. With CFD results, we designed, built and flew two near resonance flapping wing MAVs.

To test the flapping wing mechanism, a test cell was made to house the prototype and the load cell. When measuring the aerodynamic forces produced in the experiments, it was found that thrust was constantly generated, while lift was periodic in nature following a sinusoidal trend. It was found that lift is predominantly generated on the downstroke, with negative lift being generated on the upstroke. It was found out that the thin wing has both lift and thrust produced on than the PET film and thick cambered wing. Flexible wing generated higher velocities, frequency, lift and thrust. In observing the wing angle motion, it was found out that the lift occurs most when the wing is at $0°$ and $-10°$, while negative lift at $30°$ and $45°$.

The design sections of prototypes 2 and 3 have been discussed and evaluated the conceptual designs. There were two fabrication methods that were used, laser cutting and 3D printing. Although it seemed that the 3D printing was a better fabrication method as it allows for more complicated design it does has its limitations in the area of melting point and breaking strength. The third prototype could withstand the high frequency flapping and near resonance amplitude as designed. With remote control, the third prototype was able to take off, climb, cruise and land in flapping mode successfully.

AUTHOR CONTRIBUTIONS

In this paper, Woei-Leong Chan did the experiment on thrust measurement and material selection. Woei-Leong Chan also constructed the first flapping wing model. Sutthiphong Srigrarom made the second flapping wing model and wrote this paper. Notwithstanding, both authors worked together, therefore, the contribution and credit are equally shared.

REFERENCES

1. Shyy, W.; Lian, Y.; Tang, J.; Vileru, D.; Liu, H. *Aerodynamics of Low Reynolds Number Flyers*, 1st ed.; Cambridge University Press: New York, NY, USA, 2008.
2. Groen, M.A. PIV and Force Measurements on the Flapping-Wing MAV DelFly II. Master's Thesis, Delft University of Technology, Delft, The Netherlands, 2010.
3. Ellington, C.P.; van den Berg, C.; Willmott, A.P.; Thomas, A.L.R. Leading-edge vortices in insect flight. *Nature* **1996**,*384*, 626–630.
4. Jones, K.D.; Platzer, M.F. Design and development considerations for biologically inspired flapping-wing micro air vehicles. *Exp. Fluids* **2009**, *46*, 799–810.
5. Breugel, F.V.; Regan, W.; Lipson, H. From Insects to Machines. *IEEE Robot. Autom. Mag.* **2008**, *15*, 68–74.
6. Ellington, C.P. The aerodynamics of hovering insect flight III Kinematics. *Phil. Trans. R. Soc. B* **1984**, *305*.
7. Dickinson, M.H. The effects of wing rotation on unsteady aerodynamic performance at low reynolds numbers. *J. Exp. Biol.* **1994**, *192*, 179–206.

8. Sane, P.S. The aerodynamics of insect flight. *J. Exp. Biol.* **2003**, *206*, 4191–4208.

9. Dickinson, M.H.; Lehmann, F.-O.; Sane, S.P. Wing rotation and the aerodynamic basis of insect flight. *Science* **1999**,*284*, 1954–1960.

10. Sayaman, R.; Fry, S.N.; Dickinson, M.H. Unsteady mechanisms of force generation in aquatic and aerial locomotion.*Am. Zool.* **1996**, *36*, 537–554.

11. Azuma, A. *The Biokinetics of Flying and Swimming*; Springer: Tokyo, Japan, 1992.

12. Chapman, R.F. *The Insects: Structures and Function*, 4th ed.; Combridge University Press: New York, NY, USA, 1998.

13. Alexander, D.E. *Nature's Flyers: Birds, Insects and the Biomechanics of Flight*; The Johns Hopkins University Press: Baltimore, MD, USA, 2004.

14. Ellington, C. The Novel Aerodynamics of insect flight: Applications to micro air vechicles. *J. Exp. Biol.* **1999**, *202*, 3439–3448.

15. Lehmann, F.-O.; Pick, S. The aerodynamic benefit of wing–wing interaction depends on stroke trajectory in flapping insect wings. *J. Exp. Biol.* **2007**, *210*, 1362–1377.

16. Ennos, A.R. The kinematics and aerodynamics of the free flight of some diptera. *J. Exp. Biol.* **1989**, *142*, 49–85.

17. Van den Berg, C.; Ellington, C.P. The three-dimensional leading-edge vortex of a hovering model hawkmoth. *Philos. Trans.* **1997**, *352*, 329–340.

18. Warrick, D.R.; Tobalske, B.W.; Powers, D.R. Aerodynamics of the hovering hummingbird. *Nature* **2005**, *435*, 1094–1097.

19. Dial, K.P.; Tobalske, B.W.; Peacock, W.L. Kinematics of flap-bounding flight in the zebra finch over a wide range of speeds. *J. Exp. Biol.* **1999**, *202*, 1725–1739.

20. Young, A.D. *A Review of Some Stalling Research*; Her Majesty's Stationery Office: London, UK, 1952; p. 6.

21. Hong, Y.S.; Altman, A. An experimental study on lift force generation resulting from spanwise flow in flapping wings. In Proceedings of the 44th AIAA Aerospace Sciences Meeting and Exhibit, Reno, Nevada, 9 January 2006.

22. Bradshaw, N.L.; Lentink, D. Aerodynamic and structural dynamic identification of a flapping wing micro air vehicle. In Proceedings of the 26th AIAA Applied Aerodynamics Conference, Honolulu, HI, USA, 18–21 August 2008.

23. Fry, S.N.; Sayaman, R.; Dickinson, M.H. The aerodynamics of free-flight maneuvers in Drosophila. *Science* **2003**, *300*, 495–498.

24. Jongerius, S.R.; Lentink, D. Structural analysis of a dragonfly wing. *J. Exp. Mech.* **2010**, *50*, 1323–1334.

25. Kawamura, Y.; Soudal, S.; Nishimoto, S.; Ellington, C.P. Clapping-wing micro air vehicle of insect size. In *Bio-Mechanisms of Swimming and Flying*; Kato, N., Kamimura, S., Eds.; Springer Verlag: Berlin, Germany, 2008.

26. Kesel, A.B. Aerodynamic characteristics of dragonfly wing sections compared with technical aerofoils. *J. Exp. Biol.* **2000**, *203*, 3125–3135.

27. Pornsin-Sirirak, T.N.; Tai, Y.C.; Ho, C.H.; Keennon, M. *Microbat-A Palm-Sized Electrically Powered Ornithopter*; NASA/JPL Workshop on Biomorphic Robotics: Pasadena, CA, USA, 2001.
28. Usherwood, J.R.; Ellington, C.P. The aerodynamics of revolving wings I–II. *J. Exp. Biol.* **2002**, *205*, 1547–1576.
29. Hu, H.; Kumar, A.G.; Abate, G.; Albertani, R. An experimental investigation on the aerodynamic performances of flexible membrane wings in flapping flight. *Aerosp. Sci. Technol.* **2010**, *14*, 575–586.
30. Keennon, M.; Klingebiel, K.; Won, H.; Andriukov, A. Development of the nano-hummingbird: A tailless flapping wing micro air vehicle. In Proceedings of the 50th AIAA Aerospace Science Meeting, Nashville, TN, USA, 9–12 January 2012; pp. 6–12.
31. Krashanitsa, R.Y.; Silin, D.; Shkarayev, S.V.; Abate, G. Flight dynamics of a flapping-wing air vehicle. *Int. J. Micro Air Veh.* **2009**, *1*, 35–49.
32. Manuel, M. *Design of a Flapping Wing Mechanism*; Semester Project of Autonomous Systems Lab (ASL), Swiss Federal Institute of Technology Zurich (ETH): Zurich, Switzerland, 2009.
33. McMichael, J.; Francis, M. *Micro Air Vehicles—Toward a New Dimension in Flight*; Defense Advanced Research Projects Agency: Arlington County, VA, USA, 1997.
34. Blake, C.H. More data on the wing flapping rates of birds. *Condor* **1948**, *50*, 148.
35. Shyy, W.; Lian, Y.; Chimakurthi, S.K.; Tang, J.; Cesnik, C.E.S.; Standford, B.; Ifju, P.G. Flexible wings and fluid-structure interactions for micro-aerial vehicles. In *Flying Insects and Robots*; Springer Verlag: Berlin, Germany, 2010; pp. 143–157.
36. Yang, L.J.; Cheng, C.M.; Chiang, Y.W.; Hsiao, F.Y. New flapping mechanisms of MAV "golden snitch". In Proceedings of 8th International Conference on Intelligent Unmanned Systems (ICIUS), Singapore, 22–24 October 2012.
37. Image of Single Gear Crank Design—University of California Biomimetic Millisystems Lab Ornithopter Project. Available online: http://robotics.eecs.berkeley.edu/~ronf/Ornithopter (accessed on 1 April 2014).
38. Festo: SmartBird, IMAGE of SmartBird Ornithopter Inspired by Herring Gull. Available online: http//www.festo.com/cms/en_corp/11369.htm (accessed on 1 April 2014).
39. Flapping Flight—How Birds Fly. Available online: http://www.n6iap.com/ornithopter/howbirdsfly.html (accessed on 1 April 2014).
40. Lentink, D.; Jongerius, S.R.; Bradshow, N. The scalable design of flapping micro-aerial vehicles inspired by insect-flight. In *Flying Insects and Robots*; Springer Verlag: Berlin, Germany, 2010; pp. 185–205.

Index